NATURAL DISASTER RESEARCH, PREDICTION AND MITIGATION SERIES

MELTDOWN: CLIMATE CHANGE, NATURAL DISASTERS AND OTHER CATASTROPHES – FEARS AND CONCERNS FOR THE FUTURE

NATURAL DISASTER RESEARCH, PREDICTION AND MITIGATION SERIES

The Phoenix of Natural Disasters: Community Resilience
Kathryn Gow and Douglas Paton (Editors)
2008. ISBN: 978-1-60456-161-6

Natural Disasters: Public Policy Options for Changing the Federal Role in Natural Catastrophe Insurance
U.S. Government Accountability Office
2008. ISBN: 978-1-60456-717-5

Solar Activity and Forest Fires
Milan Rodovanovic and Joao Fernando Pereira Gomes
2009. ISBN: 978-1-60741-002-7

National Emergency Responses
Paul B. Merganthal (Editor)
2009. ISBN: 978-1-60692-354-2

Earthquakes: Risk, Monitoring and Research
Earl V. Leary (Editor)
2009. ISBN: 978-1-60692-648-2

Hurricane Katrina: Impact, Recovery and Lessons Learned
Nessa P. Godfrey (Editor)
2009. ISBN: 978-1-60692-478-5

Cyclones: Background, History and Impact
Terrance G. LaBeau (Editor)
2009. ISBN: 978-1-60692-064-0

Cyclones: Background, History and Impact
Terrance G. LaBeau (Editor)
2009. ISBN: 978-1-60876-711-3
(Online Book)

Indigenous Knowledge and Disaster Risk Reduction: From Practice to Policy
Rajib Shaw, Anshu Sharma and Yukiko Takeuchi (Editors)
2009. ISBN: 978-1-60741-574-9

Indigenous Knowledge and Disaster Risk Reduction: From Practice to Policy
Rajib Shaw, Anshu Sharma and Yukiko Takeuchi (Editors)
2009. ISBN: 978-1-60876-674-1
(Online Book)

Meltdown: Climate Change, Natural Disasters and other Catastrophes – Fears and Concerns for the Future
Kathryn Gow (Editor)
2009. ISBN: 978-1-60876-153-1

NATURAL DISASTER RESEARCH, PREDICTION AND MITIGATION SERIES

MELTDOWN: CLIMATE CHANGE, NATURAL DISASTERS AND OTHER CATASTROPHES – FEARS AND CONCERNS FOR THE FUTURE

KATHRYN GOW
EDITOR

Nova Science Publishers, Inc.
New York

For permission to use material from this book please contact us:
Telephone 631-231-7269; Fax 631-231-8175
Web Site: http://www.novapublishers.com

LIBRARY OF CONGRESS CATALOGING-IN-PUBLICATION DATA

Meltdown : climate change, natural disasters, and other catastrophes--fears and concerns for the future / editors, Kathryn Gow.

p. cm.

Includes bibliographical references and index.

ISBN 978-1-60876-153-1 (hardcover : alk. paper)

1. Climatic changes--Physiological effect. 2. Climatic changes--Psychological aspects. 3. Climatic changes--Social aspects. I. Gow, Kathryn.

QC903.M45 2009

577.2'2--dc22

2009025202

Published by Nova Science Publishers, Inc. ✢ New York

THAW

At last the bucket now full
I dropped in the apple
Displacement did show
in resultant overflow.
Density's dish and Empirical Law
climate now warming icy thaw.

As pride goeth before destruction
a haughty spirit before a fall
and though the snake's tempting apple
of knowledge foresaw
thoughts of excess production,
my ego's musings led to the deduction:

While Rodin's statue still holds pride of place
methinks with Artic's landless mass
some takeup ensures less envisaged harrass.

(Frances Murphy, Gold Coast, Australia)

CONTENTS

PREFACE

Why the word Meltdown in the title of the book? With the world heating up, bushfires wiping out whole communities, money markets and economic systems collapsing, mining operations replacing quality farming land, factory chemicals poisoning the waterways, the natural environment being destroyed, and whole societies being displaced, we are indeed witnessing a meltdown. People are now very concerned and some are afraid for their futures.

Does the human race, or at least sections of the populations in different countries of the world hold beliefs about, and attitudes towards, social and ecological issues such as climate change and futurist scenarios that are apocalyptic? In a completely different vein, are they prepared to take action about their environmentally unfriendly behaviours? Are all the natural disasters that have beset the world in the past decade an indicator that the world is about to end, particularly coupled with famine, war and pestilence, and lately the breakdown in the global economic systems, all having been prophesised by different seers and religious leaders?

This book is timely and in some ways timeless; the issues discussed within its pages are matters that are of interest to all people across the world and really across time. In this book, there are a number of chapters that focus on the theoretical positions and cognitions about fears and concerns for the future, in different segments of the world's population. There are other chapters that describe nature's situation as it is today, with water shortages, threats of sea-level rise, loss of forests, habitats and wildlife in various parts of the globe. These chapters demonstrate the complexities involved in attempting to understand which aspects relate to climate change, which aspects are distinct from climate change, and indeed which aspects were already in existence, but have been, and will be, exacerbated by climate change influences. The four basic elements of life - fire, water, earth and air – are covered by contributions on bush fires, floods, drought, water shortages, and air pollution.

Another focus of the book is to balance the concentration on fears and concerns with problem solving and mitigation strategies, thus leaving the reader with the sense that we can utilise more control and take more action when and where it is needed. Several of the chapters demonstrate that while there are complex variables involved, it is possible to think about the consequences of our actions and then to take stock and act sensibly and determinedly to create a much better future scenario for the world than the one we are constantly being presented with through the media and government reports. Because, as the United Nations Foundation (2007) reminds us: The challenge is to avoid the unmanageable and manage the unavoidable.[1]

[1] United Nations Foundation. (2007). Confronting Climate Change: Avoiding the unmanageable and managing the unavoidable. Executive Summary, February. www.confrontingclimatechange.org.

ACKNOWLEDGMENT

All images are copyright Tim McCrorey 2009 and are provided for use in this one book only, and for advertising the book. They are not to be used for any other unrelated purpose without prior permission from Tim McCrorey.

Part I:

Conceptualising Fears and Concerns about Future Catastrophes

Contributed by Tim McCrorey, Toowoomba.

Photo: A Portent of Disasters Coming our Way.

In: Meltdown: Climate Change, Natural Disasters…
Editor: Kathryn Gow
ISBN 978-1-60876-153-1

Chapter 1

SHOULD WE BE WORRIED ABOUT THE FUTURE?

Kathryn M. Gow
Queensland University of Technology, Australia

INTRODUCTION

Natural Disasters: Frontstage

While Australia and China battled with the searing droughts, the power of the Tsunami demonstrated that something in the world's core appeared to have gone awry, and thousands of people were drowned by tidal waves as high as skyscrapers. Then when everyone was lulled back into the security of thinking that nature had demonstrated its might and power sufficiently, powerful cyclones ripped the county of Burma apart and to show equality of intention, then turned it fury on China and shook the homes and lives of its people upside down with severe earthquakes. Nature's fury continued to demonstrate its might through tornadoes, cyclones, powerful wind and rain storms, earthquakes, deluges, floods and wildfires. "Katrina" was one catastrophic event which showed that where there should have been sufficient warnings and emergency preparedness and response in a wealthy country such as the USA, these were either not sufficient or not activated in time to save lives, property and infrastructure.

With a heightened emphasis on climate change, people started to wonder if, and some were convinced that, the frequency and severity of these natural disasters were harbingers of things to come in a world that now seemed out of balance.

The Financial Screen Saver: Backstage

The concentration on concerns about natural disasters and saving the natural world of the future distracted the populations' attention away from the economic world stage, at a time when our economic security was being unravelled backstage. The majority of people in the developed world were not worried about their employment opportunities or concerned about the safety of their savings. While the advent of a recession was not unexpected, people were not prepared for the fast financial slide (not even a roller coaster) all the way down to the lowest point in 60 years. Like an amusement park thriller breaking loose from its cables, the USA's financial system crashed into others, causing severe damage to the financial machinery of many other countries.

There is the possibility, however, that the collapse of the current financial systems may lead to remodelling, not just of our economic and financial machinery, but also of our current ways of living with, and working in, the natural environment. This may occur in spite of the probability that little funding in various nations will be allocated to preventing, mitigating or adapting to climate change effects.

Climate Change: Waiting in the Wings

What do we make of the statement that 'the ice caps are melting and sliding into the ocean'? The stories about the melting ice and snow have been published now for more than 50 years, along with the threat of tidal waves and environmental destruction on a very large scale.

There are many questions to be asked. For instance, is it possible that climate change is just one more adjustment that people have to take in their stride, or is it that climate change is a risk of apocalyptic proportions? In reality, what exactly is causing climate change? How much of what we have been hearing about climate change is hype? How much of it is fact and how do we determine the differences and measure the actual risk factors? What are the components and weightings in the equations that we need to know about, in order to make informed decisions about our futures?

Who can we trust? Who stands to make the most out of the climate change debate and who stands to lose, if it is true? If on the other hand, it is totally false, or partially true and partially false (a bit like the false memory recall debate), who in each case then stands to gain or lose?

Weart (2008) is one of the investigators who goes to great lengths to trace the history of the findings leading up to the present day emphasis on climate change; as a scientist, he brings to the table detail mindedness combined with the ability to link ideas together. He assists us in some way to dissembed the trees from the jungle; he separates out for us the component parts of this hard- to- grasp concept of climate change.

As the Editor of this book, I needed to gain an overview of the whole picture, but I could only do this in truth by delving into many of the component parts. The first thing I noticed in the media reports and the short reports on the internet was the mixing up of concepts with details. Or taking Edward Hirsch's (1988) guidance about understanding the science of language, things that were actually a subset of a class were being taken as the class itself; or in statistical jargon, one aspect of the total variance of the subject to be measured was in fact being taken for the totality. The terms global warming and climate change were used as if they had exactly the same meaning.

It seemed to the ordinary citizen that suddenly natural disasters, which had been with us all our lives, were now evidence of climate change; solutions to the projected scenarios of climate change were embroiled with money making business schemes such as carbon trading. Reports on climate changes meant that suddenly water catchments and supplies would have to be privatised in order to 'fix' climate change. While drought had been with us a long time, suddenly it was new and different and to be expected forever. Therefore the small lot farmers should leave their properties immediately, so that very large scale primary producers or new timber growing companies (promising all kinds of tax incentives) could buy out their plots cheaply and turn much of our vast land into huge money making enterprises where small towns would no longer exist, or where the land (valued for its intrinsic value by both farmers[1] and indigenous peoples alike) would be mined to such a depth and extent that black soot would sail across the pristine green mountains and clog the waterways, thus choking and poisoning the wildlife and causing breathing difficulties in nearby towns and cities.[2]

Language and Classification Problems

The dialogue on climate change has been hampered, not just by the array of languages across the world and the different socio-economic standing of first and emerging countries, not just by greed and protectionist policies within and across regional blocks, but by the use of a jumble of terms without definition and explanation that confound and confuse the general population. The media, which disseminates the scientific findings and projections to the world, has been instructed to condense information to simple language structures and key points of interest to keep the interest of their readers/viewers/listeners.

Nevertheless, the media has a major role to play in saving the world, if indeed it needs to be saved, because without them in many cases, we would know a lot less about what is going on in distant parts of the globe. Without the media in its many forms, we would probably not have as many action groups forming across the world to save the whales, force chemical companies to stop polluting the air and waterways, and to make sure that logging companies do not continue to destroy the remaining forests on this planet with napalm-like chemicals.

So how much have the media's readers, viewers, and listeners really understood about the information portrayed by the media and internet reports? In interviews conducted by Kristina Searle and others by Kathryn Gow (see Chapter 3), it became evident that the majority of interviewees did not understand what climate change was and even when they could enumerate a whole range of environmental concerns, they had no facts on which to base their views. For example, the interviewees could not cite any statistics concerning the increased temperature locally or globally over the past century, although they felt that it had been getting much hotter in the past four years.

Are we in a state of denial when it comes to climate mitigation measures? From their work with focus groups in Switzerland in 2001, Stroll-Kleeman, O'Riordan and Jaeger attempted to ascertain why people would not take responsibility for mitigating climate change either individually or collectively, and determined that people used denial to avoid creating cognitive dissonance, so that they did not have to change their comfortable lifestyles. Even though they stated that they were concerned, the authors considered that they blamed other agencies, and doubted if any individual effort, that they might make, would be efficacious.

---

[1] Paige Taylor, Exposed: Green myths against farming. The Australian, April, 10, 2009. http://www.theAustralian.news.com.au/story/0,25197,25319011-11949,00.html.

[2] Even editors have to admit to being passionate about the things that matter to us in life, as Stephan Lewandowsky and Elizabeth Tindle remind us at the close of the book.

Another paper emanating from Sweden by Sunblad, Biel and Garling (2007), investigating climate change and affective risk judgements, revealed that the actions people took after they had assessed the probability of serious negative consequences, rested on their knowledge of both the causes and consequences of climate change, especially with respect to health matters. Interestingly, while women were more worried, they were equally in line with men when it came to assessing risks. Across continents, there were also differences; for example, Kage (2007) reported on a survey that determined that Europeans were more concerned about climate change than Americans.

Joe Brewer (2008) went even further and explained how fear might be being used to quell any action on climate change. This point takes us right back to the earlier questions that I posed: Who stand to gain and who stands to lose in the promulgation of preparedness and mitigation strategies concerning climate change scenarios?

Overview of Contents

There are three parts to the book and they are labelled: (1) Conceptualising Fears and Concerns about Future Catastrophes; (2) What are Some of the Effects of Climate Change and Natural Disasters? and (3) Can We Afford to Take Chances with our Future? The following overview is organised by themes that are related in different ways and hence the introduction to the chapters transverses the material according to those themes, rather than simply introducing each chapter sequentially.

In her chapter (19) about *Serious Fears across Cultures about Collapse of the World Order,* Elizabeth Tindle (with Waveney Croft) challenges us to take stock of the Malthusian collapse of our civilization. Written before the world financial crash which commanded front stage in the last quarter of 2008, the thoughts and opinions of this writer seem to have encapsulated the ruminations and reflections of many people in the western world after the financial slide took hold.

So why did the financial world go 'pear shaped', or as Paul Wildman challenges in his chapter (20) (*Three Economic Tsunamis: Should we have seen them Coming? A Boundary Rider's Perspective)* was the evidence there all along that the economy could not continue in such a vein? Do you agree with his approach to economic recovery, even though that might not lead to the 'high flying' economy we have just seen toppled? One thing is for certain, and that is that you will experience Paul's waves of thoughts about these serious matters, as you read his words. The style of language, in Chapter 20, has been deliberately published in a non-academic vein, as an invited contribution, in order to break through our set ways of thinking and acting. It may well be that the world will have need of more bush mechanics, as Paul occasionally describes himself.

Was globalisation a factor in what is now viewed by many as a play about 'the emperor's new clothes' [3] - the creation and continuance of the fake world of finance and the lie that every person in the world could have everything, if only he/she could borrow enough money? Will we now be returning to the slogan 'think global, but buy local', as the economist Paul Wildman would advocate? Regardless, it does appear that the force of globalisation means

[3] The Emperor's New Clothes" was one of Hans Christian Andersen's fairy tales. For background, see http://en.wikipedia.org/wiki/The_Emperor's_New_Clothes.

that 'we are all in this together' and unfortunately when a team of horses is yoked together to win the race, and one or two horses falls over, so does the whole team.

Mark King draws parallels between globalisation and climate change as phenomena, the most obvious of which, he says, is their effect on virtually every human being on the planet. In Chapter 18, *Globalisation and Climate Change,* he asks how climate change will affect globalisation, because globalisation is linked to power, politics and development. Mark demonstrates concern for third world countries (which were reclassified as emerging countries when the world economy was apparently booming) and while there are positive aspects of globalisation, he wonders if the same can be said about the projected climate change effects, if they eventuate as predicted.

But is the climate change threat real? In their chapter (2) *Climate Change as Challenge: Can Human Resilience and Resourcefulness Provide Solutions?* Marek and Andrzej Celinski team up to approach this question from both the perspectives of public perception and of scientific fact. Giving us an interesting overview of what humanity has struggled with in the past and survived, they argue that there is a great amount of resourcefulness and resilience in humanity that can be harvested, if the projected IPCC reports (ranging from 1990 to 2008), about the threats of the impacts of climate change, eventuate.

What if the world is really running out of potable water? Are there sufficient alternatives to naturally available spring water in times of drought and water shortage? Howard Guille and Hannah Evans believe that there are alternatives and highlight one Australian State's attempt to build a water grid across a large area in South East Queensland. Their chapter (17) *The South East Queensland Water Grid: Using the Drought to Reengineer the Market* follows the introduction of water saving measures during a severe drought that almost paralysed Queensland, in which the capital city and many other cities and towns were either close to running out of water, or had run out of water, during the long years of drought. Should public water sources be privatised? Read what the authors have to say about this contentious issue.

One of the saddest outcomes of long and severe droughts is that water holes, ponds, dams and waterways dry up; and as Arthur Georges[4] says: 'empty dams means no turtles'. While quite a lot of attention has been drawn to the plight of sea turtles in the past 15 years, little interest has been directed at the death of millions of the smaller sized land turtles. Arthur Georges and John Roe, who have spent many years studying turtles in their various forms and locations throughout the world, have addressed the possible impact of climate change on land turtles in their chapter (11) on *Responses of Freshwater Turtles to Drought: The Past, Present and Implications for Future Climate Change in Australia.*

If people are really so disturbed or distressed about the future, because of pessimistic scenarios about climate change, wars and terrorism, would they be less concerned about their own lives than the likelihood that there will not be enough food for everyone? Indeed, would they dream about these things more often? In their chapter (5) *Do Global Concerns affect Dream Content more than Personal Concerns?* Sandra Sacre and colleagues don't find quite what they thought they would. Their findings, however, confirm the previous research which highlights the fact that people generally are much more concerned about their own immediate problems than those on the larger world stage.

[4] Arthur Georges, cited by Rosslyn Beeby. Turtles under siege as drought bites. *The Canberra Times,* Monday January 29, 2007. "How long have our turtles got?".

From the viewpoint of Hong Kong, Lida Lee, in her chapter (4) *Providing Scaffolding for Students in Light of Concerns over Climate Change: An Educational Perspective,* sets out some of the proactive approaches in educating children about the future and climate change, and it is worth remembering that we need to be able to do this in a way that enables, not disables, the future leaders of our world. As Chris Rapley (2007) advises: "So that has to be the positive message: We are all trying to make the future better than it would otherwise be. Knowledge and evidence help us do that. And we have to be careful not to make people so complacent that they don't do anything, or so worried that they just paralyse themselves. We have to find that middle path".[5]

As we speak about the leaders of the future, will the leaders of the past be the best people to guide us into the future, with all its unknown parameters and immense challenges? In *A Third Watershed: The Capacity to Adapt and Solve Future Dilemmas Confronting our Planet* (Chapter 21), Ian Plowman lays out some hypotheses about leadership styles and when, where, and with whom those leadership styles are most effective. Perhaps you can discuss with others what type of styles you think are the best for different roles in the future, given the additional financial disaster that has now overtaken the world. Plowman, based on historical events, doubts that the leadership, shown at most levels of powerful political and industrial machinery, will be facilitative, even adequate, for the problems facing the world now and in the future.

Poetry can often make a connection with people that mere prose cannot, and when Frances Murphy muses about what is happening with the world, we are forced to pause and think, as her ontological message seeps into our consciousness. Anna Lee Mraz Bartra in Chapter 22, *It's Now or Never: The Importance in Raising Ecological Conscious Awareness of The World's Climate Change,* gives us a passionate address on saving the environment in both developed and underdeveloped countries, and compares what has happened to the natural habitat in two countries in the southern hemisphere - Mexico and Australia - over the past few decades. It is interesting to note that few of the reports on climate change allocate any considerable amount of space to the natural environment or wildlife conservation and protection. Anna Lee stresses the importance of sustainable ecology for the survival of both animal and human species.

Throughout the text, Tim McCrorey contributes his photographs about birdlife to remind us of the beauty of nature and what we stand to lose if we do not protect our natural resources. Closely linked to the ecological theme are the thoughts of Ann Moir-Bussy who in Chapter 6, *Ecological Crises: A Psychological Perspective and The Path to Transformation,* takes an ontological perspective about humankind's spiritual connections with the earth. The absence of such perspectives in reports or tomes on climate change, or in ecological treatises on vanishing or endangered species, is not surprising in industrialised and technologically entrapped communities; but will it take the extinction of most of what we have in 'Paradise' to bring people back into balance with the earth? One of the keys is held by indigenous peoples around the world, but we have not yet reached a sufficient level of intellect to realise this powerful resource. Workineh Kelbessa acknowledges such indigenous intelligence in climate change preparation by different peoples in Ethiopia, Tanzania, Mozambique and Zambia (see Chapter 15). However, and Workineh Kelbessa would agree, we should not think

[5]http://www.sciencepoles.org/index.php?/articles_interviews/professor_chris_rapley_taking_stock_on_climate_change/&uid=990&type=15.

that we can 'mine' this, as we have every other resource; it will come at a price: humility, gratitude and integrity; sharing and caring and respect.

The most likely outcomes of global warming and associated climate effects of that change are addressed in several chapters; this theme is then extended to other physical health and psychological distress effects in two chapters; one (Chapter 3) that deals with the psychological factors by Kristina Searle and Kathryn Gow is titled *Exploring the Psychological Aspects of Risk, Fears and Concerns about Climate Change*, and the second, which focuses on both physical and psychological health, is by Grace Wai-man Ip in her chapter (7) on *The Physical and Psychological Challenges of Environmental Threats and Climate Change.*

Deluge is a word that is well known to people who have studied the Christian bible and there has been a renewal of interest about deluges in the current literature on climate change; it is now associated with natural disasters of emergency proportions. In the distant historical, indeed perhaps mythical literature, we are told by several writers, including Alfred de Grazia (1983), that the expression 'raining cats and dogs'[6] is not all that impossible, because the old terminology surrounding deluges did indeed include objects such as frogs and rats and not just blocks of ice, that fell from the sky at a distance point in the past; however discussion on that point would take us away from the immediate earthly issues at this moment.Those of us who live in highly vegetated rural areas can vouch for the fact that after heavy rain, all types of insect and frog life overflows, and if not falling from the sky certainly leaps across the roads in dancing vaudeville style; however a recent documentary did reveal that masses of fish fall from the sky during tornadoes.[7]

The topic on deluges finds a place in this book in two chapters: as one of the accepted effects of climate change in the chapter (10) *Can We Anticipate More Heatwaves, Wildfires and Deluges?* by Kathryn Gow, and as a forerunner of floods and backbuilding problems in the chapter (12) written by Francine Pritchard titled: *Can Understanding Backwater Effects Help us to Prepare for and Manage Deluges?* The recent natural disasters that have combined forces to burn, flood, snap and flatten large areas of Australia in 2008/2009 may be repetitions of patterns across time, or they may be harbingers of the types of climate change events portrayed in the work of the IPCC. One thing is certain, and that is we must not let our guard slip; Australia was lulled by the dullness and deadness of drought for so long, that it had forgotten that floods had ever occurred.

Drought is certainly an accepted outcome of increased temperatures and lack of precipitation, and much of the world (both developed and underdeveloped) has experienced recent droughts, some of devastating proportions. Kathryn Gow takes us for a walk through the heartbreak of drought in rural communities and the importance of water for all, in Chapter 16 *Drought in Rural Areas: Not Just the Absence of Water*. Africa is well known for its droughts of despair and hopelessness over the past century; but few in the western world seem to really comprehend the complicated nature of the sagas of starvation and disease, pestilence and poverty that accompany these drought heartbreaks. Workineh Kelbessa understands and in his chapter (15) on *Climate Change Impacts and Planning in Africa,* he leads us through some of the confounding factors that impede and prevent the restoration and

[6] Editor's expression, not that of de Grazia.
[7] *Perfect Disasters*. ABC DVD. www.abcshop.com.au

rebuilding after drought, and make of the philosophy of sustainability a 'dirty word'.[8] My interpretation of his writing is that Kelbessa is calling for 'a fair go' for Africa and spells out what the loss of Africa's contribution would mean to the rest of the world.

It is interesting that it is the rural and farming areas that suffer the most initially from droughts - both short and prolonged droughts; whether that be from the loss of income, the shortage of food and water, or the breakdown of relationships in marriages, families and communities, and then the slow collapse of whole townships and their surrounding provinces. Chapter (8) *Squeezing Relationships Dry: The Impact of Drought on Australian Farmers' Intimate Relationships* by Zoe Pearce takes a look at one of the unaddressed issues following drought; the consequences of the stress during and after drought, that leads to the break-up of relationships between husband and wife through forced separation initially and then through financial, physical and emotional collapse.

If the world is heating up, and some countries, such as Australia, have already registered higher temperatures than other countries in the past century, what can be done to assist residents to prepare for more bushfires? In their chapter (13) on *Bushfire Disaster Preparedness: A SEQ Rural Survey,* Francine Pritchard and colleagues returned to a rural area which had suffered devastating fires in late 2004/early 2005 and found out what the residents were doing to prepare for future wildfire events. While the results can be utilised to assist in preparing rural residents for similar future seasons of wildfire threat, the 2009 February firestorms that destroyed people, animals, properties and infrastructure in Victoria have shown us that the fearful might of fire and heat still needs a great deal of study, reviews of preparedness and bushfire action, as well as changes to pertinent legislation.

In the early 1950's, as children, we were taught that tidal waves would wipe out coastal areas and that the sea would run as far inland as it was flat, and that only those people who lived on high hills or mountains would escape the gigantic waves that were to come in the future. So knowledge about the effects of climate change are not new, as can be verified by searches on the internet; indeed one city council has been planning for sea level rise since the 1980's and Peter Hastings and Iraphne Childs speak about their research in a seaside town in their chapter (14): *Adapting to Sea-Level Rise: Lessons for The Future from Redcliffe, Queensland.* The authors track the problems that were identified and then addressed by this city council, across more than two decades; their findings may assist other local councils in preparing for sea-level rise.

One of the more sensible approaches to take to 'tackling' climate change is to look at the whole picture, then each component part and seek to address that part, while keeping in mind the whole picture. Concern about air pollution has been around for many decades and a lot of media coverage has translated the scientific research about the effects on our health of such pollution, whether it be from car exhausts, factory emissions, chemicals spills, smoke from wildfires or nuclear leaks. In Chapter 9, Adrian Barnett and Craig Hansen ask and answer the next question in terms of the bigger picture: *How Might The Health Effects of Air Pollution Change when The Planet gets Warmer?* While we have known about some of these pollution effects for several decades now, what the authors do is to help us focus on the range of emissions that are the subject of discussion in the global warming and climate change discourse.

---

[8] Editor's expression, not those of Kelbessa.

At the close of the book, we include a short commentary by Stephan Lewandowsky titled: "How did we Blow it on Climate Change?"[9] reminding us about the sacrifice that may need to be made for the generations to follow, so that they, too, can enjoy Mother Earth. And finally lyrics from a child, aged 6; Jordy Roberts demonstrates that the youngest generation cares greatly about the planet in his poem (which he has also set to music): *How could we live without These?*

However our prevention, intervention, adaptation, and mitigation strategies need to be sound. Professor Ross Garnaut (2008)[10] was reported as warning that hasty action, implemented to lessen the effects of global warming, could lead to the end of the economic boom and could even have the same economic effects as the Great depression, if not managed properly; a warning that is particularly relevant in what he calls "knee jerk reactions" by politicians. However, since then, we have plunged into the global financial crisis and no-one can blame the worst monetary collapse since the Great Depression on climate change mitigation efforts; nevertheless, ill conceived 'shackling and compressing' policies can hasten the demise of both industry and communities.

Action Stations: Will the Curtain Come down or Go up Again?

What if the sceptics are right[11] and each generation should simply be preparing for each new decade, because, over time, history has demonstrated its might and power through climate variability? Then, that gives us some hope and some time to plan; because then we can really sit down and start preparing for the changes before they come, by utilising the best scientific intellect in the world and the best innovators, who can work with the everyday resourceful inhabitants of the earth, and the energy of the young who will inherit the earth. Regardless of the causes of our ecological problems, we need to address each one carefully and effectively, and as quickly as possible.

CONCLUSION

If the book can be utilised to stimulate discussion in families, communities and work places, as well as in research centres, then it will have made a significant contribution to assisting people to think about, and sort through, a range of factors influencing our actions, as well as better understanding our interconnectedness with the natural world, and as David Suzuki (Suzuki, 2003; Suzuki and Knudtson, 1992) says: we are all interconnected; all of nature is one.

In essence, we can no longer act as if what one person, or community, or country, does is of no concern to other peoples and countries; we must begin to understand, as our indigenous peoples across the world and across time have learned, that Mother Earth must be protected, if we are to survive and live in harmony; and thus Elizabeth Tindle's poem *Save our Planet* at

---

[9] This piece was originally published in the *West Australian* (2007) and is reproduced here with permission.

[10] Rural Weekly, p. 27. Southern Edition, Friday May 9, 2008. Queensland rural newspaper.

[11] Verity Edwards. Climate sceptics ready to storm heaven with earth's geological history. The Australian. 23.04.09 http://www.theaustralian.news.com.au/story/0,25197,25372986-11949,00.html.

the close of the book speaks directly to our hearts and souls. As Nicola Marcus (2009) warns us, none of us should continue to allow such destruction 'on our watch'.

From a scientific viewpoint, we must share and disseminate research from different scientific fields to plan for the contingencies that these changes may bring about; from a social stance, we are going to have to learn to adapt, if the majority of the changes are outside the influence of human kind. Something much bigger may be at play here, and it is time we found out what is really going on in our planet.

ACKNOWLEDGMENTS

I would like to thank Monique Miotto for her work and patience as the Editorial Assistant in finalising this book, with its many iterations.

REFERENCES

Beeby, R. (2007). Turtles under siege as drought bites. How long have our turtles got? *The Canberra Times,* Monday, January 29.

Brewer, J. (2008). Climate and the Psychology of Loss. Rockridge Nation, 2008. *Washington Post.http://www.rockridgenation.org/blog/archive/2007/11/06/climate-and-the-psychology-of-loss*

de Grazia, A. (1983). *The Lately Tortured Earth: Exoterrestrial Forces and Quantavolutions in the Earth Sciences.* US: Metron Publications.

Hirsch, E.D. (Jr.). (1988). *Cultural Literacy. What every American needs to know.* New York: Random House.

Kage, B. (2007). Survey finds Europeans more concerned about climate change than Americans. Natural News.com. January 17. http://www.naturalnews.com/021444.html

Markus, N. (2009). On Our Watch. *The Natural World.* Melbourne: Melbourne University Press.

Nilsson, A., von Borgstede, C. and Biel, A. (2004). Willingness to accept climate change strategies: The effect of values and norms. *Journal of Environmental Psychology, 24* (3), 267-277.

Stoll-Kleemann, S., O'Riordan, T., Jaeger, C. C. (2001). The Psychology of Denial Concerning Climate Mitigation Measures: Evidence from Swiss Focus Groups. *Global Environmental Change*, 11, 107-117.

Sunblad, E-L., Biel, A., and Garling, T. (2007). Cognitive and affective risk judgements related to climate change. *Journal of Environmental psychology, 27* (2), 97-106.

Suzuki, D. (2003). *A lifetime of ideas.* Vancouver: Greystone books.

Suzuki, D. and Knudtson, P. (1992). *Wisdom of the elders: sacred Native stories of nature.* New York: Bantam Books.

Weart, S. W. (2008). *The discovery of global warming* (2nd ed.). Princeton, USA: Harvard University Press.

Contributed by Monique Miotto, Victoria.

Photo. Mankind and Nature Existing in Harmony.

In: Meltdown: Climate Change, Natural Disasters… ISBN 978-1-60876-153-1
Editor: Kathryn Gow

Chapter 2

CLIMATE CHANGE AS CHALLENGE: CAN HUMAN RESILIENCE AND RESOURCEFULNESS PROVIDE SOLUTIONS?

Marek J. Celinski[1] and Andrzej R. Celinski[2]
[1]Workplace Safety and Insurance Board, Ontario
[2]York University, Toronto, Ontario, Canada

ABSTRACT

Naturally occurring catastrophes need to be differentiated from those that result from human activities which can be modified, or their negative impacts reduced, through awareness of the issues involved and via implementation of appropriate measures. In this chapter, we will focus on climate change as a human made disaster. It is an extremely challenging issue from the moral, spiritual, existential and emotional perspectives forcing people to examine their not fully conscious assumptions about themselves, their lifestyles, legal systems and their consumption habits. Furthermore, it forces nations to find a way to recognize the common interests of humanity as a whole and the ways in which international co-operation may become effective in dealing with climate change and other environmental challenges. Climate change, irrespective of its actual origin or threat to the survival of humanity, is a very powerful metaphor for enhancing international co-operation by appealing to the fundamental values and interests of humanity. Historic examples of collective efforts, sacrifices and profound changes in the habitual way of life in the name of human solidarity, that triggered human resilience and resourcefulness, offer hope that the current challenges are also manageable. The chapter ends with futuristic scenarios that may potentially develop, based on current decisions and on continuation of present trends.

Keywords: climate change, resilience, resourcefulness, solidarity, environmental challenges

INTRODUCTION

There is a growing body of research-based evidence that a host of continuing long-term processes cause global climate change and result in various undesirable socio-economic and health-related consequences. McMichael (1993) stated that:

In the past:

> The composition of the Earth's atmosphere has changed much over the past four billion years, particularly in response to evolution of human life...whereas those changes occurred on a grand time scale, human activities have very recently introduced further, rapid changes. Polar ice cores show that emissions of various greenhouse gases into the earth's atmosphere have increased substantially since the early industrial revolution 200 years ago, and markedly so since 1950. Methane concentrations have doubled since around 1800. Carbon dioxide concentrations have gone up by almost one-third, with well over half of that increase occurring since the 1950's (p. 137).

Currently:

> The recent, brief, energy-intensive phase of human history has entailed a hundred-fold increase in the global use of energy since around 1800. This rate of increase has surged since about 1950 causing an approximately hundred-fold increase in the rate of annual global production of carbon dioxide (pp. 138-139).

In the future:

> The main anthropogenic greenhouse gases are carbon dioxide, methane (sometimes called "marsh gas"), nitrous oxide and the entirely synthetic chlorofluorocarbons (CFCs)...Within 50 years, however, methane could become the main greenhouse gas because, as the climate warms, the voluminous peat-like permafrost in the sub-arctic tundra may release much more methane (pp. 137-138).

Similarly, the Union of Concerned Scientist (2005), in various reviews of this issue, recognizes the increased production of carbon dioxide due to burning fossil fuel and deforestation, and methane from agricultural activities, as creating a heat-trapping shield over the earth which is regarded as contributing significantly to global warming.[1]

The possible natural disasters caused by climate change include floods, heat waves and drought. Hulme (2003) explored the implications of abrupt climate change on society focusing on the United Kingdom and northwestern Europe. One consideration raised by Hulme refers to abrupt climate change reflected in extreme weather conditions, or accelerated unidirectional change in climate that require development of a strategy for likely long-term consequences. An increasing number of recent publications suggests that there is a fundamental relationship between the impact of climate change on the environment and both physical and mental health (Beniston, 2002; Morrisey and Reser, 2007; Sartore, Kelly and Stain, 2007).

---

[1] [See Chapter 9 for a further elaboration on the history of emissions and climate change.]

Kovats (2004) stated that climate hazards such as floods and heat waves would continue to increase in intensity and frequency as global climate change intensifies. Already it is known that heat waves are associated with significant excess mortality in the United Kingdom, especially with respect to elderly people. Kovats also believes that our capacity to address the health impacts of such events is limited and he postulated a need for research to improve the effectiveness of interventions in the near term, as well as to assess the adaptation strategies that would address climate change in the long term. Kovats considers that global warming creates unique and complex risks principally with respect to our planet's health, which are directly linked to human health and, in psychological terms, to mental health.[2]

From the Italian perspective, global warming has been found to have a link to an enhanced risk of suicide, as presented in data from Italy extending from 1974 to 2003 (Lentini and Maugeri, 2007). It was stated that for males, abnormal increases in monthly average temperatures was associated with higher monthly suicide rates from May to August, while the rates were lower from November to January. For Italian females, the links between temperature and suicide are less consistent. These findings provide a statistical example that increased temperatures and heat waves, presently associated with global warming, have adverse consequences on human mental health. Lentini and Maugeri (2007) postulate a need for communities to adjust to temperature changes by implementing public health interventions that would play an important part in preserving the wellness of the general population.

Curtis, Kvernmo and Bjerregaard (2005) studied the interrelationship between environment and societal changes presenting, as an example, life in a small community in Greenland where changing environmental conditions affected employment opportunities in the fishing industry to such an extent that the size of the population decreased dramatically. They further illustrate the relationship between rapid social, cultural and economic changes, and the health of the population linking psychosocial stresses with the transition to a new job, or migration.

Does the General Public Care?

From an evolutionary perspective, Sherman (2002) commented that natural selection puts a premium on thinking short term and having an emotional commitment to immediate living space and to a limited and cooperative group of relatives and friends. To describe the consequence of such an attitude, Sherman quotes Wilson (2002) who states that "We are innately inclined to ignore any distant problem not yet requiring examination" (p. 1467).

Sherman further comments that, for the most part, humanity has not yet grasped the magnitude of global warming and has not seriously addressed its long-term consequences. From the everyday perspective, the western family unit "cannot comprehend responsibility beyond their grandchildren" (p. 1467). Sherman agrees with Wilson (2002) in his recommendation that concern over environmental issues must start in the neighbourhood and attention should be directed to the problems that face the local community such as littering, pollution or forest destruction.

---

[2] [Grace Ip expands on the effects of climate change on health in her chapter (6) in this book.].

Regarding public perception and awareness of various environmental risks, as assessed by the French community (Fleury-Bahi, 2008), there are differences between country versus town residents and whether perceived risk is viewed as immediate or long-term. The risks which are difficult to conceptualize and which contain both high uncertainty and long-term consequences, such as climate change and loss of biodiversity, are perceived as less dangerous to one's self by the inhabitants of the town (as opposed to the inhabitants of the countryside) than more concrete and immediate hazards such as chemical spills. Such a difference in perception means that only technological and chemical hazards would be likely to trigger some environment oriented behaviour, whereas other risks (i.e., global warming) are perceived as quite distant (geographically or timewise), and as such would not require direct personal involvement.

Ignoring more distant or difficult to conceptualize problems is just one example of psychological factors, related to setting priorities and breaking with traditional thinking and ways of life, which would prevent individuals putting effort and money into solving the global warming crisis.

Regardless how distant or still uncertain the various consequences of global warming may be, there is a need to increase awareness of the potentially dangerous outcomes, if the current lifestyle related environmental trends are left to their natural course. The best that we can do at the moment is to focus on the already available measures that may help in preventing some of the long-term negative consequences of global warming; some ideas on other preventative actions are outlined in the concluding sections of this chapter.

Frumkin et al. (2008) specified the principles by which societies need to be guided regarding the consequences of climate changes due to global warming. One of these principles is prevention. Frumkin et al. listed among prevention those measures which, on one hand, slow down climate change and stabilize or reverse its course, and on the other encourage efforts at adaptation and preparation for the anticipated effects of climate change, thereby reducing the associated health burden. Climate change is expected to result in heat waves causing respiratory diseases related to increased ozone and pollen formation, such as asthma, allergies, bronchitis, etcetera and in drought causing water and food shortages, malnutrition and waterborne diseases. An extreme weather event may cause mass population movement and international conflicts. The set of practices that deal with these issues are collectively known as public health preparedness and risk management with systematic ongoing efforts to identify and reduce risks to health. Economic issues in public health planning also need to be taken into consideration as climate change will bring enormous costs including health care costs; therefore, mitigation and adaptation efforts may be far less costly if undertaken sooner, rather than decades into the future. Frumkin et al. stated that the cost of procrastinating may far exceed the cost of timely actions in both economic terms and health terms.

In order to focus on prevention, we emphasize that it needs to be made known clearly to the public that the negative impacts on the environment are the result of human activities which are under people's voluntary control. Therefore, as nations, groups and as individuals we have a responsibility to take care of the environment, as it affects all of us to some degree. It is human made disaster which implies tremendous moral consequences and yet it is the most insidious, as our routine daily activities do not make us fully aware of potential wrongdoing. The wrong doing is basically our lifestyle which is part of the culture and the socioeconomic system that requires promoting consumption to survive; the undesirable outcome, however, is that continuously increasing use of energy for production results in excessive industrial pollution.

Is Industrial Growth the Only Option?

Throughout history, especially in Western civilizations, one of the most spectacular developments that has occurred is based on people's realization that instead of being helpless victims of circumstances and subject to impacts from factors which are seemingly beyond our control (such as physical and biological laws, societal restrictions and our own physical and psychological weaknesses), people do have the capability to choose a path that could lead to changes in the socioeconomic reality of their lives, that better satisfy their basic needs. Rather than leaving things to chance or to the mercy of more powerful others, individuals and societies have developed more and more effective strategies that allow them to act with confidence that their action will likely bring desirable outcomes. The strategies change over time. In ancient times, the primary way of achieving desirable results was through collective efforts either by forcing (i.e., slavery) or encouraging others to work towards common goals; a period of Enlightenment brought into humanity's consciousness an even more efficacious way of achieving desirable results, which was through acquiring and utilising knowledge and reason, thus allowing individuals to perceive the world as an understandable and manageable place.

Subsequently, the industrial revolution, which started in Western Europe, opened people's lives to new opportunities and experiences that otherwise would not have been possible. Production of various goods helping people with daily living (i.e., availability of food and clothing through mass production, new means of transportation, such as railway and steamships, the discovery of quick ways of communication through telegraph, telephone and radio) not only satisfied people's aspirations for a better life, but created new needs and demands; furthermore, for those who were in the position to deliver the desired goods and to satisfy demands, new ways of amassing fortunes were created, rather than through plundering or stealing land. These days we benefit from even further accomplishments that have been made in all of these areas and specific advances in air travel and the internet (information highway) give us an impression that we live in a "global village".

As technological advancement provided better health care which decreased mortality, promoted growth in populations and created opportunities for satisfying people's needs by producing more with less cost and effort, a spiral of growth started: there were more people who needed work, other than in traditional occupations, such as in agriculture or trades, but new technology-based occupations in various industries could only sustain their usefulness, if their products were in demand. Therefore, people's consumption had to continuously expand to either better satisfy their basic needs or to create new and more sophisticated wants. Such a process worked quite well for most of the 20th Century for the majority of Western European countries, as well as for North America, Australia and New Zealand. However, there are natural limits on growth that are caused basically by two factors: the availability of resources and the impact of consumption on the environment.

Since its inception in 1968, the Club of Rome[3] raised disturbing questions about the impact of industrial growth on the environment and published a report in "The Limits of Growth" (Meadows et al., 1972) which outlined the consequences on the lives of future generations, associated with: (a) the depletion of resources; (b) the probable impact of limited resources; and (c) the increasing population. The observed temporary drastic increase in the price of oil in 2008 is likely a harbinger of things to come. Nevertheless, even a short period of price increases caused a change in the way in which people make decisions about purchasing their vehicles and the means of transportation they choose to utilize. These events and decisions brought the North American automotive industry to the brink of bankruptcy.

From a social and psychological perspective, the situation that developed over the last two centuries represents the conflict between the lifestyles that we have managed to achieve through rapid industrialization and availability of goods which, on one hand, has a damaging impact on the environment, but, on the other hand, provides employment and satisfies the needs of people who are the purchasers of these products.

In attempts to address this dilemma, primary consideration has to be given to those industries which are energy-intensive and which are also producing pollutants that ultimately create global warming. Depending on the analysis of which industries are particularly responsible for contributing to global warming, there would be a pressure on this industry to regulate itself and/or to be regulated. If current trends continue, various industries in North America would be affected even further, requiring major restructuring in years to come. The reliance on cars as the means of transportation created tremendous growth in the related sectors that provide raw material, steel, build bridges and roads, and provided employment opportunities to communities established around factories. If the automobile industry, a major economic force in North America, is compelled to implement dramatic changes, it will have serious consequences for other related industries.

There is growing tension between the interests of the industries that are primarily responsible for global warming and the need to contain the negative impacts of their productivity on the environment. Governments will have to balance the impact of reduced industrial activity on the general economy against the costs of social and economic consequences resulting from unemployment. It is one of the primary responsibilities of any government to create favourable conditions for the economic survival of those who depend on hired labour. At the same time, the governments will be pressed to introduce mandatory measures that to some degree decrease productivity and/or the degree of pollution. As this occurs, we will be living in an increasingly restrictive political environment in which various activities that contribute to pollution or destruction of the environment would be heavily regulated. Moreover, if such policies create significant increase in unemployment, the governments will have to face potential social unrest and significant budgetary deficits caused by supporting the most affected groups. With the global financial crisis, such policies will need to be tempered with great wisdom.

Another example of the probable impact of environmental concerns on the present lifestyles in the industrialized world is with respect to suburban communities. At least in North America, where vast land is still available, suburban life has become a very desirable way of living. However, it requires means of transportation to and from the workplace. Unless other means of transportation are created, people will still have to depend upon cars to

[3] http://www.clubofrome.org/eng/home/.

travel to their employment sites. An undesirable outcome of individuals using their own cars for transportation has been the environmental pollution and rapid expansion of the cities that consume forests and farmland. Eventually, the increased commuting distance will make suburban life less affordable and desirable, and expansion will have to be stopped, unless suburbs create industries and economic opportunities of their own. If local employment is not available, this could create higher unemployment in many sectors, such as construction, and hasten the disintegration of suburban communities that have spread too far away from industrialized areas. Rural areas are also very sensitive to price rises in oil; if these prices rise, it may cause many small town communities which are situated at a substantial distance from large towns and cities to fail and close down.

The need for lesser reliance on individual means of transportation and for limiting expansion of big cities will likely be reflected in new legislative policies and business planning which take into consideration the environmental consequences of increased pollution and the destruction of the natural environment. If implemented, these will result in changes to our current lifestyle.

The issues that concern North Americans, Europeans or Australians are not, however, shared to the same degree by people in developing countries who aspire to achieve the standard of living already available in developed countries. China and India are by far at the top of the list of those developing countries which pollute the environment; Zissis and Bajoria (2008) chart China's costs of growth and industrialization in terms of water pollution, desertification and greenhouse gases.

It is likely that the developing countries would consider it to be unfair, if international pressure is put on their respective governments to prevent them from achieving a better standard of living for their people. There remains an obvious question as to who should sacrifice more, in order to create better conditions for everybody. Yet another issue will be how to deal with people who, as a result of droughts or floods or tornados, become displaced or deprived of opportunities to survive. If this happens in industrialized countries, chances are that there would be means to rebuild and feed people who have lost their jobs and homes. There are, however, already millions of people in Africa who are affected by famine and political infighting; they are deprived of sufficient help, and left to starve and die in dire conditions.

Patz, Engelberg and Last (2000) noted that "The Sahel region in sub-Saharan Africa is particularly vulnerable to the effect of drought caused by increasing desertification, which has been exacerbated by over cultivation and overgrazing. In 1973, 100,000 people died as a result of drought in the Sahel. Prolonged and more frequent periods of drought associated with climate change could have widespread consequences for population health in the Sahel and other parts of the world threatened by desertification. Drought-induced wildfires can cause direct injury and have the potential to affect air quality. Fire smoke carries a large amount of fine particles that exacerbate cardiac or respiratory problems" (p. 285).

These considerations behove us to ask to what degree countries and communities would be able to cope on their own and how much international cooperation would improve their plight, considering that the need for assistance will increase far beyond the current level.

Hulme (2003) commented that the limits of adaptation in some sectors or regions may be reached and the cost of appropriate adaptive behaviour may be large, but strategies have to be developed on the basis of predicted long-term unidirectional changes in climate. The degree to which the world economies and governments can prepare themselves for such outcomes, on a national or international basis, needs to be evaluated and procedures must be introduced to cope with the consequences.

It is quite probable that when the threat becomes evident, to prevent a catastrophe, more restrictive measures would have to be introduced at national and international levels. The prospect is that the United Nations would need to play a greater role in developing policies, contingency plans and procedures and to coordinate international cooperation to assist regions struck by natural disasters which jeopardize peoples' physical and/or economic survival. The alternative is chaos and the 'invasion' of intact regions by those who become seriously affected and/or displaced.[4]

Climate Change and People's Resilience and Resourcefulness

Resilience: A Sense of Identity and a Will to Survive

On the positive side, climate change, to the degree that reflects humanity's aspiration for better living, is also a metaphor that should inspire a collective effort at containing its negative impact on the environment and our ability to survive. Resilience refers to an individual's ability to withstand pressure and traumas without suffering serious long term personality disintegration and/or emotional psychopathology such as anxiety or depression, and without developing long term undesirable maladaptive reactions. The concept of resilience is gaining recognition as an important quality of mind and character (Celinski and Pilowsky, 2008; Rutter, 2006; Zautra et al., 2008). According to Celinski and Pilowsky (2008), it is based on a sense of inner cohesiveness of cognitive and emotional faculties that are associated with efficacy to achieve certain goals in accordance with how people understand themselves, and their role in the promotion of their personal development and in the further evolution of humanity. For the first time in human history, there is a global threat to humanity at large. The threat crosses national boundaries, but it is possible to predict which geographic regions may be more or less affected. It also crosses various cultural and religious barriers; on one hand, this may cause more divisions between people, as there is the likelihood that some regions will be more affected than others; on the other hand, because there is a probability that every place on earth may in the long run be affected to some degree, international co-operation of unprecedented proportion may develop.

The resilience and resourcefulness model follows principles derived from clinical observations and rehabilitation of individuals affected by traumas and life adversities; we believe that the diagnostic and recovery model that we have found clinically useful may also be applicable to the global issues that require mobilization of resources and seeking and finding solutions.

The major components of the resilience model (see Celinski and Pilowsky, 2008) take into consideration the perceived (or reasonably expected) severity of the trauma and an individual's self-confidence in the ability to handle stressful situations and adversity of life for the sake of emotional stability and continuity; resilience further requires proactive

[4] [Workineh Kelbessa speaks about these issues in his chapter in this book.].

engagement, rather than withdrawal. Additional traits are associated with the people's acceptance of themselves as imperfect and vulnerable individuals, but also with a belief that we are all valued people whose lives are worth living, regardless of our objective circumstances and personal conditions. Therefore, resilience requires the discovery of the personal meaning of life and having a belief that our efforts matter, and also accepting that suffering or sacrifices, either for self-advancement or for the common good, are indispensable parts of life and the condition for progress.

The resilience model further postulates that, whereas reality must be assessed from the point of view of danger, the ultimate response should be inspired by a positive bias that the situation is understandable and potentially manageable, and that our efforts will bring desirable outcomes. Such an assumption is sometimes more 'wishful thinking' than 'realistic judgment' but, at the same time, it gives people hope and directs efforts at finding solutions. People are asked to draw on their past experiences and evaluate how they have successfully overcome various adversities in life which, in turn, prepared them for facing the current situation with a confidence that solutions will be found, as they have been found in the past. In essence, when there is adversity that threatens our lives, physical integrity, way of life or well-being, the optimal response is to focus on momentary survival and on hope in the ultimate will of humanity to continue its existence, and in its wisdom to use individual and collective abilities to solve even the most difficult problems.

The most relevant situation from the past that resembles current and future threats to survival may be found in the period between 1948 (the beginning of the "Cold War") and the collapse of communism in the Soviet Union in 1991. Every day during this period, the world faced major consequences in terms of nuclear war that threatened the survival of humanity (or at least survival of the nations that potentially would be the target of the attack which limited the scope of the disaster). In that period, artificial divisions across Europe and in other parts of the world were created between neighbouring nations and even within the nations (e.g., Germany, Vietnam and Korea). The strategy at that time was to survive (i.e. continue regardless of the circumstances) and to contain or minimize the negative impact created by the threat of the nuclear exchange.

Let us take into consideration the perspective which had to be taken during the Cold War period; even though it was not entirely consciously formulated by millions of people, they were aware, at least to some degree, of the disastrous consequences from a nuclear attack. Nevertheless, they remained hopeful that by committing themselves to the continuation of life and being inspired by values such as democracy and freedom of expression, economic prosperity and human rights, gradual changes behind the "Iron Curtain" would be possible. Eventually, human rights presented as the innate and unalienable characteristics of human existence, brought peaceful changes in Middle and Eastern Europe, and transformed these nations in accordance with the best humanistic traditions. In Europe, this tradition has its roots in the great thinkers of ancient Greece who formulated principles of rational thought. Another important source of inspiration that contributed to Western civilization came from Roman Law that protected individual possessions and made Roman citizens equally subject to legal standards. Probably most noteworthy was the impact of Christianity on the social, political and cultural evolution that reflected the necessity of caring for others, as the principle of social justice and foundation of social order. While these values linked the nations of Europe, there were other factors that created deep divisions, related especially to rivalry between the major European powers and to their tendency to dominate and expand.

By referring to common values which should overcome deep historic divisions, the European Union was established. In the founding declaration that was signed by historic enemies, France and Germany, there is an explicit reference to human solidarity. On May 9, 1950, French Foreign Minster Robert Schuman (Schuman, 1950) proposed that the coal and steel industries of France and West Germany be placed under a common higher authority. The realization of this proposal led to the creation of the European Coal and Steel Community (ESCS) which is the forerunner of what is now the European Union (EU). It is specifically stated in the declaration that:

> Europe will not be made all at once, or according to a single plan. It will be built through concrete achievements which first create a de facto solidarity. The coming together of the nations of Europe requires the elimination of the age-old opposition of France and Germany. Any action taken must in the first place concern these two countries.
>
> With this aim in view, the French government proposes that action be taken immediately on one limited, but decisive, point: it proposes that Franco-German production of coal and steel as a whole be placed under a common High Authority, within the framework of an organization open to the participation of other countries in Europe.
>
> The pooling of coal and steel production should immediately provide for the setting up of common foundations for economic development as the first step in the federation of Europe, and will change the destinies of those regions which have long been devoted to the manufacture of munitions of war, of which they have been the most constant victims.
>
> (Schuman, 1950)[5]

The declaration created an opportunity for France, Italy, West Germany and the Benelux countries (Belgium, Netherlands and Luxemburg) to share strategic resources in order to make war, not only morally unacceptable but also materially impossible, and to build a lasting peace in Europe. The ultimate outcome of this initiative was the creation, on the 18th April, 1951, of the European Coal and Steel Community (ECSC), the first of the European communities and the predecessor of the European Union.

Contemporarily, the hostile trends that dominated European life for so many centuries have generally changed and evolved into a decisively more co-operative political and economic union.

The example of post World War II Western European successful integration, inspired by the common European tradition and ideas of democracy, liberalism, human rights and economic freedom united people in the spirit of common values and led to the creation of the Solidarity Movement in Poland. The changes followed a shipyard worker's strike in Gdansk in August, 1980. By the beginning of the 1990's, the former Soviet Union did not exist and the former Central and Eastern European satellites regained their independence hoping to join NATO and the European Union (these goals have already been achieved for the vast majority of these countries). The Solidarity Movement was based on an appreciation of life and on the individual's and the collective's responsibility for betterment of the human condition from psychological, social and economic perspectives. It drew its primary strength from the sense of unity with other people who shared a similar plight and who were prepared to offer mutual support, and even to sacrifice their lives for such a good cause. At a time when the common

[5] http://www.robert-schuman.eu/declaration_9mai.php.

adversary was clearly recognized, the Solidarity Movement embraced more than 10 million people in a population of less than 40 million.

It is amazing what this movement achieved, in a peaceful way, in spite of the logic of socioeconomic and political conditions which made people prone to believe that they would remain unchanged, regardless of what people wish, think and do. The Soviet Union, having been a "guardian" of the communist ideology and the status quo, at that time still enjoyed the position of a world superpower and possessed the military capability that had, in the past, been used to squash any aspiration for democratic changes in Hungary (1956) and Czechoslovakia (1968). In a similarly peaceful way, India regained its independence from Great Britain on August 15, 1947 which unfortunately was followed by violence between Muslims and Hindus; however, later on, after having achieved their freedom, they have progressed to global power status.

It is perhaps too early to expect that, in the near future, a potent movement to save lives on our planet will emerge on a similar scale against the logic of the day, lifestyle comfort, and societal and individuals' priorities. A French study (Fluery-Bahi, 2008) refers to natural catastrophes associated with climate changes as still being at a low level of people's awareness and as not yet representing such a degree of threat that sacrifices and action would be required. To mobilize people to act more responsibly for the sake of their environment, any future more efficacious environmental movements need to appeal, on a mass scale, to a deep appreciation for life in general and to emphasise the uniqueness of individual life, as a product of evolution from biological to spiritual levels, and stress the individuals' roles and responsibilities to contribute even in a small way. Such a movement should also inspire confidence in the manageability of the problems, based on solidarity with all the people and creatures on the planet.

As it is quite likely that various geographic regions and nations, and certain social groups may be affected more than others, a notion of human solidarity is particularly important for the success of future scenarios that potentially may offer solutions to problems on a global scale. For this purpose, people have to be convinced that they have good reasons for working together. In this respect, an appeal to ethical norms and cross cultural social sensitivities may be helpful as discussed now.

Frumkin et al. (2008) refer to the three concepts which guide public health ethics: utilitarianism, liberalism and communitarianism. *Utilitarians* postulate that the net sum of people's well being, especially when future generations are considered, will likely increase, if the health impact of climate change is controlled. *Liberal* analysts would postulate a right to a healthy environment and would therefore support policies and practices that prevent environmental degradation. *Communitarians* would argue that climate change undermines the requisite conditions for intact social order based on moral and legal principles and will cause all people to suffer. This latter concept is closest to what we propose as a more unifying idea of human solidarity which refers to people's mutual support for defence of common values and for the continuation of humanity, while facing global and serious adversity.

Frumkin et al. argue that there is a need for a proper data gathering system to monitor climate change and the accompanying environmental and disease risks, while identifying and investigating these problems through the health system. A key focus should also be on informing, educating and empowering people about health and other issues associated with climate change.

In our opinion, all of these strategies require implementation not only across regions, but internationally. From a social and psychological perspective, making any positive impact with respect to global warming requires a special collective state of mind that does not have any parallel in history. Never in recorded history could humanity, as a whole, be potentially affected in a similar manner; consequently, never before has the prospect of the dangers that we face required such a degree of social awareness, sense of responsibility, international cooperation and sharing of both human and material resources. It is indeed a very crucial moment for humanity, because this time, the enemy which unites us is not another nation of oppressive ideology, but is clearly within us. It is our lifestyle and our desire to continue with things as they are because they give us, at the present time, a greater level of comfort.

Situations can only be changed, if the principles which operated for an extended period of time and brought a certain degree of stability to the world, or parts of the world, are recognized as no longer leading to the same desired consequences. The continuation of the status quo is associated with a prospect of economic collapse and social disorder. We should accept anxiety caused by such a perspective as the inspirational force from which we either develop a new awareness of ourselves, or otherwise short-sightedness and hence the promotion of national or group interests will prevail. This new sense of responsibility puts a tremendous burden on individuals and society with respect to who will endure the costs and to what degree. In order to achieve any progress in this respect, we need to create meaning and clear purpose for our efforts (Langford, 2002) that will make the sacrifices that must be made more bearable. Langford quotes Maddi (1967) who defined existential neurosis as the chronic inability to believe in the truth, importance and value of anything one does or can imagine doing. Langford postulates that facing danger can provide meaning in that focusing on certain aspects of risk and provision of some solutions can give purpose to life.

Thus, the current situation challenges our passivity associated with what psychologists understand as the concept of external locus of control; sometimes we do not act, because we believe that powerful others or external forces determine our behaviour (Bandura, 1986; Rotter, 1966). This may lead to denying self-responsibility by stating that individuals cannot do much because of the enormity of the issues. Langford (2002) distinguishes those who are willing and keen; they both show a high degree of personal responsibility and belief in their obligation to do the right thing personally with individual action. The fear of uncertainty causes the willing to lose some motivation for taking action based on their beliefs, but the keen focus on using collective efforts to bring about change in both business and government. Overall, in Langford's study, fear of death and threats to survival were assuaged by a belief in communal efficacy linked to the possibility of social change based on education and political pressure. Langford concluded that only by providing people with a genuine chance to understand, have hope, and belief in the possibility of instigating change, will communication strategies effectively inform the public about various environmental hazards. Otherwise, fear of the unknown, alienation, helplessness and reactions to this state of mind will always win the day.

Resourcefulness: The Skills to Succeed

Another concept derived from an individual perspective, but which needs to be utilized on a mass scale, is that of personal resourcefulness (Celinski and Gow, 2005) which is comparable to material resources (Hobfoll et al., 2003).

Analyzing further the psychological aspects of global warming from the individual perspective, one can benefit from clinical and rehabilitation practice in which reference to positive psychology and resourcefulness prevents people from developing psychopathology and also enables people to recover, when they are already affected by adversity. The proposed model of resourcefulness refers to the multi-level concept of freedom (i.e., a personal ability to generate, in an autonomous way, some causative action that provides an impetus for the individual or external world) to provide some solutions, depending on the level of autonomy, awareness of the issues and the levels of expected control while keeping in mind that the intention is to achieve certain well delineated outcomes. Rollo May (1981) quoted Bertrand Russell as saying that "in the realm of the mind there are no limits" (p. 62).

In dealing with various difficult situations in life, we may utilize various levels of freedom that corresponds to autonomous activities; these represent various states of mind (Celinski, 2003, 2004) and abilities to act in a chosen way. They include:

- *Helplessness:* Perceived total inability to change an unfavourable subjectively experienced condition or objective circumstance into a more desirable one.
- *Procedural Freedom:* Minor variations in otherwise instinctual, habitual or stereotypic behaviour.
- *Intentional Freedom:* An ability to generate ideas and wishes without implementation.
- Sub-Optimal Freedom: Exercising freedom on a level less than optimal to one's peers', thus resulting in underachievement.
- *Optimal Freedom:* Accomplishing results in a manner currently viewed as most optimal, by utilizing typical approaches in accordance with societal norms and established traditions.
- *Sovereign Freedom*: Autonomy to choose goals and implement them which may be different or contrary to the prevailing political or cultural views. On this level, the underlying principle is to change the existing relations with the group that is chosen for comparison, so that one's own activity better reflects an individual's objectives, sometimes against the restrictions imposed by norms, opinions or established procedures.
- *Absolute Freedom:* A sense of total autonomy and power over certain aspects of objective reality. The underlying principle is to change the nature of the process involved, so that it better suits an individual's objectives (an example may be merging plants' and animals' genetic materials which does not occur in the natural environment, or utilizing technological processes that repeatedly lead to the same predictable and desirable results).
- *Ultimate Freedom:* Doing good things because they are good and thus inspirational and rewarding by themselves, without expecting immediate results or rewards (i.e. engaging in research, education, diet, exercise, etc.). A metaphor for this is "planting

seeds and not knowing what will grow out of it" (i.e., size of yield and/or how the crop could be useful).

- *Victimization*: This has double meaning; if associated with helplessness, it may result in a state of emotional numbing and "paralysis." By contrast, having nothing more to lose and being freed from the usual bonds and preoccupations pertinent to the other levels of freedom, such a state of mind may trigger extraordinary acts of courage, flights of imagination, creativity, generosity, kindness, inner peace, and so on.

Based on this conceptualization, there is a potential for addressing the global warming crisis in diversified and creative ways. For example, the optimal level of freedom may refer to improving the current utilization of energy so that it produces the least possible amount of pollution. In this context, frugality is considered (Sachs, 1999) as the primary solution for the expansive utilization of resources and the need for continuous growth as the means to combat poverty.

The sovereign way of approaching global warming would be to overcome routine and stereotypic ways, currently commonly accepted, by introducing alternatives, an example being less reliance on environmentally detrimental energy sources by introducing alternatives (i.e., wind and solar power). In this way, we not only utilize natural resources, new technologies and industrial potential in a more efficient and frugal way, but we address energy problems frequently against the well established status quo. This may also go against the interests of certain industrial sectors.

Absolute freedom means changing the nature of the processes involved in the problems we intend to solve, in order to maximize the desired outcome. To act on this level, it is necessary to change the principles of industrial productivity (i.e. expansion and profit), people's appetites for continuous improvement of their personal standard of living, as well as the nature of the physical processes involved.

Proposed solutions from these perspectives would require that an economy based on expansion and growth (being the essence of the free market) be replaced by the concept of production for only necessary sustenance, and that peoples' appetite be curtailed to a level consistent with reduced production quotas which probably may only be achieved on a global scale through autocractic or totalitarian measures. Applying the concept of absolute freedom to technology postulates that only those processes be developed and utilized which, by their very nature, would not substantially contribute to further pollution; an example is hydrogen power which produces energy with water as its output.

The introduction of such measures would obviously require that citizens give their governments greater legislative authority and funds, while governments would do the same for international institutions, to make it possible for these measures to be implemented.

However, we must also remember that ultimate freedom and the prospect of victimization will act as inspirational forces to propose even more daring, creative and innovative solutions that we cannot even imagine at this time. It has always happened at a time of crisis, be it political, economic or social, that extraordinary ideas emerged and charismatic individuals presented solutions that would invigorate and unite people for the common good, in spite of sacrifices that needed to be made. Let us believe that this will also be the case, with respect to environmental protection, and hopefully avert global catastrophe.

As an alternative possibility, let us consider some imaginary futuristic scenarios as an extension of the argument that global warming is potentiated (or primarily caused) by human based activity and that, if it continues, it will result in undesirable social and psychological consequences; these would jeopardize further development of humanity, in accordance with its best traditions and/or it will cause disintegration of the present social order and revert to previous developmental stages.

Scenarios for the Future

The dire scenarios outlined below represent likely consequences of global warming; these should trigger anxiety that, in turn, should lead to positive action (rather than cynicism, withdrawal and helplessness).

1. Economic consequences associated with increased health costs, people not being able to work, shortages of food and potentially social disorder due to spread of illness, loss of farmland, and lower industrial and agricultural productivity, unemployment and forced migration.

2. Migration which is out of control, due to droughts or floods in some areas or due to infection and economic disaster as a result. This may also create social and political instability, and potentially start either aggressive or defensive wars.

3. Threat to democracy, as change in climate will likely necessitate introduction of restrictive and drastic measures related to people's movement, economic activity, food, goods and services rationing (i.e., medical), and restriction on political and personal freedoms, as a way of coping with social discontent and protests. These drastic measures may be required, both at the national or international levels, and they would threaten the achievements especially of the prior 50 years, with respect to broadening democracy and enhancement of human rights that have increased the sense of human value and dignity, and that have opened a variety of choices and opportunities for individuals. Economic opportunities, combined with democracy, that allow for people's mobility, independence and self-responsibility will be reversed, if the vast majority of the population is made much more dependent on government policies, social agencies and are restricted in their choices.

These are pessimistic outcomes, but, nevertheless, are possible. Awareness of these issues should give us all sufficient impetus to introduce preventative strategies which would slow down, stop, or reverse the negative environmental changes.

CONCLUSION

The potential for 'big scale' disaster, caused by global warming, that could be precipitated or augmented by who we are as people and what we do to satisfy our aspirations, needs and wants, poses an existential question: is our own nature programmed for self-destruction, or we are able as humanity, to prevent these scenarios from happening through appealing to what represents a higher awareness of ourselves and our social history, and an appreciation of the role that humanity plays in the world's evolution? Prevention requires tremendous investment of hope, effort, economic resources and personal resources, along with political will stemming from the belief that humanity, as a whole, is worth its defence and its future continuity. Furthermore, we should act with a hope that we have the intellectual and emotional capabilities to cope with such a challenge. Ultimately, awareness of these issues forces us to consider the most profound existential and ethical issues which humanity has ever faced: what will be the cost of our survival, in terms of economic prosperity, comfort and traditional ways of thinking and living? What should we defend and retain at any cost, and what can we compromise on? What do we value in life and want to retain and improve?

From the authors' perspective, in this chapter, we attempted to outline those human qualities (solidarity, resilience and resourcefulness) that may give us hope, confidence and measures that enable us to face and deal successfully with the environmental challenges facing the world.

References

Bandura, A. (1986). *Social Foundations of Thought and Action: a social cognitive theory*. Eaglewood Cliffs: Prentice- Hall.

Beniston, M. (2002). Climatic Change: Possible Impact on Human Health. *Swiss Medical Weekly*, 132, 332-337.

Celinski, M.J. (2003). Cultural Experience, Naturalistic Hypnosis and Therapeutic Outcomes. *Hypnos*, 30, 75-84.

Celinski, M.J. (2004). Freedom as a Mental State. *Hypnos*, 31, 189-200.

Celinski, M.J. and Gow, K.M. (2005). Trauma Clients: How Understanding Disintegration Can Help To Restore Resourcefulness of the Self. *Australian Journal of Clinical and Experimental Hypnosis,* 33, 195-217.

Celinski, M.J. and Pilowsky J. (2008). Measuring Resilience to Psychological Outcomes of Natural Disasters: More Questions than Answers. In K. Gow and D. Paton (Eds.), *The Phoenix of Natural Disasters: Community Resilience* (pp. 73-86). New York, N.Y.: Nova Science Publishers Inc.

Club of Rome. http://www.clubofrome.org/eng/home/

Curtis, T., Kvernmo, S., and Bjerregaard, P. (2005). Changing Living Conditions, Lifestyle and Health. *International Journal of Circumpolar Health*, 64, 442-450.

Fleury-Bahi, G. (2008). Environmental risk: perception and target with local versus global evaluation. *Psychological Reports*, 102, 185-193.

Frumkin, H., Hess J., Luber G., Malilay J., and McGeehin, M. (2008). Climate Change: The Public Health Response. *American Journal of Public Health*, 98, 435-445.

Hobfoll, S.E., Johnson, R.J., Ennis, N. and Jackson, A.P. (2003). Resource Loss, Resource Gain, and Emotional Outcomes among Inner City Women. *Journal of Personality and Social Psychology*, 84, 632-643.

Hulme, M. (2003). Abrupt Climate Change: Can Society Cope? *Philosophical Transactions: The Royal Society of London, 361*, 2001-2021.

Kovats, R.S. (2004). Will climate change really affect our health? Results from a European assessment. *Journal of British Menopause Society*, 10, 139-144.

Langford, I. H. (2002). An Existential Approach to Risk Perception. *Risk Analysis*, 22, 1, 101-120

Lentini, P. A. and Maugeri, M. (2007). Global warming possibly linked to an enhanced risk of suicide: Data from Italy, 1974-2003. *Journal of Affective Disorders*, 102, 19-25.

Maddi, S. (1967). The Existential Neurosis. *Journal of Abnormal Psychology*, 72, 311-325.

May, R. (1981). *Freedom and Destiny*. New York, N.Y.: W.W. Norton and Co.

McMichael, A. J. (1993). *Planetary Overload: Global Environmental Change and the Health of Human Species*. New York, Cambridge University Press.

Morrisey, S.A. and Reser, J.P. (2007). Natural disasters, Climate change and mental health considerations for rural Australia. *Australian Journal of Rural Health*, 15, 120-125.

Meadows, D.H., Meadows, D.I., Randers, J., and Behrens, W.W, III. (1972). *The Limits of Growth. A report for the Club of Rome's Project on the Predicament of Mankind.* New York: Universe Book.

Patz, J.A., Engelberg, D., and Last, J. (2000). The Effect of Changing Weather on Public Health. *Annual Review of Public Health*, 21, 271-307.

Rotter, J.B. (1966). Generalized expectancies for internal versus external locus of control. *Psychological Monographs*, 80 (1), whole issue.

Rutter, M. (2006). Implications of Resilience Concepts for Scientific Understanding. *Annuals of New York Academy of Science*, 1094, 1-12

Sartore, G.M., Kelly, B., and Stain, H.J. (2007). Drought and its effect on mental health. How GPS can help. *Australian Family Physician*, 36(12), 990-993.

Sachs, W. (1999). *Planet Dilectics. Exploration in Environment and Development*. Halifax, N.S.: Fernwood Publishing.

Schuman, R. (1950). Robert Schuman's May 9, 1950 Declaration. (http://www.robert-schuman.eu/declaration_ 9mai.php)

Sherman, D. (2002). Time and tide wait for no man. *British Medical Journal*, 325, 21-28.

Union of Concerned Scientist (2005). http://www.ucsusa.org/global_warming

Wilson E.O. (2002). *The Future of Life*. London: Little, Brown.

Zautra, J., Hall, J.S. and Murray, E., and the Resilience Solutions Group. (2008). Resilience: A new integrative approach to health and mental health research. *Health Psychology Review*, 2(1), 41-64.

Zissis, C. and Bajoria, J. (2008). China's Environmental Crisis. *The Council on Foreign Relations* (updated August 4, 2008). http://www.cfr.org/publication/12608/

Contributed by Tim McCrorey, Toowoomba.

Photo: Nature's Built-in Resilience.

In: Meltdown: Climate Change, Natural Disasters… ISBN 978-1-60876-153-1
Editor: Kathryn Gow

Chapter 3

EXPLORING THE PSYCHOLOGICAL ASPECTS OF RISKS, FEARS AND CONCERNS ABOUT CLIMATE CHANGE

Kristina Searle and Kathryn Gow
Queensland University of Technology, Australia

ABSTRACT

Theories about risk taking and decision making may explain why many people do not take action on climate change scenarios, but there are a range of other factors that would need to be taken into consideration in order to persuade people to change their behaviours in relation to environmental preservation and wastage of resources such as water and energy. Reports indicate that some people express great concern and are angry and distressed about the future in general and about climate change issues in particular. In this chapter, we outline some of the theories about beliefs and emotions that might help us to understand who are worried and why. Comments are included from a series of exploratory interviews that were conducted as a component of research that aimed to investigate the psychological impact of climate change. At the end of the chapter, we also report on the preliminary findings of a survey on psychological perspectives about fears and concerns for the future among a group of university students.

Keywords: climate change, future fears, risk theory, psychological impact; environmental concerns

INTRODUCTION

For me it is the ability to say with any confidence, how much and how quickly the ice from the great ice sheets (Greenland and the Antarctic) will melt and slide into the ocean.
(Professor Chris Rapley, 2007)[1]

Risk and Change

Risk society (U. Beck, 1992) would suggest that climate change and environmental disasters are side effects of industrialisation and modernisation that have led to a collective uneasiness and social anxiety. Public opinion polls indicate that this is true. Research also indicates the existence of an apocalyptic sensibility in relation to environmental beliefs that Tolsma (1968) would argue can be traced to existing pathology.

Within the Australian community, there has recently been a call to focus on global climate change issues in light of the relative neglect shown towards what has been referred to as one of the greatest challenges of our time (Breckler, 2007). With public opinion polls and the media indicating concern within the general population and, for some, a rising concern about global warming, it appears that further investigation is warranted.

It may well be that individuals' reactions to significant problems such as climate change scenarios are similar to those about other major dilemmas about which they feel powerless to do anything that could avert the course of the promulgated disaster. The size, severity and global nature of the catastrophe may render them powerless, as has the felt powerlessness (Gow, 2002a, 2002b) that ordinary citizens of the world endure regarding their inability to assist in situations where millions of people have died from famines in Africa and Asia, and holocausts in countries such as Germany, Poland, Rwanda, Cambodia and Russia.

The lack of plans and goals, the absence of logical strategies for problem solving and clear solutions, in relation to what is being portrayed in the media as insurmountable problems, can only heighten the anxiety of people who believe that: (1) the world is coming to an end based on the media reports; (2) spiritually, this is inevitable as it has been foretold through sacred scripture or prophecies; (3) from a moral perspective, the earth and its people are being punished for wrong doings following interference with nature's ecosystems; and/or (4) as humans, there is nothing that we can do to prevent the inevitable negative outcomes of global warming, whether man-made or by planetary or geological causes.

In light of Beck's (1992) ideas about Risk Society, concern for climate change can be conceptualised as a crisis of modernity. Media headlines such as "Pulling back from the brink of destruction", "Apocalypse soon" and, from Time magazine, "Be worried. Be very worried" (Castles, 2006) convey a sense of panic about how climate change is impacting the Earth. Despite the collective uneasiness these statements reflect, little research has been conducted regarding the psychological impact of climate change within the general population. However, public opinion polls and surveys suggest that climate change is of great concern, more so than terrorism and other global issues; however these took place before the 2008/2009 global financial crisis. Prior to that time, research on fearful future outcomes also

[1]http://www.sciencepoles.org/index.php?/articles_interviews/professor_chris_rapley_taking_stock_on_climate_change/&uid=990&type=15.

highlighted environmental issues and hinted at the existence of an 'apocalyptic sensibility' (J. Gow and Leahy, 2005).

Risk Society

During the late twentieth century, new social anxieties have emerged in advanced industrialised countries that include environmental, nuclear, chemical, biological and medical issues (Ungar, 2001). Beck (1992) conceptualized these social anxieties as a result of what he termed a risk society. According to Beck (1992), industrialisation and modernisation has led to side effects that produce global risks and the potential for catastrophe. These risks are very complex in their causation. They are the product of human decisions and have led to advances in science, technology and economic gain. They are also undetectable to our physical senses, unpredictable and incalculable (Ungar, 2001). A risk society therefore produces wide spread worry and uneasiness that, at the extreme, contributes to a perception of a catastrophic society that is doomed by impending disaster (Welch, 2006).

Concern for climate change and current environmental issues can be conceptualised as a crisis of modernity where the boundless growth of demand clashes with the limited capacities of the natural environment (Gow and Leahy, 2005). Catastrophic risks, in the form of global warming and increased natural disasters, are perceived as a result of the very processes that create modernity and affluence (Gow and Leahy, 2005). For example, scientists propose that through industrialisation, a process which has necessitated the burning of fossil fuels, modern society has released thousands of tons of carbon dioxide and other man-made gasses into the atmosphere leading to an increase in the concentration of gases that trap heat and as a result, an increase in the warming of the earth's surface (Australian Greenhouse Office, 2007).

The idea that climate change and the issues surrounding climate change have resulted from advances in technology and the demands of modern living was indicated in several interviews that were conducted as a part of a study undertaken by the first author. Comments from university students reflected a sense of interference with, and abuse of, nature for the sake of material gain and technological advancement; for example:

- I think climate change is a sign that man has been interfering with nature too much and that we haven't balanced our resources properly and now that we're starting to run out of things, everything is going to go haywire.
- I think people have become too focused on what money they can make, on all the products they can buy, and they didn't think about how pollution would affect the world.
- Humans have destroyed the environment for their own benefit.
- I think we really abused what we had (in nature) and I think the human race has become very lazy; we use cars and have introduced things (new technology) that takes short-cuts and nowadays we're in a perpetual trap and we don't know how to do anything without it. And I think everyone knew years ago what the effect would be and they didn't stop, so it's sort of their own fault. I think we deserve everything climate change throws at us.
- The environment is slowly deteriorating, we're ruining it.

Written comments from the survey reported later in the chapter support these views with references to climate change occurring as a result of “modern convenience”, “over consumption”, “modernisation”, “industrialisation”, “human greed”, “abuse of nature”, “human development”, unnecessary methods of production”, “overuse and misuse of natural vital resources”, and “man’s need to control nature”.

It is widely believed by environmental scientists that greenhouse gas emissions are the main factor contributing to climate change since the 1950’s (Australian Greenhouse Office, 2007). As a result, research indicates that temperatures in the second half of the 20th century are likely to have been the highest for the past 1300 years, with 11 of the past 12 being the warmest since reliable measurements began in 1860 (Australian Greenhouse Office, 2007). Further, there has been an increase in heat waves, warming of deep oceans, fewer frosts, a retreat of glaciers and sea ice, and a 17 cm rise in the sea level during the 20th century which has impacted on many species of plants and animals (Australian Greenhouse Office, 2007).

For example, Alaska’s climate has warmed around 2 degrees Celsius since the 1950’s, leading to a 14% retreat in sea ice and a 60% thinning of the ice (National Assessment Synthesis Team, 2001). Permafrost melting has caused erosion and landslides and warming has been accompanied by forest disturbances and increased insect attacks (National Assessment Synthesis Team, 2001). Further, some scientists argue that climate change is leading to more dangerous consequences, such as the collapse of the West Antarctic Ice Sheet, which would cause an increase in sea levels by up to two metres and lead to the forced migration of millions of people (Leiserowitz, 2005).

Many of these specific environmental events mentioned above were also identified in our study where participants were asked to explain how they conceptualised climate change. Rather than offering precise or accurate definitions explaining the scientific mechanisms at work, concrete examples were offered that related to current changes in climate, the environment and lifestyle. These specific environmental events were generally viewed as leading to greater social problems, such as mass migration, increased food prices and war over vital resources, as well as more extreme climatic events in the future.

For example, for these participants, climate change was seen as being about:

- Global warming, the ozone layer, extreme weather patterns, ice caps melting and flooding. And if the population keeps growing at the current rate, it’s going to be overpopulated in a really small area because of the flooding that’s going to happen. It’s going to put a lot of pressure on our resources and its going to hurt economically. All the temperatures are going to change, so there will be droughts that are even worse than what we’re going through at the moment. Everything relating to nature will just get blown out of proportion compared to what it is now. Everything will be ten times worse.
- Climate change is about the temperature of the world changing, the sea’s rising, too much pollution and the ozone layer depleting. The weather is all different, summers are coming a lot earlier or they’re delayed. What’s hot should be cold and what’s cold should be hot. There are cyclones, you hear about in America or wherever else in the world, unexpected ones, they come out of nowhere... I have an image of the earth becoming a desert; there will be no food, and all the species will be dying from a chain reaction.

- Climate change is about extremely hot weather and water restrictions. In the future, you won't be able to have a pool anymore and you won't be able to have water fights; they will have to cut school hours back because kids won't be able to handle the heat; it's going to be so hot that it's going to dry up all the water, there will be no beaches left. This heat will have an impact on animals, like polar bears and penguins which have lost their breeding grounds already, because the icecaps are melting and now they are dying every year.
- Climate change is related to the hole in the ozone layer from industry and all the pollution they cause. We have more heat coming in, the icecaps are melting, sea levels are rising and all the ecosystems are changing because the environment is different.

Consistent with comments from the interviews, many students in the written survey referred to climate change as the effects of global warming resulting from increases in greenhouse gasses and carbon emissions which were associated with a range of natural events. These included: ozone depletion; increases in temperatures and extremes in temperatures; ice-caps melting; a rise in sea levels: submergence of entire islands and flooding of coastal areas; natural disasters such as tsunamis, hurricanes, cyclones, tornadoes, violent storms, droughts and floods; pollution including air and water quality; acidic rain, lack of rain and rivers drying up; extinction of animal species; suffering of animals and polar bears dying; great forests dying out, desertification and erosion; acidic oceans, warming of sea temperatures and bleaching of tropical reefs.

These events were closely associated with a range of social problems such as food and water shortages, limited fishing stocks, reduced biodiversity and reduction in all natural resources, increased health costs, increased rates of skin cancer, more disease and loss of human life, famines and plagues, economic effects, increased cost of living and lower standards of living, mass migration, refuges and dire consequences for poor communities and third world countries, changes in governments, political instability and wars for control of resources.

In contrast to the majority of participants, a small minority of students perceived climate change not to be a threat or even a reality. A sense of deception is reflected in references to "over-exaggeration", "outrageous claims" and "vested interests", with climate change being accused of being a "farce", "smoke and mirrors", "a marketing tool" and "a sensationalised scare tactic".

Exposure to Climate Change

Climate change affects the global population to varying degrees. Some individuals experience environmental events directly, such as farmers whose land and livestock is affected by drought or those who, for example, experience personal injury or damage to homes through flooding or hurricanes. Other individuals experience a form of low exposure, such as city dwellers who change their lifestyle to accommodate water restrictions, and those who endure record high or low temperatures. In addition, individuals experience indirect exposure to ecological events through media images of violent weather, special reports on the latest natural disaster being linked to climate change and "fear mongering" by some reporters who broadcast sensationalist headlines about events that are not linked to climate change, but are packaged in such a way that the viewers will perceive the information as such.

Research has shown that it is not only direct exposure, but also low and indirect exposure, to events that can cause distress for individuals. For example, several researchers found strong associations between anxiety and Post-Traumatic Stress Syndrome (PTSD), and exposure to television coverage after the 9/11 attacks on the World Trade Centre (Marshall et al., 2007). Specifically, two months after 9/11, probable PTSD was found in 4.3% of the adult general U.S. population who were indirectly exposed to the attacks through media images (Schlenger et al., 2002). It appears plausible then, that while the percentages are small in relation to PTSD in these global incidents, certain individuals could become vulnerable to feeling differing levels of distress as a result of direct or indirect exposure to climate change effects.

The power of indirect exposure to climate change through the media, and in particular, television images and movies, was commonly reflected in remarks made during our study as demonstrated by the following comments:

- When I saw these things on TV, the catastrophes, the erosion, the ozone layer, the heat, that's when I started realising, gee, this is true. This is when I felt it myself.
- It has made me really sad sometimes when I've seen something on the news, about global warming for example, I'll think about it and I get really upset.
- That was one of the things in the movie ("An Inconvenient Truth"), they showed footage of the animals; it's a trick by them to try and get an emotional response but it worked... I got really upset about the animals down in Antarctica, because if the ice caps melt, they're going to have no home; all these animals are going to become extinct.
- I get anxious and worried when they talk on TV about how global warming is going to lead to the end of the world.

Risk Perception

Certain types of risks and the way they are perceived are more likely to have intense psychological consequences that go beyond harming direct victims. Slovic (1987) proposed two aspects of an event that affect its impact by influencing an individual's appraisal of the risk: dread risk and unknown risk (Marshall et al., 2007). Dread risk encompasses the uncontrollable and catastrophic aspects of a hazard, such as those involved in the use of nuclear weapons or nuclear power, because of its potential for accidents. Unknown risk encompasses aspects of a hazard that include it being unobservable and not knowing or understanding when or how one is exposed to it. Unknown risk can be seen in examples from the 1980's that included the introduction of DNA technology, the use of microwave ovens and water fluoridation.

Terrorism is considered to possess both high levels of dread risk and unknown risk. It is conjectured that, for many individuals, climate change would also be regarded as having both high levels of dread risk and unknown risk, because it can be gauged as uncontrollable, leading to catastrophic ends, and also because the specific mechanisms are largely unobservable and not fully understood. This latter point can be demonstrated by comments taken from social commentaries such as "I try to read up about climate change, but basically I'm in the dark. Science baffles me…" (Castles, 2006) and also from the following comments written within our study: "It concerns me that people do not see the link between what they do and the effects of climate change, and I think it's because these links are not directly visible"; "I'm not sure what or who to believe"; and "Climate change makes me feel uneasy, I am unsure of what the real effects of climate change will be."

Marshall et al. (2007) proposes that for the majority of people who witness catastrophic media images, pathological distress does not result, because there is little perception of threatened death or injury to the viewer. However, in response to the media presentation of 9/11, a small minority of US residents felt sufficiently threatened to produce a response of fear, helplessness and horror in regard to future terrorist attacks (Marshall et al., 2007). It can be construed that these viewers appeared to have engaged in catastrophic secondary appraisals which were then perpetuated by the implication of a significant ongoing threat (e.g., future terrorist attacks and the potential for war) that could bring additional harm to them and others (Marshall et al., 2007).

Media coverage of climate change issues shares similarities with that of terrorism in some ways. Just like terrorism, climatic events are unpredictable and occur anywhere at any time. Climate variations are reported to be occurring on a large scale with adverse global impacts and reports of worldwide earthquakes, tsunamis, hurricanes, droughts and wildfires occurring at a greater rate than at any other time in recorded history. Such climate change events are unpredictable, as extreme weather patterns and natural disasters strike without warning; and graphic representations about these catastrophic events are captured and replayed in the media. Individuals, environmental groups and the media promote warnings that if nothing is done, our planet is heading for grave consequences - the seas will rise, extreme weather conditions will occur and mass natural disasters will increase and millions of people may die or be displaced, as a result of these effects.

So could it be possible that the nature of climate change and its presentation in the media fills some individuals with fear and uncertainty to the extent of inducing anxiety and depression at a pathological level?

Climate Change Concern

Although there has been little research conducted in the psychological literature, distress related to climate change has been indicated by public opinion polls and surveys. For example, a poll of 3,032 Australians, conducted by Coredata and NEWS.com.au in January 2007, indicated that climate change was the number one concern for 48% of Australians, ahead of any other world or national issue such as poverty or terrorism. Another Coredata survey in May 2007 (Abbott, 2007a) found that 41% of Australians believed that there was little that could be done to combat the effects of climate change, thus indicating a generalised sense of helplessness. Further, a Lowy Institute poll (2007) found that 55% of Australians were 'very worried' about climate change, rating it ahead of all external threats to Australia such as 'unfriendly countries developing nuclear weapons', 'Islamic fundamentalism' and 'international terrorism'. Again these polls took place before the effects of the global financial crisis had become evident.

Uncertainty and worry related to environmental concerns can also be observed in children. The Australian Childhood Foundation (Tucci, Mitchell, and Goddard, 2007) surveyed 600 children, aged between 10 and 14 years, and found that 44% were worried about the impact of climate change, 52% were scared that there would not be enough water in the future and 43% were concerned about air and water pollution, compared to 36% of children who were worried about terrorism. In addition, British supermarket chain Somerfield conducted a survey of 1150 children between the ages of 7 and 11 and found that half of them were anxious about global warming and often lay awake worrying about it (Friedlander, 2007). The most feared consequences were flooding, poor health, the submergence of entire countries and the welfare of animals (UK News, 2007).

Anxiety related to climate change can be seen in environmental action and support groups. In early 2007, the 'Climate Change Despair and Empowerment Roadshow' (Rainforest Information Centre, 2007) toured the east coast of Australia addressing the "hopeless despair that many feel" with the use of symbolic exercises such as 'The Mourning', where participants were invited to express deep feelings of anger and grief at what is happening to the Earth. Climate change distress is also reflected in the media with mainstream newspaper articles advising readers to "try a bit of CBT" (cognitive behavioural therapy) when they feel global warming panic take hold, and be alert to thinking errors such as catastrophizing, over-generalisation, polarised thinking and inappropriate filtering (Castles, 2006).

While the media and public opinion polls indicate concern about climate change, very little has been written about it in the psychological literature. Much of the psychology based research has focused on risk perception. For example, Sunblad, Biel and Garling (2007) investigated risk judgements related to climate change in 621 Swedish residents and found that knowledge of the causes and consequences of climate change was able to predict probability estimates of negative consequences and also worry about negative consequences. Langford (2002) used semi-structured interviews and focus groups to assess subjects' beliefs about climate change and found that responses fell into four categories; denial, disinterest, doubt, and engagement. Those in the denial group did not consider that climate change was important and reflected low interest and concern for the issue. Those in the disinterest group displayed a generally fatalistic view of life reflecting a belief that external forces were beyond their personal control and denying any responsibility or involvement (Langford, 2002), while those in the engagement group felt empowered, and showed high levels of concern and information gathering, believing that commitment and effort could create a better world (Langford, 2002).

From a consequentialist versus ethical evaluation perspective, emotional reactions to environmental risks were investigated by Böhm (2003). Four types of emotional reactions were examined: prospective and retrospective consequence-based emotions (e.g., fear and sadness respectively), and self and other-related ethics based emotions (e.g., guilt and anger respectively). The single most intense emotion found in regard to specific environmental risks was worry. Further, each emotion type, apart from retrospective feelings, was able to be predicted by specific cognitive judgements.

An examination of reactions to specific environmental events has highlighted the negative impact on mental health in specific populations. For example, in 2005, Rogers-Clark conducted personal interviews with rural Australian men and reported that conditions in country areas, particularly where the drought had been ongoing, had an impact on men's mental health, such as inducing feelings of isolation, depression and suicidal ideation. Also, according to transcripts involving 98 individuals and 23 organisations which reported on the outbreak of Ovine Johne's Disease (OJD) in rural Victoria, individuals reported experiencing trauma, stigma, a sense of failure, loss of identity, diminished self-esteem and family discord which led to grief, depression, PTSD and suicide (Hood and Seedsman, 2004).

While it appears that specific environmental events have led to negative impacts within directly affected populations, few studies have examined the general impact of climate change on mental health. Environmental concerns have, however, emerged in several studies concerning fearful future outcomes. For example, Gow and Leahy (2005) conducted focus group interviews with 100 residents and then interviewed by telephone 300 residents in the Hunter region of NSW; they concluded that respondents were fearful of a range of apocalyptic scenarios. When asked about possible events that could affect Australians negatively in the next 50 years, the most common concerns involved environmental crisis, invasion of Australia, disintegration of moral values, nuclear war, plague or poverty and unemployment (the interviews were conducted prior to September 11, 2001).

Gow and Leahy (2005) detected a distinction between concerns regarding environmental problems that were identifiable, as opposed to concerns that these problems would lead to a future apocalyptic event. That is, environmental problems, such as species extinction and threats to native forest, were regarded as a warning of unexpected and disastrous events that may happen at some time in the future. Further, the fear of a looming environmental catastrophe was associated with a belief in multiple crises and was indicative of a broader 'apocalyptic sensibility' (Gow and Leahy, 2005).

Consistent with Gow and Leary's research, comments from interviews in our study also reflected a sense that the Earth and mankind is heading towards an impending environmental catastrophe and great changes will be required for future survival, for example:

- In my children's lifetime, I think a (major environmental) catastrophe will happen.
- It's going to get to the stage that when something does happen, it's going to be a massive disaster that will affect us really severely.
- I think that the world as we know it will come to an end; the civilization that we've come to know will end. I think that we will get to a point where we will have to stop driving for example; we're going to get to a point where we have to give up our lifestyle; we're going get to a point where we have to change.
- The world is going to end, not end altogether, but it's slowly going (to die out) .... it will be too hot to live in, plants will die, it will be uninhabitable with all the pollution... it can't be stopped, I don't think it can be stopped.
- In the future it will be catastrophic.... things have started to happen that indicate how bad it's getting... it will just get worse and worse.

Apocalyptic Sensibility

Apocalyptic sensibility has been referred to by Viane (1963, as cited in Tolsma, 1968) who introduced the term the 'Cassandra Complex' after the story of Cassandra who foretold the fall of Troy. The Cassandra Complex is explained as a pathological, negatively charged future expectancy or fateful expectancy. Tolsma, in his 1968 paper on doom expectancy, distinguished between neurotic and psychotic forms of doom expectancy. Neurotic doom expectancy entails a general fear of the future and correlates with negative experiences in childhood or in the critical phases of life. Psychotic doom expectancy can be further divided into several categories: depressive; paranoid; degenerative; schizophrenic; and psychasthenic and compulsive neurotic (Tolsma, 1968).

The depressive doom-sayer group includes people who suffer from depressive syndromes and who hold gloomy ideas about the future. For these people, the future is dismal, threatening and black, and these ideas are often accompanied by sensations of anxiety. The paranoid doom-sayer category includes people who feel that they are threatened by the surrounding world. For these people, disaster can be avoided, if only the conspiracy ends and evil influences are eliminated. The degenerative doom-sayer group includes people who hold religious ideas with the central theme being about the destruction of the world. An expectancy of salvation in the future is therefore linked to negative attitudes towards the existing time zone. The schizophrenic doom-sayer holds bizarre negative future perspectives such as chaotic delusions of destruction. Finally, the psychasthenic and compulsive neurotic doom-sayer group includes people who demonstrate a pronounced element of personal insecurity and fear of failure.

Thus Tolsma (1968) is suggesting that doom expectancy results from pathological causes. In life, people are continually confronted with unpredictable and inconceivable risks that lead most people to be concerned. For those who are in a state of agitation, however, that concern may lead to apprehension, and apprehension may lead to an expectancy of doom (Tolmsa, 1968). This viewpoint is supported in a study of psychiatric patients leading up to the millennium and the anticipation of the Y2K bug. Strous et al. (2000) found that 51.9% of patients (with a diagnosis of schizophrenia, schizoaffective disorder, postpartum depression, major depression with and without psychotic features, and personality disorders) showed concern and anxiety regarding the turn of the century, in comparison to only 12.8% of controls. Further, research revealed that there was a strong correlation between individuals suffering from PTSD after September 11 and those who possessed a co-morbid diagnosis or a pre-diagnosed depression or anxiety disorder (Neria, Gross, and Olfson, 2006; Silver, Holman, McIntosh, Poulin, and Gil-Rivas, 2002).

Vulnerability Factors

While it appears that co-morbid mental illness increases the likelihood of psychological distress in the face of a trauma or ongoing threat and the anticipation of disaster, research findings have highlighted several additional factors that lead to vulnerability. These have included personality factors such as trait anxiety, future anxiety, coping style and coping self-efficacy, and intolerance of uncertainty, as well as the nature and extent of prior traumatic experience, religiosity, gender and other demographic variables (Goldstein, Dudley, Erickson, and Richer, 2002; Morrissey and Reser, 2007; Silver et al., 2002). However, little has been written about the psychological impact of climate change and the specific vulnerability factors that lead to climate change distress. Further investigation of the role of future anxiety, intolerance of uncertainty, gender and religiosity will contribute to the understanding of how this global issue is affecting society and which individuals are susceptible to the resultant distress.

Future Anxiety

The research literature indicates that both trait anxiety and future anxiety are associated with an interpretive bias for future events; that is, highly anxious individuals are more likely to perceive stimuli in a threatening fashion (Rapee, 1997; Zaleski, 1996). For example, several studies have found that students scoring high on trait anxiety are unrealistically pessimistic about examination-related events (Butler and Mathews, 1987; Eysenck and Derakshan, 1997). Further, Goldstein et al. (2002) found that trait anxiety was a predictor of Y2K anxiety, in that anxious individuals were more likely to overestimate the probability and severity of negative consequences of the Y2K computer problem. Future anxiety is described as a personality characteristic that is highly correlated with trait anxiety (Zaleski, 1996). It has been defined as "a state of apprehension, uncertainty, fear, worry and concern of unfavourable changes in a more remote personal future" (Goldstein et al., 2002, p. 165) and is also associated with pessimistic predictions about global problems faced by humanity (Zaleski, 1996). It seems likely then that individuals possessing future anxiety might perceive information about current ecological concerns in a more threatening and worrisome manner.

Intolerance of Uncertainty

Intolerance of uncertainty can be seen as a cognitive bias that influences an individual's perceptions, interpretations and responses to uncertain situations (Dugas, Schwartz, and Francis, 2004). People who find it difficult to tolerate uncertainty tend to avoid ambiguous situations, because they find them stressful and upsetting (Dugas et al., 2004). Research has shown that individuals with high levels of intolerance of uncertainty show more concern about ambiguous situations and make more threatening interpretations of ambiguous situations than those with lower levels of intolerance of uncertainty (Dugas, Hedayati, and Karavidas, 2005). Further, intolerance of uncertainty is a strong predictor of the tendency to worry (Dugas et al., 2004). Therefore, due to the uncertain nature of climate change, it would seem likely that persons who are intolerant of uncertainty should also experience greater distress about climate change. Comments from our study indicate that "future uncertainty", "increased unpredictability" and "not knowing the long-term consequences" were amongst participants' greatest concerns in regards to climate change.

Gender Differences

Women generally report more depression, anxiety and fear than men. Women have also been found to be more anxious when facing ambiguous events such as nuclear war and Y2K computer problems. In relation to climate change, Sunblad et al. (2007) investigated cognitive and affective risk judgements and found that even though men and woman did not differ in their predictions of the probability of serious negative consequences occurring as a result of climate change, women were more worried about the consequences. Our pilot study found that significant gender differences ($F(1,76) = 5.94$, $p<.05$) were evident with respect to total climate change distress; that is, the average score for males was 8.7, while the average for women was 12.9.

Religiosity

Religious beliefs have been associated with a reduction in anxiety in many populations, such as individuals those with HIV (Jenkins, 1995) and university students (Plante, Yancey, Sherman, and Guertin, 2000). Spiritual values have also been shown to be negatively related to future anxiety in that religious commitment appears to reduce pessimistic attitudes towards the future (Zaleski, 1996). However, due to claims that climate change will lead to the destruction of the earth, those who hold religious beliefs may associate Armageddon-like consequences to concepts of climate change thus increasing their levels of distress. Tolsma (1968) encompasses this notion with the idea of the degenerative doom-sayer who holds an expectancy of salvation in the future that is linked to negative attitudes towards the existing world. Research on Y2K anxiety has also shown such a link. For example, Goldstein et al. (2002) found that strength of religiosity was a predictor of Y2K anxiety which the researchers attributed to the apocalyptic undertones of Y2K speculation. Religiosity may therefore influence climate change distress from several perspectives.

Although specific comments relating to religion were not invited in the current study, several remarks were written in survey forms indicating a strong personal connection between religion and climate change. An apocalyptic undertone was demonstrated in the following comment, "Climate change is the non-Christian interpretation of what is stated in Revelations 4 in the Bible". Others saw climate change as a form of punishment from God as demonstrated by this participant's beliefs: "Because of my religion, I believe climate change is a result of people messing with everything and it is going to get worse. And what I mean by messing with everything is not just in regards to nature, but also doing wrong things towards each other such as sinning, bad behaviour, and crimes etc."

While some participants claimed not to belong to a traditional religion, their comments reflected a sense of spirituality in nature and encompassed more Eastern concepts such as karma. For example, views from one of the interviewees) follow:

- I'm not really a religious person but I find a real beauty in nature and that's why I feel so strongly about it. When everything looks really bad in my life, there is always something that is beautiful in nature. Nature reflects how we try to explain what we are. People have all these ideals, but I don't actually see them in people, I see them in nature. ... And I think what we are doing to nature is all going to come back around and get us. I think the

> world overheating, tidal waves and volcanoes erupting, I think these are just ways the world is getting back at us, sort of like revenge, like karma. It is as if whatever we do to the world, it is going to get us back.

While these comments reflect personal meanings between climate change concepts and religion or spirituality, it is not clear whether this connection increases or reduces feelings of distress about climate change.

Concerns of University Students about Climate Change

In 2008, the authors undertook a pilot study with students at a University in Queensland, Australia (41% were males and 59% were females; 86% were aged 25 and under) and an analysis of preliminary data revealed the following interesting results. [For the full research findings on the larger cohort of students, see Searle and Gow, in preparation]

In making comparisons about their felt concerns in relation to climate change over time, more then two thirds were more concerned now than previously. Sixty eight percent were now more concerned about climate change compared to three years ago and 78% were more concerned now than 5 years ago. The latter percentage remained constant comparing "now" to 10 years ago which may have been influenced by the younger age of the majority students who were in the 18-25 year age group.

Other participants who were interviewed about a range of climate change issues (see Appendix) also indicated a growing concern which, for many, was precipitated by learning about climate change through television programs, magazine articles, newspapers and documentaries such as the Al Gore movie *An Inconvenient Truth* which is shown throughout many schools in Australia. These participants commented:

- I knew about climate change before I watched the Al Gore movie, but I never really took an interest or was concerned about it till then. It really opened my eyes to things I'd never looked at before.
- I was forced to watch that Al Gore movie at school and that was the first thing I saw that got me actually thinking about it.
- I first became concerned about climate change when I was a young child after I read an article in the National Geographic (magazine) when I was waiting to go into the dentist. It was about how the Murray River was drying up and I remember thinking, "how can you let a river dry up?"
- I first became aware of climate change about four years ago. I noticed every summer had been getting hotter and hotter, and then I saw all these programs on TV that kind of opened my eyes and it just went from there.

Apart from the media, other major influences in increasing students' concern about climate change were cited as being parents and grandparents, like-minded people who were sought out on internet sites, and school subjects such as science and environmental courses.

When the participants were asked to rate their feelings and cognitions about Climate Change Issues (as outlined in Table 3.1), we were able to discern distinctions between the different types of emotions and cognitions.

Table 3.1. Cognitions and Emotions about Climate Change Issues: Percentage Endorsed

Label	%	Label	%
Concerned	57.6	Powerless	23.1
Tense	20.5	Sad	20.5
Worried	35.9	Helpless	25.6
Anxious	24.4	Stressed	15.3
Depressed	12.8	Angry	33.3
Hopeless	11.6	Scared	23.1

The percentages in the columns represent the responses that endorsed the item as being an issue for "a good part of the time" or "most of the time". An additional category of "applying to some degree" would bring the percentages in all items considerably higher (higher by 25 – 45%).

With respect to the emotional reactions, the statistics tend to indicate that one fifth (or more) of the respondents felt tense, powerless, sad, helpless, angry and scared. The low figure on stress may be due to the fact that students would tend to be aware that their own personal stressors associated with income, adequate and stable housing and academic success were of more immediate concern to them.

Significantly, the lower percentages on depression, hopelessness and helplessness seem to indicate that most of those students surveyed had either not given up on saving the environment, or that they thought that there was nothing that an individual could do about it, or that it was not an issue that was of great concern to them, at that time in their life and point in history.

Of the participants who were interviewed (chosen because they had expressed a higher level of concern than most), they expressed a mixture of negative reactions when thinking about climate change. Cognitions and emotions such as anger, sadness, fear, worry and anxiety are reflected in the following comments:

- I feel really angry at people, because something could have been done.... At the same time, I feel an immense sadness because we got lazy and we tried to change all this stuff and we thought we were more important than everything else. We've taken control of the earth and destroyed it.
- I'm sad that we haven't taken enough care of the world... and it's sad to me that these unexpected environmental disasters are happening. Mostly I'm scared, if something happened here, I'm scared about how it would end, or if I would die from it or how it would affect my family....
- I get anxious and worried when they talk on TV about how global warming is going to lead to the end of the world.
- I have a mixture of anger and fear. I get angry because we know how to be smart with resources, but we just don't do it.... And I fear that my family won't have an earth to live on... I worry about my future and my family's future.

Within the surveyed population, the university students gave some indication about the nature of their environmental beliefs and concerns (note that the figures in the right hand column of Table 3.2 represent the percentage of students who endorsed the statements about the environment either mildly or strongly).[2]

In analysing what their endorsements of these items might mean, the highest endorsements (83.3% to 66.6%) were accorded to the rights of nature, the balance of nature and the importance of not upsetting the ecology (see first six items). The next 9 items appear to be grouped under the rights of humans, as distinct from nature's rights; it is thus interesting that the percentage endorsements slide downwards in terms of agreement with the items.

From item 10 to item 15, the lower percentage endorsements (all under 50%) indicate that generally the majority of our respondents (over 50%) do not believe that humans have rights, or control, over nature. Items 7, 8 and 9 demonstrate the quandary that human beings are in; firstly the earth does seem to have plenty of resources, but given that the earth is overpopulated, therefore perhaps the earth may be under-resourced to cope with the growing number of people on earth. The last two items (14 and 15) give a very clear message that the respondents (75.6%) did not believe that the ecological crisis is exaggerated and that nature can no longer cope with the effects of industrialisation (83.3%).

What is clear from this pilot study is that the majority of students demonstrated a humility in admitting that humans will not (and probably never will) control nature. They were divided in their views on whether or not the earth had plenty of natural resources left and whether there were too many people to be sustained by the environment.

Table 3.2. Beliefs and Concerns about the Environment

Item No.	Specific Beliefs and Concerns	% Agreement
1	Humans are still subject to nature's laws	83.3
2	Humans are abusing the environment	80.8
3	Interfering with nature leads to disaster	79.5
4	Plants/animals have equal rights to live	78.2
5	If continue, soon major ecological catastrophe	74.4
6	Nature's balance delicate and easily upset	66.6
7	Earth has plenty natural resources if we learn to develop them	65.3
8	World is reaching limit of population it can support	51.3
9	Earth has very limited room/resources	51.2
10	Humans have rights to modify natural law	47.4
11	Human ingenuity will ensure earth is not unliveable	34.7
12	Humans were meant to rule over nature	33.4
13	Humans will control nature by learning how it works	33.3
14	Ecological crisis greatly exaggerated	24.4
15	Nature's balance can cope with industrial nations	16.7

Source: The wording of the items from the New Ecological Paradigm scale are condensed in the table. (For the full scale, see Dunlap, Van Liere, Mertig and Emmet Jones, 2000.) Content of NEP scale reproduced here with permission.

[2] There were 2 agreement categories (mildly or strongly agree); 2 disagreement categories and 1 unsure category.

It is interesting to note the variations in the beliefs about resources (see items 7 and 9). The beliefs may be based on facts, or they may be based on un-researched ideas. There does seem to be some disconnection evident in these beliefs. This would be worth following up, as it could be that in their thinking (perhaps in the thinking of most of us), that the concept of resources for humans to use is different from the concept of the natural world; that is wildlife and habitat may be held as one distinct aspect of life and resources such as coal, oil, and gas (energy sources) as another. It is true that as non-indigenous humans, we are not aware of the interconnectedness between humans and animals and the total natural environment in which we live and work.

We do know that some people care greatly about wildlife and take great care not to harm them; others protect the habitats of fauna and lay down before bulldozers to save forests across the world. Still others go to great lengths to save beached whales and protect the oceans' sea life. However generally, the majority of people do not take action to preserve the environment for its own sake, nor value it in terms of its role in the provision and preservation of human life, which we do not yet fully appreciate. Apparently, there is little understanding of the inherent sacrifice in nature proven by Professor George Ellis in South Africa who won the Templeton Prize in 2004 for his work on religion and science.

Interviews with other professionals and people in the community led us to understand that with respect to cognitions and emotions about climate change concerns, some people in different parts of society are moved to take action to protect various parts of our natural environment because they "feel deeply" about the loss of certain species, animals, habitats, forests, waterways and coral reefs. However, our interviews have shown us that while people can feel bad, ashamed, and passionate about protecting the environment, in most instances that activation lasts only a few minutes or at the most a few hours or a few days, as demonstrated by the following comment:

- I always feel motivated to do something, but it never lasts very long. I might think about it for a couple of days - like I'll take the bus or I'll walk to work today, but then it'll start raining, so I think I'll have to drive or otherwise I'll get wet and then I'm going to be late; and then it (my intentions) only last for a short time.

And some comments indicated a complete lack of activity:

- I guess I don't do more than I should, because I don't see anybody else doing it; so I think what can I do as one person?
- I try, I think I should do something, but I don't always get around to it.

Others have found that taking action has in fact reduced their distress levels:

- I find it (being active about climate change) is a way I can relieve my stress, because I personally can do something. I can't really do anything to change other stressful events in my life, but I can go and turn all the lights off in my house.

The types of activities these students engaged in with the aim of combating climate change revolved mostly around conserving energy and water. These included turning off lights and electrical appliances when they are not in use, using energy efficient light-bulbs, printing on the reverse side of paper, bucketing instead of hosing the garden, using buckets to collect waste-water, using timers to have shorter showers, car pooling and catching public transport. But, for the majority of these students their motivation to conserve energy came from financial incentives, rather than climate change being the impetus for their actions. This is also demonstrated in the following comments from the interviewees:

- I try to car pool and catch the train, but my motivation is probably 70% financial. You feel better about it, but you're going to do it anyway.
- My motivation comes from conserving energy and finances; mostly saving money, but hoping this will save the world too.
- What I do is kind of related to petrol prices and kind of related to reducing the petrol emissions. I'll catch the bus, even though it will take me 20 minutes longer... but honestly, it's 90% about saving petrol and 10% about reducing my carbon footprint.

For many students, financial concerns are a challenge of daily living, but it also appears that this lack of individual motivation and accountability towards taking action to combat climate change may come from a sense of blame directed towards the government and big business leaders. Many of the following remarks reflect an expectation that it is these larger bodies, in addition to countries in the developing world, who should be taking the primary responsibility for preventing negative climate change effects now and in the future:

- The government is not doing things to prevent problems in the future. The government needs to get industry to look at alternative ways to produce products where they can cut down on their gas emissions.
- I blame mainly industrial sort of people... Industrial businesses are creating the fossil fuels to construct so much; and at the same time, I blame the government because they allow it, and they just pretend that it's not doing anything, but it is.
- I hope that corporations and the government are doing what they can to slow the process down ... big corporations with their big factories definitely play a big part (in the problem).

This attitude was reflected in comments from the written survey with references to "political short-sightedness", "governments who won't move quick enough", "lack of government initiatives", "lack of action by big industry", "government loopholes that allow big business to abuse the planet" and "the government's refusal to make tough decisions". Several of the participants also made references to the role of developing nations by expressing concern about: "the developing world adopting the mistakes of the industrial world"; "countries like China producing so much and not being responsible for the pollution"; "the poor habits of the western developed countries being repeated, at an even bigger scale in developing countries"; "the rapid rate at which China is growing and its seeming total disregard for the environment" and "the rapid industrialisation of third world countries".

It should be stated that the survey and interviews were conducted just before the release of the Garnaut report on climate change in 2008 (Garnaut, 2009) which was followed quickly by the Green paper on Carbon Emissions (Carbon Pollution Reduction Scheme, 2008). Both reports were widely publicised in the media and the then new Prime Minister of Australia, The Honourable Kevin Rudd, reinforced the seriousness of the Climate Change Agenda, putting into place major reforms to reduce carbon emissions and to reduce the impact of other climate change influences on Australia and in turn on other parts of the world.

There is no doubt that Australia had been lagging behind many other developed countries in the actions it had taken to address the Kyoto protocol and to take stock of the impact of the encroachment of construction and mining and agriculture on the natural environment. Perhaps it was the long relentless drought that brought home the stark truth that without water, there could be no vegetables, fruit wine, meat, or even wool. Without rain, there might be no drinking water, no dams to quench the thirst of the stock and no rivers to nurture the natural environment and its creatures.

After the advent of the disastrous bushfires in Victoria in February 2009, with record high temperatures occurring in a land bereft of rain for so long, there was no doubt left in the Australian psyche, after watching the scenes of death and destruction on television and hearing the stories from some of the survivors, that the concept of climate change might just hold some seeds of truth (see Gow, 2009). The world was indeed hotting up and one part of the country in the south (Victoria) burned for weeks, as parts in the north were flooded by deluges that moved across the state of Queensland.

CONCLUSION

When we commenced the research, we were of the opinion that many people were worried about the future, not just in terms of climate change, but also in relation to the environmental issues that have been negatively affecting our health and lifestyles for some decades. We were already aware that for most people, natural disasters generally held no dire fear, unless one had been exposed to a traumatic experience in a storm, flood, bushfire, cyclone, earthquake or tornado. [However, it must be kept in mind that the severe floods, bushfires and cyclones that beset Australia in 2008/2009 had not struck at the time of the research.]

What became clear to us was that changing behaviour, in spite of one's experiences and fears, was a much more difficult situation, and that people would only change their behaviour, if they were firmly convinced that it would affect themselves personally and significantly. Otherwise, they could simply dissociate from the call to action to save the world, as it was not relevant or not real, or was outside their sphere of control.

These findings about lack of change or difficulty in changing behaviours, in terms of preparation for, and mitigation of, hazards, confirm those of the disaster management literature and point to a need to make climate change scenarios as applied as possible to individuals in their local circumstances and to bring the discussions back across the time line to the decades nearer those who are alive on this threatened planet now.

Thus talk about 2070 is of interest to people, but half of the population will be dead by then and simply asking them to think of their grandchildren will not work in terms of behaviour change. Using the "drag and drop" feature across the timeline (from 2070 back in time to 2020) might work a lot quicker and a lot more effectively in terms of changing beliefs and behaviours. If one needs to "drag and drop" the year 2050 to 2015, in order to bring people face to face with serious issues in their own life time, then presenting them with non conflictual data and pictures, sounds and tactile interactions with the projected negative scenarios will work, as long as the action plan is logical, well thought out and people can be convinced that it will indeed achieve the desired outcomes.

REFERENCES

Abbott, J. (2007a). *Australians Warm to Government Action on Climate Change*. Retrieved 20.09.07, from http://www.newsdigitalmedia.com.au/inc/docs/mediareleases/Media%20 Release- climateChange-FINAL.pdf

Abbott, J. (2007b). *Climate Change the No. 1 Concern for Australians*. Retrieved 20.09.07, from http://www.newsdigitalmedia.com.au/inc/docs/mediareleases/070117_Climate_ Change_media_release.pdf

Australian Greenhouse Office. (2007). Is the Earth's climate really hotting up? *Information from Department of Environment and Heritage: Frequently Asked Questions*.

Böhm, G. (2003). Emotional reactions to environmental risks: Consequentialist versus ethical evaluation. *Journal of Environmental Psychology, 23*(2), 199-212.

Beck, U. (1992). *Risk Society: Towards a New Modernity*. London: Sage.

Breckler, S. J. (2007). Science Directions: Rising to the challenge. *Monitor on Psychology, 38*(4), 28.

Butler, G., and Mathews, A. (1987). Anticipatory anxiety and risk perception. *Cognitive Therapy and Research, 11*(5), 551-565.

Carbon Pollution Reduction Scheme. Green paper. (2008, July). Canberra: Department of Climate Change. http://www.climatechange.gov.au/greenpaper/index.html

Carleton, R. N., Norton, M. A. P. J., and Asmundson, G. J. G. (2007). Fearing the unknown: A short version of the Intolerance of Uncertainty Scale. *Journal of Anxiety Disorders, 21*(1), 105-117.

Castles, S. (2006). Cool head needed in times of heat and confusion. *Taranaki Daily News,* p. 4.

Dugas, M. J., Hedayati, M., and Karavidas, A. (2005). Intolerance of Uncertainty and Information Processing: Evidence of Biased Recall and Interpretations. *Cognitive Therapy and Research, 29*(1), 57-70.

Dugas, M. J., Schwartz, A., and Francis, K. (2004). Intolerance of Uncertainty, Worry, and Depression. *Cognitive Therapy and Research, 28*(6), 835-842.

Dunlap, R. E., Van Liere, K. D., Mertig, A. G., and Emmet Jones, R. (2000). Measuring endorsement of the new ecological paradigm: A revised NEP scale. *Journal of Social Issues, 56*(3), 425-442.

Eysenck, M. W., and Derakshan, N. (1997). Cognitive biases for future negative events as a function of trait anxiety and social desirability. *Personality and Individual Differences, 22*(5), 597-605.

Friedlander, J. (2007, 15 May). Out and about beats fear. *Sydney Morning Herald, The,* p. 12.

Garnaut, R. (2009). *The Climate Change Review.* Cambridge, UK: Cambridge University Press.

Goldstein, S. B., Dudley, E. A., Erickson, C. M., and Richer, N. L. (2002). Personality traits and computer anxiety as predictors of Y2K anxiety. *Computers in Human Behaviour, 18*(3), 271-284.

Gow, J., and Leahy, T. (2005). Apocalypse probably: Agency and environmental risk in the Hunter region. *Journal of Sociology, 41*(2), 117-141.

Gow, K (2002a). Trauma counselling in Cambodia (Part 1): A socio-cultural narrative. Forum, Newsletter of the Australian Critical Incident Stress Association Inc. September.

Gow, K. (2002b). Trauma Counselling In Cambodia: Part 2: Be Prepared To Descend Into Hell. Forum, Newsletter of the Australian Critical Incident Stress Association Inc. October.

Gow, K. (2009). The long way home. ABC News online. Australian Broadcasting Corporation. Tuesday February 10. Accessed 06.03.09. Http://www.abc.net.au/news/stories/2009/02/10/2486966.htm

Hamilton, S. B., Keilin, W. G., and Knox, T. A. (1989). When thoughts turn toward nuclear war: Stress responses, coping strategies, and the importance of trait anxiety in moderating effects on 'mental health.' *Journal of Applied Social Psychology, 19*(2), 111-139.

Hood, B., and Seedsman, T. (2004). Psychosocial investigation of individual and community responses to the experience of Ovine Johne's Disease in rural Victoria. *The Australian Journal of Rural Health, 12*(2), 54-60.

Jenkins, R. A. (1995). Religion and HIV: Implications for research and intervention. *Journal of Social Issues, 51*(2), 131-144.

Koenig, H., Parkerson, G. R., Jr., and Meador, K. G. (1997). Religion index for psychiatric research. *American Journal of Psychiatry, 154*(6), 885-886.

Langford, I. H. (2002). An existential approach to risk perception. *Risk Analysis, 22*(1), 101-120.

Leiserowitz, A. A. (2005). American Risk Perceptions: Is Climate Change Dangerous? *Risk Analysis, 25*(6), 1433-1442.

Lovibond, P. F., and Lovibond, S. H. (1995). The structure of negative emotional states: Comparison of the Depression Anxiety Stress Scales (DASS) with the Beck Depression and Anxiety Inventories. *Behaviour Research and Therapy, 33*(3), 335-343.

Lowy Institute. (2007). Australians Speak on Foreign Policy: Media Release.

Marshall, R. D., Bryant, R. A., Amsel, L., Suh, E. J., Cook, J. M., and Neria, Y. (2007). The psychology of ongoing threat: Relative risk appraisal, the September 11 attacks, and terrorism-related fears. *American Psychologist, 62*(4), 304-316.

Morrissey, S. A., and Reser, J. P. (2007). Natural disasters, climate change and mental health considerations for rural Australia. *The Australian Journal of Rural Health, 15*(2), 120-125.

National Assessment Synthesis Team. (2001). *Climate Change Impacts on the United States: The Potential Consequences of Climate Variability and Change: Foundation*. Retrieved 25.09.07, from http://www.usgcrp.gov/usgcrp/Library/nationalassessment/foundation.htm

Neria, Y., Gross, R., and Olfson, M. (2006). Posttraumatic stress disorder in primary care one year after the 9/11 attacks. *General Hospital Psychiatry, 28*(3), 213-222.

Plante, T. G., Yancey, S., Sherman, A., and Guertin, M. (2000). The association between strength of religious faith and psychological functioning. *Pastoral Psychology, 48*(5), 405-412.

Rainforest Information Centre. (2007). Climate Change Despair and Empowerment Roadshow.

Rapley, C. (2007). Taking Stock on Climate Change. http://www.sciencepoles.org/index.php?/articles_interviews/professor_chris_rapley_taking_stock_on_climate_change/anduid=990andtype=15

Rapee, R. M. (1997). Perceived threat and perceived control as predictors of the degree of fear in physical and social situations. *Journal of Anxiety Disorders, 11*(5), 455-461.

Rogers-Clark, C. (2005). *Tough times for men in rural Australia.* Retrieved 20.09.07, from http://www.usq.edu.au/newsevents/news/tough+times.htm

Schlenger, W. E., Caddell, J. M., Ebert, L., Jordan, B. K., Rourke, K. M., Wilson, D., et al. (2002). Psychological reactions to terrorist attacks: Findings from the National Study of Americans' Reactions to September 11. *JAMA: Journal of the American Medical Association, 288*(5), 581-588.

Schottenbauer, M. A., Rodriguez, B. F., Glass, C. R., and Arnkoff, D. B. (2004). Computers, anxiety, and gender: An analysis of reactions to the Y2K computer problem. *Computers in Human Behavior, 20*(1), 67-83.

Searle, K. and Gow, K. (in preparation). Climate Change Distress.

Silver, R. C., Holman, E. A., McIntosh, D. N., Poulin, M., and Gil-Rivas, V. (2002). Nationwide longitudinal study of psychological responses to September 11. *JAMA: Journal of the American Medical Association, 288*(10), 1235-1244.

Slovic, P. (1987). Perception of risk. *Science, 236*(4799), 280-285.

Storch, E. A., Roberti, J. W., Bravata, E., and Storch, J. B. (2004). Psychometric Investigation of the Santa Clara Strength of Religious Faith Questionnaire--Short-Form. *Pastoral Psychology, 52*(6), 479-483.

Strous, R. D., Shtain, M., Oselka-Goren, H., Lustig, M., Stryjer, R., Zerzion, M., Baruch, Y., Chelben, J. (2000). Anticipatory reactions of psychiatric inpatients to the year 2000. *Journal of Nervous and Mental Disease, 188*(11), 786-788.

Sundblad, E.-L., Biel, A., and Gärling, T. (2007). Cognitive and affective risk judgements related to climate change. *Journal of Environmental Psychology, 27*(2), 97-106.

Tolsma, F. J. (1968). Doom expectancy from a psychopathological point of view. *Psychiatria, Neurologia, Neurochirurgia, 71*(5), 435-446.

Tucci, J., Mitchell, J., and Goddard, C. (2007). Children's fears, hopes and heroes: Modern childhood in Australia. The Australian Childhood Foundation. Retrieved 15.09.07 from http://www.childhood.org.au

UK News. (2007). *UK Kids are Greener than their Parents.* Retrieved 20.09.07, from http://www.lifestyleextra.com/ShowStory.asp?story=MI2334379Dandnews_headline=uk_kids_are_greener_than_their_parents

Ungar, S. (2001). Moral panic versus the risk society: The implications of the changing sites of social anxiety. *British Journal of Sociology, 52*(2), 271-291.

Welch, M. (2006). Seeking a Safer Society: America's Anxiety in the War on Terror. *Security Journal, 19*(2), 93-109.

Zaleski, Z. (1996). Future anxiety: Concept, measurement, and preliminary research. *Personality and Individual Differences, 21*(2), 165-174.

Contributed By Ray Norton, Melbourne.

Photo: Is This What Lies Ahead For Us Now?

APPENDIX

Format for Interviews: With the Focus of "Keeping in Mind the Issues about Climate Change

How do you conceptualise the issues surrounding climate change?
What are your concerns about the issues surrounding climate change?
When did you begin to feel concerned/ distressed about the issues surrounding climate change?
Did your concern/distress about the issues surrounding climate change come on gradually, or was there a specific event that triggered your concerns? (For example, watching footage of the Boxing Day Tsunami on TV, experiencing flooding, death of an animal due to extreme weather patterns etc). If so, what was that event? How would you describe your response to that event? What concerned you most about that specific event? How long after that event did you begin to feel concern/distress? Do you ever re-experience that event in any way? (for example, recurring images, thoughts, dreams etc.)
How would you describe your response to the issues surrounding climate change? (emotional and physiological)? How would you rate the intensity of that feeling on a scale from one to ten where one is very low and ten is very high?

How often does this occur?
What events or experiences remind you of your issues surrounding climate change distress? (For example, climate related programs on TV, news events related to extreme weather events, dam levels, looking at dying vegetation/animals, experiencing record temperatures.) When you experience these reminders of the issues surrounding climate change, what is your response? (Both emotional and physiological.)
Do you ever avoid situations or experiences that reminded you of the issues surrounding climate change? (For example, you may stop watching the news, avoid conversations about it, or you may try to stop thinking about the issues surrounding climate change.)
Do you ever feel that your response to the issues surrounding climate change is excessive or unreasonable? Do you ever feel that others think that your response is excessive or unreasonable?
Do you ever find it difficult to control your worry about the issues surrounding climate change?
How have your feelings about the issues surrounding climate change impacted on your life? (Socially, occupationally, lifestyle, relationships, mental health)
On a scale from one to ten where one is very low and 10 is very high, how would you rate the extent to which your concern/distress about the issues surrounding climate change has impacted on your functioning in these areas? (Socially, occupationally, lifestyle, relationships, mental health)
Who have you shared your concerns with? What was their response?
What do you find helpful in managing your concern/distress about the issues surrounding climate change?
Does your understanding of the issues surrounding climate change have any impact on your meaning in life? How do you "make sense" of it? Do you ever think that the issues surrounding climate change are associated with the end of the world? Does thinking of the issues surrounding climate change give you a sense of a foreshortened future (For example, not expecting to have grandchildren, or not living a normal lifespan)? What do the issues surrounding climate change mean for your future?
Since becoming concerned/distressed about the issues surrounding climate change, do you ever experience the feeling of being detached, numb, dazed, have the sense that either yourself or the outside world is not real, or have memory blocks about aspects of the issues surrounding climate change that distress you? Do you have difficulty sleeping, poor concentration, feelings of restlessness, feelings of being on edge, feelings of irritability, an exaggerated startle response, less interest or less participation in activities, become easily fatigued, have muscle tension, be tearful, have feelings of hopelessness, general pessimism about the future or feelings of worthlessness ? Are you taking any recreational drugs, medications, or do you have a medical condition (including a prior mental health diagnosis) that may account for any of these symptoms?

In: Meltdown: Climate Change, Natural Disasters… ISBN 978-1-60876-153-1
Editor: Kathryn Gow

Chapter 4

PROVIDING SCAFFOLDING FOR STUDENTS IN LIGHT OF CONCERNS OVER CLIMATE CHANGE: AN EDUCATIONAL PERSPECTIVE

Lida Lee
Hong Kong Shue Yan University, Hong Kong

ABSTRACT

Taking an educational psychology perspective about preparing students for climate change scenarios, I argue the position that education can serve as a vital path to alleviate concerns and worries, among the younger generations, over the potential impact of climate change and to prepare them for meeting that challenge. In this chapter, I explore what educators might do to assist in easing students' worries about the issues, as well as how to raise their awareness and knowledge about climate change. The context is set within the local educational environment of Hong Kong, in which knowledge pertinent to climate change is typically covered in school subjects such as geography, integrated science, and liberal studies.

Keywords: climate change, educational psychology, school curriculum, Hong Kong

INTRODUCTION

Climate change has captured the spotlight around the globe. Numerous organizations, government offices, and research establishments have been set up to study or tackle issues brought on by climate change at the local, regional, or global level. Among them are, for instance, the Intergovernmental Panel on Climate Change (IPCC)[1] and the Climatic Research Unit of the University of East Anglia in the U.K.[2] Countries such as France, Japan, Italy,

[1] http://www.ipcc.ch/.
[2] http://www.cru.uea.ac.uk/cru/about/.

Britain, Canada, and the United States have been engaging in dialogue on issues such as the reduction of greenhouse gases, which is often tied to climate change. An endeavour sought by the nations is the well-known Kyoto Protocol,[3] which is a mechanism put in force in 2005. Other ongoing events included the United Nations' Convention to Combat Desertification (UNCCD)[4] and the celebration of the "World Day to Combat Desertification" on the 17th June every year,[5] the latter being an effort to raise awareness on combating desertification in nations with serious drought which is sometimes tied to climate change. The wave of concern over climate change attracted public attention recently when a number of special reports on global warming were issued in *Time* magazine (e.g., Underhill, 2007, April 16). This focus on climate change manages to touch our everyday lives at the individual level.

THE LOCAL CONTEXT OF HONG KONG

In Hong Kong, there has been an increasing awareness about the impact of climate change. According to the Hong Kong Observatory Technical Note (Local),[6] local people experienced an increase in hot nights by more than 50% during the period 1971-2000 and Lee, Leung, and Chan (November, 2006) confirm that such changes are "in line with global warming and the increasing extent of urban development in Hong Kong" (p. 6).

In fact, measures have been implemented by the local government to protect the environment in light of concern over the impact of urban development on local climate change. For instance, a campaign called the "Action Blue Sky"[7] was launched in 2006 by the local government to raise people's awareness on air quality control. Local companies and schools joined the campaign by pledging to save energy in their everyday lives. Additionally, the local government introduced the Smoky Vehicle Emission Control Programme to ensure that owners of vehicles emitting excessive black smoke may be asked to bring in their vehicles for testing.[8] Under the programme, repeated failures to comply may result in having their vehicle licences cancelled.

Moreover, coping with climate change is regarded as an important issue in a special meeting of the Finance Committee in the Environmental Bureau held in 2008.[9] Also, at the regional level, there have been efforts to strengthen environmental cooperation. For instance, Mr. Edward Yau, the Secretary for the Environment, went to Guangdong in China in 2007 in an attempt to strengthen dialogue about ways to improve air quality for the region (Environment Bureau (EB), n.d.). Furthermore, about 70 Asian cities, including Hong Kong, switched off lights on June 21, 2008 in an attempt to raise people's awareness about reducing light pollution and also to save our Earth from global warming.[10]

Given the momentum in the local context, what can educators do, for instance, about raising people's awareness about climate change and/or assisting them to cope with their concerns about the impacts of climate change? Specifically, how can teachers educate

[3] http://unfccc.int/kyoto_protocol/items/2830.php.
[4] http://www.unccd.int/convention/text/convention.php) (UNCCD, n.d.a).
[5] http://www.un.org/events/desertification/2007/ (UNCCD, n.d. b).
[6] No. 83 published by the Hong Kong Special Administrative Region Government.
[7] http://www.epd.gov.hk/epd/english/action_blue_sky/action_blue_sky.html (EPD, n.d. a).
[8] http://www.epd.gov.hk/epd/english/environmentinhk/air/prob_solutions/cleaning_air_atroad.html (EPD, n.d. b).
[9] http://www.enb.gov.hk/en/legco_matters/finance_committee/speech_08_09.html.

students for meeting real world challenges, such as climate change impacts? In the following sections of this chapter, some relevant questions are explored: Which theoretical perspective can be instrumental to our understanding more about the dynamics in play regarding students' awareness and/or concerns about climate change? What are the possible roles that teachers might play? Which instructional methods might be conducive in helping students learn more about issues related to climate change? What are the pertinent learning outcomes that would be beneficial to students?

The Social Cognitive Approach

To facilitate our understanding about the dynamics in play regarding students' concerns about climate change, psychology has much to offer. Among the various psychological perspectives, Albert Bandura's social cognitive theory is very instructive. To Bandura, the child is an active agent for him/herself. A given individual can exercise control over his/her own behaviour and make things happen when interacting with the environment. In the process, the person's intention and self-beliefs are influential. The person would initiate action to fulfill their intention or purpose and his/her self-efficacy beliefs would guide and sustain the ongoing effort. Having a sense of personal efficacy is vital, as it is a key feature of personal agency. According to Bandura (1997), self-efficacy can affect the course of action a person would take, how persistent the person is, how resilient they are when facing adversity, whether he/she would cope well or become depressed with dealing with environmental demands, etcetera. In the face of real world challenges, such as climate change, it is desirable that young generations believe that they are capable of mastering their own future. For this reason, the concept of self-efficacy is relevant to our discussion in climate change.

The notion of reciprocal causation in social cognitive theory is also helpful in understanding the dynamic interplay between human behavior and the environment. Reciprocal causation can be presented as a triangle comprising three angles or elements: the person (including beliefs, cognition, emotions and personality), the behaviour, and the environment. A given individual's behaviour can then be understood in terms of triadic reciprococity among these factors. For instance, salient changes in global climate (i.e., the environment) may reinforce our beliefs about the importance of protecting our ecology or may increase our concerns about potential natural catastrophes that may eventuate. Such beliefs or concerns may then be manifested in our observable behaviors, such as using fewer plastic bags and reducing vehicle emission which, in turn, help to improve the environment.

Also noteworthy in the social cognitive framework is the notion of modeling. Bandura (1977, 1986) postulated that children can learn vicariously from their models by the means of observational learning. Modeling is a process which involves interaction between a learner and his/her model and which influences the learner's decisions as well as their new behaviours (Bandura, 1986). Indeed, modelling is a vital path for transmitting attitudes, values and thoughts to a given individual. An implication here is that teachers and parents who want to educate young children about climate change can teach by their own examples. That is, they can provide the opportunity for children to observe them conserving electrical energy in everyday lives; the children may then come to realize that such behaviour is

[10] http://www.foe.org.hk/lightsout/en/index.html (Friends of the Earth (HK), n.d.).

important for them also. On the other hand, adults may also serve as models of indifference to the issue of climate change or environmental protection in their daily lives.

Additionally, the social cognitive theory posits that motivation is the key to activate what one has learned. That is to say, once children have paid attention to their models (e.g., teachers) and remembered information about the behaviour to be modeled, they need incentives to motivate them to actually perform what have been observed. In this regard, the classroom setting can serve as an arena for children at the school age to actualize and practise what they have learned regarding, for example, environmental protection designed to ease the concern about climate change. In this respect, what we gain from the social cognitive perspective is an optimistic picture: not only do students serve as their own agent, but also school teachers have a role to play in the context (see Dunn et al., 1994, about the effect of context). Teachers can change student behavior directly by providing incentives and serving as a model or indirectly by altering students' cognition which then leads to a change of behaviour.

THROUGH AN EDUCATIONAL LENS

After a brief introduction of a psychological theory pertinent to the issue of climate change and its relevance to educating school children, we ask what teachers might do to assist in this regard. Specifically, what roles might teachers take in raising students' awareness or easing their worries about climate change? What instructional strategies could be instrumental in educating students to meet real world challenges such as climate change?

Roles of Teachers

Despite the fact that climate change is not an independent subject in Hong Kong schools, local teachers who want to educate students about climate change can still take part through the behavioural, emotional, and intellectual dimensions of student life in schools.

If we consider the behavioral dimension, teachers can serve as models for students in the classroom. According to the social cognitive perspective, the characteristics of the model are critical if the children are to pay attention to the model or not. To be an effective model, teachers need to appear to their students as credible and enthusiastic and give correct and up to date scientific information on the issue. When students see that their teachers are credible and are already carrying out the activities that they have suggested the class do (e.g., use fewer plastic items, or save electricity by turning off appliances when idle) and witness their passion and belief that this will have positive effects on the environment, students are more likely to pay attention to teachers as appropriate role models. Such vicarious forms of learning are beneficial to students because they are a resource in the development of students' self-efficacy (Bandura, 1997).

With respect to the emotional dimension, teachers can utilize counseling-related skills for strengthening students' capacities to deal with emotional concerns over authentic issues such as climate change when they interact with students (see Davis, 2005). Teachers obviously already have a lot to do in the classroom and referring over-anxious students to school counselors is advisable. Teachers can also work in conjunction with school counselors in conducting group discussions: for example, by sharing with peers ways of mobilizing inner resources; or talking about current issues to raise students' awareness about the impact of climate change; or by giving correct information on effects and ways of mitigating the negative effects, so as to alleviate potential concerns. One interesting and simple topic is to discuss the effect of heat on health and behaviour. Research has revealed that heat can be connected to aggressive behaviour such as assaults and mental health complications (see Cohn and Rotton, 1997, Hansen, Nitschke, Ryan, Pisaniello, and Tucker, 2008), and carefully led discussion groups on the issue can be useful in raising students' awareness, thereby possibly preventing emotional-behavioural manifestations such as mood swings or aggression during heat waves which are predicted to increase in the future.

In relation to the intellectual dimension of student life in school, teachers also have much to offer. They can equip students with what they may need to know about the impact of climate change personally if they have the knowledge and skills or arrange for guest speakers who are experts on the specific topic areas. Because the sources of students' information about local environmental issues (including humans' impact on the weather, and the effect of the increase in carbon dioxide content on the atmosphere) come mainly from their geography lessons and also textbooks (Yeung, 1998), school teachers can have a role to play in educating students about climate change. In fact, recent evidence demonstrates that students learning about geography can be deepened as a result of the use of particular instructional approaches (Kwan, 2008; Yeung, 2002). In the process of teaching about climate, teachers can also help instill in students the importance of having pro-environmental behaviours, particularly around issues researched by local scholars (e.g. Lee, 1997; Yeung, 1998). Some studies have revealed a lack of pro-environment habits among students in Hong Kong. For instance, in Yeung's (1998) study, only 1.8% of the students indicated one or more pro-environment habits (e.g., horticulture, hiking); and the rate of pro-environment behaviours that involved more active personal effort (e.g. taking part in environmental protection work) were shown to be much lower than general behaviours that required only some degree of inconvenience (e.g., persuading others to be environmentally friendly, and behaving themselves as environmentally friendly consumers).

Desirable Elements of the Learning Processes

As mentioned earlier, teachers have much to give in educating students about climate change. Yet, before addressing the question of how to better teach students about the issue, we first need to know which elements are conducive to student learning.

It appears that student collaboration is essential. People-relating skills are vital, because students are now living in an increasingly complex global village. Issues and problems such as climate change cannot be solved merely at the individual or even the local level. Rather, mutual dialogue and ongoing cooperation among parties concerned are needed. Local educators appear to have such foresight as recently forums discussing education in relation to globalization have been held. Indeed, classrooms are sociocultural contexts in which learning occurs and in which students learn to cooperate with their fellow classmates. Vygotsky (1978) states that social interactions with peers are beneficial for learning which in turn enforces positive child development.

While interacting with peers, students may come to a better realization about, for example, what climate change means and what its ramifications may entail. Interactions with peers can stimulate the learner's cognitive development by inducing cognitive conflicts, which result in changes of the existing knowledge schemas as suggested by Jean Piaget. For example, students may originally have some misconceptions, or even be indifferent toward the negative effects of climate change, but they may change their stance after acquiring new information about climate change from their peers. Students are not merely passive receivers of information; they have the capacity to self-reflect and act as their own agent (Bandura, 1989). Critical reflection on the part of the learner is thus an essential element in the learning process. With regard to climate change, it is desirable that students should be able to reflect on the phenomenon and its relevant issues, develop insights about what changes their behaviours may bring to the environment, and decide on any actions they may want to take, instead of just staying at the level of passively receiving factual information about climate change. Such heightened concern and ability to notice what is happening in the local and global environment is akin to Gardner's (1983) notion of naturalistic intelligence.

Given the intricacies of issues about climate change, innovative ways to manage them would likely be needed. To prepare younger generations for meeting unprecedented challenges in the future, teachers need to arrange for learning activities so that students can be exposed to new experiences via creative problem-solving, which is also a desirable component in the learning process. In geography lessons, for example, teachers can plan for students to undertake in-class exercises which require students to experiment with different methods in solving problems around the theme of climate change. In the process, students may need to compare, for instance, random searches, heuristics (such as means-ends analysis), and algorithms and then decide for themselves which is the best approach in investigating and solving the problems at hand.

Lastly, educators generally hope that students will be able to apply what they have learned in the classroom to their everyday lives. This coincides with the long held ideal for education as a social function (Dewey, 1916). Transferability of learning is among the most desirable elements of learning. Certainly, genuine encounters with real world challenges, such as climate change, do not stay within the confines of the school. Each student's awareness and ability needs to be able to transfer their acquired knowledge and skills beyond the classroom setting. As such, enterprising teachers could tailor classroom activities accordingly, so as to prepare students for taking part in constructive dialogues and brainstorming potential solutions on climate change. Obviously teachers need to be given the resources and the time to add these additional activities to their already full schedules.

Relevant Instructional Approaches

Within the educational context of Hong Kong, knowledge pertinent to climate change is normally covered in school subjects such as geography, integrated science, and liberal studies. In terms of instructional strategies relevant to teaching about climate change, the didactic approach which is focused on transmission of facts to students is commonly used. However, this instructional approach is not ideal, as it is rather weak in promoting students' independent thinking and positive behaviour (Marsden, 1995, cited in Yeung, 2002). Now I attempt to put forward a few other approaches: problem-based inquiry, computer-mediated learning, and collaborative learning.

Problem-based Learning

Problem-based learning (PBL) is a method adopted in medical schools from the early 1970s and it represents a shift from teacher-centred to student-centred learning (White, III, 2001). PBL consists of "carefully selected problems... related to the objectives of learning in which the learner acquires the critical knowledge and skills of problem solving, self-directed learning strategies and team participation" (Stimpson, 2004, pp. 161-162). The benefits of PBL for students lie, not only in the acquisition of content knowledge, but also in training in critical thinking skills, communication, and solving authentic problems (Kiernan, Murrell and Relf, 2008; Stimpson, 2004). Research has shown that students learn better when they are asked to do something instead of merely read textbooks (Mueller and Fleming, 2001).

In problem-based learning, students are normally assigned to groups of eight to ten, in which a teacher plays the role of a tutor in helping students work towards solving a problem. With respect to teaching climate change to students, educators can anchor the classroom learning experience within a well-defined case or problem that is relevant to climate change. The purpose and value of the learning activities must be made clear to students in the beginning. Also, the task needs to be an authentic one; the best ones being those that are situated in the local context (e.g., "What can we as school children do about the potential impacts brought by climate change on our everyday lives?" "How we can tackle the excessive use of plastic bags by local people?). In the process, students are not merely asked to study the facts or information which their geography or science texts may provide, but rather to engage in scientific discourse about, for example, any initial solution paths they may have and to test their initial ideas as well as seeking alternative solutions. The learning outcomes can be beneficial; for instance, geography students studying in a secondary school were found to have obtained more in-depth knowledge, as well as problem-solving skills, in environmental learning through the use of PBL (Kwan, 2008).

To make the best out of PBL projects, students are encouraged not to limit sources of information to textbooks and readings assigned by teachers, but to make use of pertinent information such as greenhouse effect, global warming, and global sea levels available from the Hong Kong Observatory.[11] Examples include a user-friendly website called the "Energyland"[12] (and a newsletter called "EnergyWits"[13] for raising public awareness about heat and energy-saving issues. Moreover, teachers can ask students to undertake research projects under supervision on potential impacts brought about by climate change; such as the rise in temperature, rainfall, visibility, sea level, cloud amount, and tropical cyclones. For senior school students, teachers can engage them in learning about climate change and its ramifications at the national and global level. Students can be asked to do project reports that may utilize information made available by, for example, the Australian Bureau of Meteorology,[14] the Intergovernmental Panel on Climate Change (IPCC) Assessment

[11] http://www.weather.gov.hk/abouthko/logoexplain/logoexplaine.htm.
[12] http://www.energyland.emsd.gov.hk/eng/new/index.htm.
[13] http://www.emsd.gov.hk/emsd/eng/welcome/index.shtml.
[14] http://www.bom.gov.au/climate/change/.

Reports,[15] and the "Climate Change Information Kit" provided by the United Nations Environment Programme (UNEP).[16]

Furthermore, it would be most desirable if teachers are ready to step in and facilitate student discussion in the process. When needed, teachers may help to alert students to thinking through relevant issues and questions, by prompting through questions such as: "Is there any evidence for you to support the concern over climate change?" "What are the potential impacts of climate change?" "How are phenomena such as the greenhouse gas effect related to climate change?" "What happens if we ignore climate change?" "How can we decide on feasible solutions to the problem of climate change?" "What are the strengths and weaknesses of each solution?"

Computer-mediated Learning

Teachers can take advantage of the interactive features in learning afforded by computers in this Internet Age. Computers and multimedia devices can serve as teaching/learning tools that help to stimulate or maintain children's motivation to learn. An example of such educational tools is a fun zone called "EnergyWits",[17] designed to raise people's awareness and teach them about energy-saving, which can be linked to climate change.

While using computers, students can gain more control over their own learning. For instance, they can search online information anytime at their own pace and can use tools such as *Flash* to create interactive artifacts for themselves. Besides, learners in the computerized learning environment reported that they enjoyed learning and acquired new skills and knowledge (MacDonald, Stodel and Chambers, 2008); they were also found to have higher level of satisfaction and also more improved learning performance than those in the traditional environment (Chou and Liu, 2005). Other than assisting student learning at the individual level, computer tools also support learning in group settings (known as computer-supported collaborated learning). In the latter context, collaborative tools such as video conferencing, co-authoring tools, and voting tools may be used. Students can also work on projects in the asynchronous or synchronous access mode. In addition, students can also take the opportunity to learn about how to conserve energy when using computers.

Given the fact that students in Hong Kong are not living in an environment marked by natural phenomena such as desertification, drought and ozone layer damage, teaching about climate change using textbooks alone makes education seem like spoon-feeding. Some students may not be able to sense the relevance or importance of global climate change for their everyday lives. Yet, when coupled with simulation games on climate change afforded by computers, local students can get a "real" sense of what the climatic phenomenon may bring to them. The learning outcome would be even more fruitful when students are asked to conduct follow-up activities, such as group investigations or discussions on the pertinent issues.

[15] http://www.ipcc.ch/ipccreports/assessments-reports.htm.
[16] http://www.unep.org/.
[17] http://www.energyland.emsd.gov.hk/eng/energy/prin_forms_six.htm.

Collaborative Learning

Teachers can plan for students to experience collaborative learning, which is sometimes referred to as a social contract among peers or between peers and the teachers (Dillenbourg, 1999) and which leads to creation of knowledge (Caplow and Kardash, 1995, as cited in Mueller and Fleming, 2001). Not only does collaborative learning serve the social needs of students, but it also has psychological significance as it can be tied to creativity (Wegrif, 2005) and can help child development to proceed from the inter-psychological to intra-psychological plane (Vygotsky, 1962, 1978). In addition, such a student-centred approach parallels a current trend in the teaching of geography (Stimpson, 2005).

In the classroom, teachers can assign or encourage students to form groups for undertaking collaborative projects. For example, they may be asked to examine and report on a related phenomenon such as holes in the ozone layer. In the process, different members in a given group may have the chance to take up different roles. Instead of focusing on a climatic phenomenon, teachers may also choose to design project activities applied to a certain locality. For example, the teacher may ask students to study how India has managed to fight climate change in its national policy. The project may require students to first conduct an information search about this nation, which may perhaps be quite familiar to some people here, and then to see what and why Indian companies are doing about investment in renewable energy or clean technologies. Not only would students gain some knowledge about potential solutions to climate change, but they may also develop an appreciation for what other nations, such as India, are doing for the common good of humanity.

Additionally, there are specific methods or strategies utilized in collaborative learning approaches. One useful way is called the jigsaw method. Depending on the teacher's instructional plan, the jigsaw method may be run differently. A simple example is for a class of twenty to be divided into four groups, with each having five members. Then, one member from each group enters a new expert team, which is composed of one member from each of the original group. The Geography teacher, for instance, may divide the reading materials into four sections (or divide a problem into four sub-parts to be solved), each of which go to one expert team for students to master (or to solve their part of the problem). Students then discuss and also devise their own strategies to teach the section of the reading material or the solution to their classmates. After that, students leave the expert team and rejoin their original grouping, in which they act as experts to teach their fellow classmates about what they have found and learned in the expert team.

Depending on the educational objectives set by the teacher, group size and composition may vary. It is desirable that the teacher monitors ongoing development of student groups and, when needed, re-assigns roles in the group in order to facilitate the process. At times, the teacher may also provide the needed resources to support student learning. An example of such resource tools is the *Handbook for the Montreal Protocol on Substances that Deplete the Ozone Layer*, which was published by the United Nations Environment Programme.

Conclusion

Climate change is of increasing concern to many people nowadays. Governments and officials across nations have been engaging in ongoing dialogue seeking to find potential solutions. Within the local context of Hong Kong, measures have been taken by the government in light of concerns about climate change. In the academic realm, professionals such as educators have their part to play also. To help prepare students for meeting real world challenges, teachers can make use of instructional strategies instrumental to teaching students about climate change. As aforementioned, instructional approaches such as student-centred collaborative learning are desirable; whereas the didactic approach alone may not serve well, especially when the instructional objectives are not merely about the transmission of facts to students.

Moreover, a myriad of other strategies can also help to enrich student learning about climatic phenomenon: for example, field trips to the Hong Kong Observatory, board displays in school on the Concern for Global Warming Day, visiting experts talking about the impact of climate change, etcetera. As such, educators have an important role to play in the process. Such roles can actually be understood within the social cognitive perspective. An underlying assumption of the theory is triadic reciprocity among the person, behavior, and external environment variables.

Notably, certain conditions and support are needed for reinforcing local teachers to help and teach students about the impact of climate change. Effective practice of instructional methods such as problem-based learning, collaborative learning and interactive learning would require the teacher to have at least a sense of the know-how, if not formal training, in being able to facilitate learning: "What kind of problems or cases should I assign to my class?" "How can I identify and/or nurture students' naturalistic intelligence?" "What group size may work better?" "What level of task difficulty would be appropriate?" "What kind of group composition is ideal?" "What can I do about group conflicts when they emerge?" Also important is the availability of computing resources and technical support from school management for backing dynamic or interactive approaches in teaching about climate change. Such institutional support is essential for conveying to teachers a sense of the value about what they are doing for students.

Despite the challenges for teachers, there are opportunities within the local Asian region. Given that climate change does not stand alone as a school subject, we have been addressing it in pertinent subjects such as geography and liberal studies, as well as in the cross-disciplinary context. For the former, my aforementioned discussion on instructional strategies is relevant. Also, with the topic of climate change being embedded in environmental education, we actually have the opportunity to expose students to even broader issues of, for example, forming environmentally friendly behaviour, having an holistic understanding of all aspects of environmental problems at the local and global levels, as well as appreciating and loving nature, which are the objectives of environmental education set in the *Guidelines on Environmental Education in Schools* (Curriculum Development Council, 1992). For the latter, local educators also have a lot of room to maneuver. For instance, we may adopt ideas from Porter's (1991, cited in Kwan, 2003) suggestions on citizenship education. With respect to teaching about climate change, we may then set our sights on helping students to understand pro-environmental behavior/attitude as the connection or relationship between them and the climatic phenomenon in question, as their being part of humanity, or as something that requires of them their active participation in the issue.

Apart from this, the Hong Kong Government has, over time, played its part in educating the public about environmental protection. There are environmental resource centres and also the green education and resource centre being set up in various locations to provide information, activities and programmes. For example, the largest environmental resource centre is the Fanling Environmental Resource Centre, which was opened in 2003 and occupies about 400 square metres. The Environmental Campaign Committee (ECC), which was founded in 1990, aims to educate children about the importance of environmental protection (Lee, 1997). There is also the Student Environmental Protection Ambassador (SEPA) Scheme held in connection with local primary and secondary schools. Outstanding student ambassadors travel overseas to learn about environmental protection and join environmental leadership summer camps.

There are challenges as well as opportunities related to problem solving in relation to climate change which occurs on all fronts, individual/societal and local/global. What is important for us is to constantly remind ourselves whether we are determined to educate the younger generations for meeting real world challenges. In respect to this, local teachers in Hong Kong have been found to have positive intentions toward promoting environmental education (Lee, 2000), which is good news. It sets a baseline from which we embark on our journey. Once we are determined, ongoing dialogue and subsequent action will follow.

REFERENCES

Bandura, A. (1977). *Social learning theory*. Upper Saddle River, NJ: Prentice Hall.

Bandura, A. (1986). *Social foundations of thoughts and action: A social cognitive theory*. Upper Saddle River, NJ: Prentice Hall.

Bandura, A. (1989). Human agency in social cognitive theory. *American Psychologist, 44*(9), 1175-1184.

Bandura, A. (1997). *Self-efficacy: The exercise of control*. New York: W. H. Freeman and Company.

Climatic Research Unit. (n.d.). *About the Climatic Research Unit*. Retrieved December 18, 2008, from the University of East Anglia, Climatic Research Unit Web site: http://www.cru.uea.ac.uk/cru/about/

Chou, S.-W. and Liu, C.-H. (2005). Learning effectiveness in a Web-based virtual learning environment: a learner control perspective. *Journal of Computer Assisted Learning, 21*, 65–76.

Cohn, E., and Rotton, J. (1997). Assault as a function of time and temperature: A moderator-variable time-series analysis. *Journal of Personality and Social Psychology*, *72*(6), 1322-1334.

Curriculum Development Council. (1992). *Guidelines on environmental education in schools*. Hong Kong: Government Printer.

Davis, T. (2005). *Exploring school counseling: Professional practices and perspectives*. Boston, MA: Houghton Mifflin Company.

Dewey, J. (1916). *Democracy and education*. New York, NY: The Macmillan Company.

Dillenbourg, P. (1999). What do you mean by collaborative learning? In P. Dillenbourg (Ed.). *Collaborative-learning: Cognitive and computational approaches* (pp.1-19). Oxford: Elsevier.

Dunn, W., Brown, C., and McGuigan, A. (1994). The ecology of human performance: A framework for considering the effect of context. *The American Journal of Occupational Therapy, 595-607.*

EPD (Environmental Protection Department). (n.d. a). *Clean air for a cool Hong Kong.* Retrieved December 16, 2008, from the Government of the Hong Kong Special Administrative Region, Environmental Protection Department Web site: http://www.epd.gov.hk/epd/english/action_blue_sky/action_blue_sky.html

EPD (Environmental Protection Department). (n.d. b). *Changing the air at street level.* Retrieved December 16, 2008, from the Government of the Hong Kong Special Administrative Region, Environmental Protection Department Web site: http://www.epd.gov.hk/epd/english/environmentinhk/air/prob_solutions/cleaning_air_atroad.html

EB (Environment Bureau). (n.d.). *Speech by SEN at the FC Special Meeting.* Retrieved December 18, 2008, from the Government of the Hong Kong Special Administrative Region, Environment Bureau Web site: http://www.enb.gov.hk/en/legco_matters/finance_committee/speech_08_09.html

Friends of the Earth (HK). (n.d.). *Dim it.* Retrieved December 18, 2008, from http://www.foe.org.hk/lightsout/en/index.html

Gardner, H. (1983). *Frames of mind: The theory of multiple intelligences.* New York, NY: Basic Books.

Hansen, A., Bi, P., Nitschke, M., Ryan, P., Pisaniello, D., and Tucker, G. (2008, October). The effect of heat waves on mental health in a temperate Australian city. *Environmental Health Perspectives*, *116*(10), 1369-1375. Retrieved March 22, 2009, from Academic Search Premier database.

Intergovernmental Panel on Climate Change. (n.d.). *About IPCC.* Retrieved December 18, 2008, from the World Meteorological Organization and the United Nations Environment Programme, the Intergovernmental Panel on Climate Change Web site: http://www.ipcc.ch/

Kiernan, M. J., Murrell, E., and Relf, S. (2008). Professional education of psychologists using online problem-based learning methods: Experience at Charles Sturt University. *Australian Psychologist, 43*(4), 286-292.

Kwan, T. (2003). Geography and Citizenship Education in Hong Kong. *International Research in Geographical and Environmental Education, 12*(10), 64-71.

Kwan, T. (2008). Environmental learning using a problem-based approach in the field: A case study of a Hong Kong school. *International Research in Geographical and nvironmental Education, 17*(2), 93-113.

Lee, J. C. K. (1997). Environmental education in schools in Hong Kong. *Environmental Education Research. 3*(3), 359-71.

Lee, T. C., Leung, W. M., and Chan, K. W. (November 2006). *Climatological Normals for Hong Kong 1971- 2000* (Technical Note (Local) No. 83). Hong Kong: Hong Kong Special Administrative Region Government.

Lee, J. C. K. (2000). Teacher receptivity to curriculum change in the implementation stage: the case of environmental education in Hong Kong. *Journal of Curriculum Studies, 32*(1), 95-115.

MacDonald, C., Stodel, E., and Chambers, L. (2008, March). An online interprofessional learning resource for physicians, pharmacists, nurse practitioners, and nurses in long-term care: Benefits, barriers, and lessons learned. *Informatics for Health and Social Care, 33*(1), 21-38. Retrieved March 22, 2009. http://www.informaworld.com/openurl?genre=journalandissn=1753-8157

Mueller, A., and Fleming, T. (2001). Cooperative learning: Listening to how children work at school. *Journal of Educational Research, 94*(5), 259-265.

Stimpson, P. (2004). Coming to grips with change: The initial training of geography teachers in Hong Kong. *International Research in Geographical and Environmental Education, 13*(2), 159-63.

Stimpson, P. (2005). A Hong Kong reflection: Similar theme, different story. *International Research in Geographical and Environmental Education, 14*(1), 83-85.

Underhill, W. (2007, April 16). This way forward. *Time*, 37-43.

United Nations Convention to Combat Desertification. (n.d.). *Desertification and climate change: One global challenge.* Retrieved December 18, 2008, from the United Nations, United Nations Convention to Combat Desertification Web site: http://www.un.org/events/desertification/2007/

UNCCD (United Nations Convention to Combat Desertification). (n.d. a). *Text of the United Nations Convention to Combat Desertification.* Retrieved December 18, 2008, from http://www.unccd.int/convention/text/convention.php

UNCCD (United Nations Framework Convention on Climate Change). (n.d. b). *Kyoto Protocol.* Retrieved December 18, 2008, from http://unfccc.int/kyoto_protocol/items/2830.php

Vygotsky, L.S. (1962). *Thought and language*. Cambridge, MA: MIT Press.

Vygotsky, L. S. (1978). *Mind in society*. Cambridge, MA: MIT Press. Wegerif, R. (2005, May). Reason and Creativity in Classroom Dialogues. *Language and Education: An International Journal, 19*(3), 223-237. Retrieved March 22, 2009, from Academic Search Premier database.

White, III, H. B. (2001). Getting started in Problem-based learning. In B. J. Duch, S. E. Groh, and D. E. Allen (Eds), *The power of problem-based learning*. Sterling, VA: Stylus Publishing.

Yeung, S. P.-M. (1998). Environmental consciousness among students in senior secondary schools: The case of Hong Kong. *Environmental Education Research, 4*(3), 251-68.

Yeung, S. P.-M. (2002). Teaching Approaches and the Development of Responsive Environmental Behaviour: The Case of Hong Kong. *Ethics, Place, and Environment,* 5(3), 239-69.

Contributed by Tim McCrorey.

Photo: Wildlife Utilising Mankind's Scaffolding.

In: Meltdown: Climate Change, Natural Disasters… ISBN 978-1-60876-153-1
Editor: Kathryn Gow

Chapter 5

DO GLOBAL CONCERNS AFFECT DREAM CONTENT MORE THAN PERSONAL CONCERNS?

Sandra Sacre[1], Kathryn Gow[1], Julie Hansen[1] and Mark Blagrove[2]

1. Queensland University of Technology, Australia
2. Swansea University, United Kingdom

ABSTRACT

At a time when global events such as natural disasters, rising oil prices, climate change and terrorist attacks seem to be prominent in our wider environment, it would not be surprising that concerns that individuals have about these matters might affect their unconscious and find their way into their dreaming. A large body of literature shows that personal concerns are frequently incorporated into dreams; although there is a dispute at the theoretical level as to whether individuals process and resolve these problems while they sleep and dream, or whether dreams merely reflect waking preoccupations. On the other hand, it could also be argued that human beings are inherently resilient and have the capacity to cope and survive by blocking out threats (to their wellbeing) that surround them in their daily lives, especially when such threats are distant or global, and hence dream incorporation should not occur. Or, while we may be worried about global events, it is also possible that these events remain less immediate, while directly personal events are much more stressful to deal with. This chapter reports on an empirical investigation into the incorporation of world wide and immediate personal concerns into dreams. In this chapter, we will explore whether greater weight is attached to personal concerns in dreams, or to global events such as 9/11, the Bali bombings, the Iraq war, fears about global warming, the worsening world economic climate, or national security concerns.

Keywords: dreams, global events, stressful events, future concerns, well being, adjustment

INTRODUCTION

We begin this chapter with an overview of the literature on the presence of waking stressors in people's dreams. Then we outline the methodology of the study that was carried out with undergraduate students in relation to comparisons between the effects of personal and global events on dreams. The findings in relation to low and high distress groups on several factors, such as life satisfaction, depression, adjustment and trauma symptoms, are then discussed.

Links between Stressful Events and Dreams

Many studies have provided evidence that stressful waking events are often reflected in the dreams of the people who experienced those events (e.g., Breger, Hunter and Lane, 1971; Brunkow, 1996; Esposito, Benitez, Barza and Mellman, 1999; Germain, et al., 2004; Kramer, Schoen and Kinney, 1984; Kroth, et al., 2004; Mellman, 2000; Mellman, Nolan, Hebding, Kulick-Bell and Dominguez, 1997; Mellman, Ramsay and Fitzgerald, 1991; Muller, 1996; Punamaki, 1997, 1998; Raymond, Nielsen, Lavigne and Choiniere, 2002; Ross, et al., 1994; Siegel, 1996; Stoddard, Chedekel and Shakun, 1996; Wood, Bootzin, Rosenhan, Nolen-Hoeksema and Jourden, 1992; Zadra O'Brien, and Donderi, 1997-1998), as well as the dream incorporation of stressful or traumatic events have been an important focus of modern dream research (De Koninck, 2000; Delorme, Lortie-Lussier and De Koninck, 2002). Most studies have focussed on waking events that directly and personally involved the dreamer. For example, studies have examined the dream content of people who experienced stressful or traumatic events, such as major surgery (Breger et al., 1971), personal trauma and assault (Brunkow, 1996; Germain, et al., 2004; Kroth, et al., 2004; Muller, 1996; Raymond et al., 2002; Stoddard, Chedekel and Shakun, 1996), war trauma (Esposito et al., 1999; Kramer et al., 1984; Mellman, 2000; Mellman et al., 1997; Mellman, Ramsay and Fitzgerald, 1991; Ross, et al., 1994); children living in threatening environments (Punamaki, 1997, 1998); exposure to wartime experiences (Schredl and Piel, 2006) or natural disasters; Siegel, 1996; Wood et al., 1992). These studies pinpointed some common dreaming patterns, such as the tendency for traumatised individuals to incorporate traumatic life events in their dreams, often recurrently, and for this incorporation to increase waking distress (Zadra et al., 1997-1998). It is also now generally agreed that dreams do reflect waking emotional concerns and preoccupations, and therefore frequently have significant meaning for the dreamer (Kramer, 2007).

There are numerous factors that influence how well and how quickly an individual will recover from the effects of a stressful or traumatic life event, or even whether or not the individual recovers at all (Wang and Amato, 2000). Some of these factors are personal characteristics of the individual, such as personality, resilience, earlier experiences, self-efficacy and coping skills (Garvin, Kalter and Hansell, 1993). Still other factors relate to interpersonal features, such as levels of social support, relationship quality, and conflict (Demo and Acock, 1996; Gray and Silver, 1990). Then there are factors that exist at a community or structural level, such as employment opportunities, financial support, government policies, and community services (Amato, 2000). The literature emphasises the importance of two types of inner resources, in addition to material resources: psychological resources and coping strategies (Andrew and Robinson, 1991; Booth and Amato, 1992; Cohen and Dekel, 2000; Demo and Acock, 1996). According to Pearlin and Schooler (1978), psychological resources are the personality characteristics that people may draw upon to create resilience in relation to threats and events, while coping strategies are the cognitions, perceptions, and behaviours that they actually use when contending with life challenges.

Several personality factors, including ego strength, boundary thickness, neuroticism and openness to experience, have been found to be correlated with dream content and distress in relation to nightmares (Berquier and Ashton, 1992; Blagrove and Akehurst, 2000; Hartmann, 1990; Hartmann, Rand and Rosen, 1998; Levin, Fireman and Rackley, 2003; Schredl, Landgraf and Zeiler, 2003). Among these, ego strength and boundary thickness have emerged as particularly strong predictors of dream content and experiences (Funkhouser, Würmle, Cornu and Bahro, 2001; Kaminer and Lavie, 1991, 1993; Lavie and Kaminer, 1996; Levin, 1989; Tonay, 1995; Watson, 2001).

Several dream theories have attempted to explain why the incorporation of stressful and traumatic waking experiences into dream content occurs, and whether such incorporation serves any actual purpose. Is it that dreaming is one way of coping when considering the possibility of further future events? It could be that the dreaming actually assists the mind to process fears in a safe haven.

It seems that, overall, the results of the research on dreams of people experiencing stressful or traumatic life events have been mixed. Some findings seem to suggest that dreams which relate to stressful events are symptomatic of greater distress, while other studies suggest that there is a relationship between dreams of the events and improvements in psychological well-being. Interestingly, the re-experiencing of trauma in dreams, as in therapy, may or may not be helpful to the dreamer, and the repression or suppression of distressing material may or may not be adaptive (Blagrove, 1996).

Most of the studies which have attempted to test the hypothesis that dreams help with emotional problem-solving have involved the study of participants who are undergoing or have undergone a real-life, emotionally significant or traumatic event. The common feature in these stressful situations and post-traumatic situations is that an external event has occurred which is perceived by the individual to be threatening to their life, livelihood, health, happiness, or person, and which is difficult to integrate with available psychological resources (Mack, 1992). According to Hartmann (1998 a, b), the dreams of traumatised people represent the clearest examples of dreams helping to deal with dominant emotions and emotional concerns. He proposed that dreams after trauma, or during stressful situations, are an especially useful starting point for examining possible functions of dreaming, since, in these cases, it is known what must be on the dreamers' minds a large part of the time. However, Blagrove (1992, 2000) argues that a relationship between stress and nightmares, or even between a change in nightmare content and recovery from trauma, does not equate to dreaming being considered as playing a causal role in that recovery.

A popular current theory of dream function is the problem-solving theory, which proposes that dreams function to help us solve waking problems, particularly those with an emotional focus. Several authors have put the emotional problem-solving or adjustment theories forth since the 1960s (e.g., Breger, 1967; French and Fromm, 1964; Greenberg, Pearlman, Fingar, Kantrowitz, and Kawliche, 1970; Hartmann, 1995, 1998a; Jones, 1962; Kramer, 1991, 1993). There have been some interesting studies, such as that by De Koninck

(1996), which have shown correlations between dream incorporations of waking problems and adaptation to those problems.

Studies investigating this hypothesis have distinguished between the influence of traumatic events and normal stressful experiences on dream content. Indeed, Schredl's (2003) model emphasises the 'emotional intensity' factor of daytime experiences, in terms of the likelihood that those events will be incorporated into subsequent dreams. Wright and Koulack's (1987) model proposed that variations in normal dream quality over time are due to different processes of adaptation to stressful events. Koulack (1993) suggested a relationship between contemporary stress and dreams under non-traumatic conditions. Another finding that was consistent with the idea that dreams are somehow implicated in the 'working through' of interpersonal difficulties was that of Nielsen et al. (2004), who found that delayed dream incorporations were more likely to refer to waking situations of interpersonal interactions, problem resolution and positive emotions. A small body of research (e.g., Barrett, 1993; Cartwright, 1991), relating to individuals who have experienced major stressful life events, is also often cited as support for the theory that dreams serve the specific function of helping us to adjust or adapt to current events.

Problem-solving theories of dreaming have been challenged by several researchers (e.g., Antrobus, 1993; Blagrove, 1992, 1996, 2000; Domhoff, 1993, 2001a; 2003; Flanagan, 2000; Hobson, Stickgold, and Pace-Schott, 1998) who have contended that there is little evidence to support a direct adaptive function of dreaming. They have suggested that dreams are more likely to be an accidental by-product of REM sleep, which may itself perform a cognitive function in relation to memory and the processing of emotional material. At the same time, Blagrove (1992) suggests that dreams, like waking thought, may relate strongly to what most concerns us at any given time, and like waking speech, dreams may represent these concerns metaphorically. Blagrove (1992) has argued that dreams "are like a linguistic translation of waking concerns and known solutions, rather than an environment in which novel attempts to solve problems may be made" (p. 205).

In his critique of dream theories, Domhoff (2003) has outlined several difficulties for emotional adjustment theories, including: people's generally low recall of their dreams; the high proportion of dreams that seem to be unrelated to waking problems; and the concurrence of traumatic dream content and reduced wellbeing in PTSD sufferers. As Revonsuo (2000) has pointed out, healing dreams ought to comfort and calm us, rather than amplify our distress in relation to traumatic experiences. Reliving such traumas and stresses does not seem to be what traumatised dreamers are most in need of (Revonsuo, 2000).

The findings of other studies (e.g., Koulack, Prevost and De Koninck, 1985; Kramer et al., 1984; Zadra et al., 1997-1998) have shown connections between dream incorporations of stressors and poor subsequent adjustment. In contrast to the findings of Cartwright (1991), Kramer et al.'s (1984) research with Vietnam veterans found an association between high recall of dreams of stressful events and poor subsequent adaptation to the events. They reported that the dreams of these veterans tended to be repetitive and disturbing, and sometimes returned when they were faced with new stressors. They suggested that, rather than serving an adaptive function, the dreams of these veterans served as an expression of their emotional preoccupations at any given time. Furthermore, Lavie and Kaminer (1991) found that concentration camp survivors were more likely to be well adjusted, if their dreams did not incorporate their traumatic experiences. Germain and Nielsen (2003) found that posttraumatic nightmare sufferers exhibited a different pattern of sleep disturbances to idiopathic nightmare sufferers, suggesting that individuals with posttraumatic stress are more aroused, as a result of nightmare psychopathology. Domhoff (1996) suggests that the phenomenon of post-traumatic dreams fits better with cognitive theories of dreaming, such as the continuity theory, since such theories hold that dreams reflect a preoccupation with unresolved problems.

The influence of global events, such as the incidents which occurred in the U.S. on September 11, 2001, on dream content and imagery have also been studied (e.g., Hartmann and Basile, 2003; Propper, Stickgold, Keeley and Christman, 2007) but less extensively. There is certainly evidence that individuals often do experience significant stress symptoms as a result of large scale global events (e.g., Marshall and Galea, 2004; Schlenger et al., 2002; Schuster et al., 2001).

Here we report on research with undergraduate university students that explored the relationship between people's personal and global concerns and their dream content. We had wondered if, even though people might be concerned about global issues, personal issues would be more likely to predominate in their dreams. The study also examined whether a relationship might exist between dreaming about recent stressful events and adjustment to these events. It was hypothesised that dreaming about stressful events would be related to poor adjustment, and that dreamers would be more likely to find dreaming about stressful events unhelpful instead of helpful. The study also included personality and coping style variables in order to more clearly explicate the relationships between dream content, personal resources and adjustment.

METHOD

Participants

Participants were recruited through announcements in lectures at a university in Brisbane. A convenience sample of 57 undergraduate Psychology and Nursing students volunteered to participate. They ranged in age from 17 to 62 years (M = 27.96, SD = 11.47). It was a relatively young group of participants, the median age being 22 years, with 54% of the participants being aged 23 years and younger. Eight (14%) were male and 49 (86%) were female, which is the expected ratio when recruiting from students undertaking these courses.

Study Design and Procedure

The study used a cross-sectional design, assessing a group of student volunteers through the use of a questionnaire and a dream diary which was kept for a period of one week. Questionnaires and dream diaries were handed out, after class, to the participants, along with informed consent packages. They were asked to complete the questionnaires before leaving and to keep a dream diary for one week, and then mail them back to the chief investigator. The questionnaire contained items regarding demographic information (age, gender); individual difference scales (ego strength, boundary thickness, level of concern about global events); dream factors (number of dreams about the distressing event, number of dreams about global events, number of dreams per week, nightmare frequency, and level of nightmare distress); a description of the most difficult or distressing recent event; a question relating to the level of initial distress about the personal event; a question relating to the level of current distress about the personal event; adjustment measures (satisfaction with life, depression, self-rated adjustment, and impact of event).

Materials

Recent Difficult or Distressing Event Questionnaire

Participants were asked to provide a brief written description of "the most difficult or distressing event or situation" that they had personally experienced over the previous couple of months. They were then asked to rate the severity of the distress that this event or situation had initially caused them personally, on a scale of 1 to 10, where 1 represented "minimal distress or upset" and 10 represented "very deeply distressing". They were then asked to rate the severity of distress which the difficult or distressing event, described previously, was currently causing them personally on a scale identical to that used to measure initial distress.

Dreams

Dream content and frequency. Dream content was measured in two ways. On the questionnaire, specific data was gathered in relation to dream content through the use of several explicit questions. In addition, participants were all asked to keep a semi-structured dream diary for a period of one week, in which they were to record daily, all of their recalled dreams. Dream frequency was assessed in the questionnaire, using a simple question asking how many dreams that the individual recalled having, on average, per week. On the questionnaire, inclusions of the recent stressor were assessed through participants being asked if they had dreamed about the personally difficult or distressing event that they had described. If they had, they were asked to approximate how many dreams they had experienced which they felt were related to this event or situation. They were then asked if the dreams had helped them in any way, and if so, to describe exactly how they had found them helpful. When their dream diaries were returned, these were also analysed by the chief researcher to ascertain the number of inclusions or references to the personally distressing event, or to global events.

Recalled dream inclusions of global events were assessed through participants being asked, following completion of the Global Events Concern Questionnaire, if they had ever dreamed about any of these events or issues, and if they had, how many dreams they had over a specified time period. In addition, they were asked to describe any such dreams.

Nightmare frequency, sleep terror frequency, and nightmare distress. Nightmare frequency was assessed using a question that directly asked participants to estimate the number of nightmares that they had in the past year. Sleep terror frequency was assessed in the same way. The differences between nightmares and sleep terrors were first explained, to ensure that participants did not confuse them. This approach was recommended by Belicki, Chambers and Ogilvie (1997). While such retrospective estimates tend to underestimate nightmare frequency, they do correlate highly with diary estimates (Wood and Bootzin, 1990). Nightmare distress was measured using an adaptation of the Nightmare Distress

Questionnaire (NDQ) (Belicki, 1992a), omitting one therapy item. While nightmare distress and nightmare frequency are significantly correlated, their correlations are sufficiently small ($r = .29$ to $r = .45$, with a mean $r = .35$) to warrant them being treated as separate constructs (Belicki, 1992 a, b; Belicki and Cuddy, 1991).

Individual Differences

Ego strength. Ego strength was measured using the MMPI-2 Ego Strength subscale (Barron, 1953). The MMPI-2 Ego strength subscale is a good measure of adaptability, resiliency, personal resourcefulness, and effective functioning, and is also a good general indicator of psychological health and adjustment (Butcher et al., 2001).

Boundary thickness. Boundary thickness was measured using the Boundary Questionnaire (Short Form) (BQ18) (Hartmann, 1998a).

Global events concern questionnaire. This Questionnaire was developed for the purposes of this study, and contained ten items. Each item consisted of a question relating to a significant global event that was likely to be causing concern to a great many people, including: affairs in the Middle East; worldwide events precipitated by the September 11th attacks; the conflict in Iraq; global warming; the Bali bombings; the global economic climate, and national security. Respondents were asked to rate the extent of distress, upset, or personal concern that they were currently experiencing in relation to these events, on a scale of one to ten, where one represented "minimal or no distress / concern" and 10 equalled "very deep distress / concern".

Adjustment

Adjustment was measured using four scales, which broadly assessed general wellbeing and adjustment to the stressful event. Two of the scales measured depression and satisfaction with life. Adjustment to the stressful event itself was measured through a self-rated Likert scale and a scale measuring the impact of the event.

Life satisfaction. Life satisfaction was considered to be an important measure of general adjustment and was assessed using the Satisfaction with Life Scale (SWLS) (Diener, Emmons, Larson, and Griffin, 1985).

Depression. To comprehensively assess wellbeing, depression was also measured, using the Depression subscale of the Depression, Anxiety, and Stress Scales (DASS) (Lovibond and Lovibond, 1996).

Self-rated adjustment. Self-rated adjustment was assessed through the use of a scale rating adjustment from one to ten. This 10 point scale measured the individual's own estimate of their adjustment to the stressful situation, with 1 representing *"not over it at all"* and 10 representing *"fully recovered"*.

Impact of event. The impact of the stressful event was measured using the Impact of Event Scale – Revised (IES-R) (Horowitz, Wilner, and Alvarez, 1979; Weiss and Marmar, 1997).

RESULTS

Participants, on the whole, reported fairly significant levels of initial distress in relation to the recent personally difficult or distressing events that they described. On a 1 to 10 scale, the mean level of *initial* distress scored was 7.17 (*SD* = 2.21). Mean *current* distress was much lower (*M* = 4.14, *SD* = 2.33), with an average 3.03 (*SD* = 2.18) points of improvement in distress since the event occurred. There were no significant correlations found (at a level of $p < .01$) between either age or gender and initial distress, current distress, or global event concern. In their questionnaire responses, participants estimated that they had an average of 4.22 (*SD* = 3.16) dreams per week, while diary recordings yielded an average of 5.10 (*SD* = 3.08) dreams over the single week of diary-keeping.

Recent Difficult or Distressing Events and Dream Inclusions

While the events scale spanned a range of stressful events, the events that were described by the participants as the most difficult or distressing that they had personally experienced over the previous couple of months ranged from starting university and study stress, to serious illness in a close family member, uncertain immigration status, and international relocation. The majority of participants (63.2%) reported in the questionnaire that they had dreamed about the distressing event, with 28.1% reporting five to ten dreams that were related to the event. In their one-week dream diaries, participants recorded an average of two dreams (*SD* = 2.37) about the event, with the maximum number recorded being ten. Of the 36 individuals who did dream of the event, only ten (27.8%) felt that dreaming about it had been helpful to them. For example, one participant had been experiencing a worrying time since her adult son had come to her about his worsening drug problem. She had the following dream which resulted in a feeling of relief:

> There was a swimming pool in the back garden and my son, P (aged approx. 6 years) got into difficulties in the water and started to drown. My son, P (as he is now – an adult), jumped in and saved his child self.

They were much more likely to report that dreaming of the event had made them feel worse or more distressed in some way. For example, one woman whose mother had recently died from cancer reported the following dream which she found upsetting:

> I had something wrong with me and complained. Mum bent down and kissed my knee (not the problem). Then she went upstairs. I wanted to hug her and was afraid she would disappear because I knew she was dead. I went upstairs, put my arms around her and hugged her.

Another participant had recently begun to talk to friends and a counsellor about sexual assault by her brother when she was nine years old, and had not yet disclosed this to her family. She had the following dream two days before Christmas and found that it made her feel very angry:

> It was Christmas day and all my family were together. We were sitting at a big table together talking. I stood up and said 'Let's raise our glasses to J. for sexually assaulting me when I was nine years old. Mum and Dad went somewhere and my older brother went to J. and stabbed him. J. fell to the ground and then I woke up.

She then had another dream where she felt guided and comforted by an angel, and this helped to clarify her decision-making with regard to telling her parents about the assault:

> The angel was saying that I had to get the tests done to prove my brothers actions before I say anything to my family. The angel said that I had to be true to myself and not put blame on myself. I was crying – the angel was saying 'be strong, everything will be okay.'

Here are some quotes from 3 of the 26 participants who all found dreaming of the distressing event less than helpful:

1. They made me more scared and paranoid.
2. I don't feel my dreams were helpful. They actually made it a little worse, but in the light of day you know that in reality, there is a light at the end of the tunnel, whereas the dreams tended to paint a very dark picture.
3. The dream did not assist in any positive way, but increased my stress and concern levels over my preparation for uni exams.

Global Events Concern and Dream Inclusions

The maximum achievable score on the Global Events Concern Scale was 100. Participants in this study scored from 20 to 87 on this scale, with an average score of 50.51 ($SD = 17.87$). Despite this relatively high level of concern regarding global events, only eight participants (14%) recalled ever having dreamed about them. In the week of dream diary recordings, only three participants (5.3%) recorded one dream each that related to a global event. As one participant put it, 'Odd, but even though I think quite a lot about the future, I don't dream about it. Thank God!' Of those eight participants who did recall dreaming about global events, all dreamed of either war or terrorist attacks, although for some, the war was from an earlier era:

> Usually the war is more reminiscent of WWII or WWI than of current conflicts. In these dreams my family and myself are usually being chased and we are trying to hide from the soldiers…no-one is actually caught or hurt by the soldiers (who are often Nazi or of Asian descent).
>
> I dreamed about the Nazi death camps and the Third Reich. I dreamed about the Holocaust.

For most, though, the events were set in present time. Some dream events took place in familiar places with familiar people:

> I was in a war zone, rubble and destruction, then I was at home and saw a huge explosion in the distance over the suburbs…some chemical weapon had been released, then a feeling of 'this is it, it's really happening, we are going to die.'

> All these soldiers were lined up at attention. In between the two lines of soldiers, was a tank following a general. He was inspecting his troops…The terrifying part was the soldiers were all men that I loved or felt close to.

Iraq and terrorist attacks featured prominently in some of the dreams. For example, one participant had the following dream that reminded her of September 11 on the night before she joined an interstate flight:

> I had a dream that the plane crashed into Melbourne airport and killed everyone on the plane and the people at the airport were waiting to collect me.

Another participant experienced dreams of terrorist events that were triggered by moving to live in London after September 11, 2001, but the dreams ceased once she returned to Australia, "probably because I feel so safe here in Australia compared to London."

> They were always pretty detailed, with my family and I running from terrorists with machine guns etc. Once, a helicopter, or some sort of war aircraft, was shot down over my backyard and burst into flames and we were in the middle of warfare.

Another participant was an ex-soldier himself who had friends serving in Iraq:

> I dream of my mates over there…sometimes I'm over there in my old role as a soldier…conducting perimeter security in the west of Iraq…also in a gun pit with friends…being watchful for the ever present threat of possible enemy offensives…seeing locals with AK47s.

Perhaps the most typical was the following dream of a 17 year old female student, who dreamed of Iraqis trying to kill her:

> I was on the front of a tank in Iraq and I was weaving between tanks while the men were screaming 'kill her' and loading their guns and shooting at me. I was almost at the 'safe zone' when I was caught by the Iraqis…I could die by a gun shot to the head or a pillow over my face…I looked him in the eyes and he put the pillow over my face. I remember being scared and I didn't want to die. I woke up thinking it was real and it took a long time before I got over it emotionally.

None of the dreams reported were about environmental threats, climate change or economic issues. They all seemed to represent more immediate physical dangers and threats to persons. There were no dreams reported of natural catastrophes such as earthquakes, tsunamis or bushfires. Rather, all reported dreams of global events related to war or terrorism in some form.

Grouping Based on Initial Distress

In order to ascertain if different patterns of dream content, adjustment, and global events concern might be evident for people with differing levels of personal distress, participants were split into two separate groups. Participants were grouped into either a Low Distress or High Distress group, based upon their rating of their initial distress in relation to their distressing or difficult personal event. Those who scored below the mean of seven on initial distress were allocated to the Low Distress category, and those who scored seven or higher (on a 1 to 10 scale) were categorised as High Distress. There were 18 participants in the Low Distress group and 39 in the High Distress group.

Differences between Low and High Distress Participants

The differences between participants in the two distress groups were assessed by conducting a series of t-tests. The results of these analyses are displayed in Table 5.1. While Low Distress participants were less likely to dream about their distressing personal event compared to High Distress participants, this difference was not significant at a probability level of p <.01. Those participants who were in the Low Distress category were significantly less likely to suffer from nightmare distress than High Distress participants.

Table 5.1. Means and standard deviations for low and high initial distress participants, and t values for differences between means

		Low Distress (N = 18)		*High Distress (N = 39)*	
Variables	*M*	*SD*	*M*	*SD*	*t*
Initial Distress	*4.47*	*1.54*	*8.41*	*1.07*	*-11.20****
Improvement in Distress	*1.69*	*1.38*	*3.64*	*2.22*	*-3.42***
Current Distress	*2.78*	*1.35*	*4.77*	*2.42*	*-3.25***
Dreams of Event	*1.28*	*1.93*	*2.92*	*2.89*	*-2.20*
Dreams per Week	*4.03*	*3.19*	*4.31*	*3.18*	*-.31*
Nightmare Frequency	*3.97*	*5.14*	*10.09*	*26.48*	*-.97*
Nightmare Distress	*7.61*	*5.61*	*13.31*	*7.37*	*-2.91**
Sleep Terror Frequency	*.44*	*1.25*	*3.95*	*15.91*	*-.93*
Number of Diary Dreams	*4.67*	*2.79*	*5.31*	*3.22*	*-.73*
Diary Inclusions	*1.17*	*1.54*	*2.38*	*2.60*	*-1.84*
Ego Strength	*34.83*	*4.73*	*30.72*	*6.05*	*2.54*
Boundary Q18	*32.00*	*8.67*	*34.18*	*8.96*	*-.86*
Global Event Concern	*52.11*	*19.32*	*49.77*	*17.37*	*.46*
Impact of Event	*21.11*	*12.62*	*38.28*	*23.12*	*-2.94****
Life Satisfaction	*24.11*	*6.01*	*22.56*	*6.82*	*.83*
Depression	*2.94*	*3.79*	*10.87*	*12.11*	*-2.71**
Self-Rated Adjustment	*7.86*	*2.61*	*6.08*	*2.47*	*2.49*

Note: Statistically significant at *$p < .01$, **$p < .005$, ***$p<.001$.

However, the two groups did not differ significantly on nightmare frequency, sleep terror frequency, or number of dreams per week. Moreover, the two groups did not differ significantly on any of the individual difference variables. General wellbeing and adjustment to the distressing event were significantly better for the Low Distress group than the High Distress group.

Table 5.2. Correlations between Dreams of Event, Diary Inclusions, Night Dream Frequency, Nightmare Frequency, Nightmare Distress, Global Event Concern, and Wellbeing and Adjustment, for the Low Initial Distress group (*n* = 18) (below diagonal) and the High Initial Distress Group (*n* = 39) (above diagonal in bold)

Variables	1	2	3	4	5	6	7	8	9	10
1. Dreams of Event		**.51***	**.09**	**.30**	**.46***	**.21**	**.65*****	**-.35**	**-.41***	**.48****
2. Diary Inclusions	.74*		**.05**	**.10**	**.53***	**.14**	**.29**	**-.12**	**-.40**	**-.28**
3. Dreams per Week	.49	.55		**.19**	**.29**	**.29**	**-.02**	**.24**	**.03**	**-.13**
4. Nightmare Frequency	.23	.35	.35		**.53***	**.34**	**.07**	**-.21**	**.05**	**-.10**
5. Nightmare Distress	.56	.51	.56	.17		**.24**	**.22**	**-.08**	**-.14**	**-.34**
6. Global Event Concern	.28	.30	.34	-.30	.22		**.07**	**-.16**	**-.13**	**-.16**
7. Depression	-.18	.04	.33	-.02	.18	.47		**.44***	**.36**	**.60*****
8. Satisfaction With Life	.05	-.22	.08	.14	.06	-.55	.29		**.22**	**.23**
9. Self-Rated Adjustment		-.28	.02	-.19	.17	-.22	-.04	-.48	-.22	**.48****
10. Impact of Event		-.47	-.40	-.62*	-.11	-.52	-.28	.20	.05	.60*

Note: Statistically significant at *p < .01, **p <.005, ***p <.001.

Correlations between Inclusions of Distressing Events and Adjustment

Correlations were examined for each of the two distress groups, and are displayed in Table 5.2. In the Low Distress group, there was no relationship between dreams of the personally distressing event and adjustment, or between diary inclusions of the event and adjustment. In the High Distress group, the patterns of relationships were different. In this group, there were significant negative correlations between: number of questionnaire-reported dreams of the personally distressing event and depression, impact of the event, and self-rated adjustment to the event. However, in neither group was there a significant relationship between number of diary inclusions of the event and depression, satisfaction with life, impact of event or self-rated adjustment.

Relationships between Global Event Concerns and Adjustment

With regard to the relationship between global event concerns and adjustment, the pattern of relationships was not different for the two distress groups (see Table 5.2). In both groups, the correlations between global event concern and depression, satisfaction with life, impact of the personal event, and self-rated current adjustment to the event were all non-significant.

Thus, it appeared that for both groups, there was no relationship between their level of concern about global events and their level of general wellbeing and adjustment.

Discussion

The participants of the current study who dreamed of a recent life event that was personally difficult or distressing did not tend to find the dreaming helpful. In fact, many described dreaming of the event as only adding to their distress. Those who experienced the most initial distress as a result of the event were more likely to have a greater number of dreams relating to the event, and significantly more nightmare distress, than those who experienced less initial distress. It is also possible that participants experiencing more distress were more inclined to remember such dreams than those coping with less distress. Nevertheless, the analyses revealed no evidence that having more dreams about the event was helpful to the emotional adjustment process.

The number of dreams relating to the event was significantly related with wellbeing and adjustment following the event for the High Distress participants only. Importantly, this correlation was a negative one, meaning that more dreams of the event were correlated with poorer wellbeing. There was a clear discrepancy between the results for recalled dreams of events in the questionnaire and inclusions that were recorded in dream diaries. Only retrospectively-recalled dreams correlated significantly with wellbeing and adjustment, while the relationship between diary inclusions and adjustment was weaker. This suggests that participants' recall may have been influenced by their perceptions of their own distress, and they may have had trouble distinguishing in their memory, between dreams and thoughts of concerns.

Some individuals are inclined to experience waking life as particularly intense and absorbing, and they may have a greater tendency to daydream (Tonay, 1993) and also experience sleep disturbances, repetitive dreams and nightmares (Belicki and Cuddy, 1991; Najam and Malik, 2003; Schredl, Ciric, Gotz and Wittmann, 2003). The wellbeing of these individuals may become subject to a feedback loop whereby their waking and dreaming cognitions serve to perpetuate one another. Often such people tend to pay more attention to their dreams, have higher dream recall, attach more meaning to their dreams, and tend to relive the details of dreams and nightmares (Pietrowsky and Köthe, 2003; Schredl, Schafer, Hofmann and Jacob, 1999). This can lead to an overestimate of inclusions of stressors in questionnaire reporting, as possibly occurred here and which has also been found in previous research (Schredl, 1999).

As this and other studies (e.g., Hall and Nordby, 1972) have found, dreams are also largely self-referent and dream events and locations are more commonly familiar than unfamiliar to the dreamer. This seemed to be the case even in most of the dreams about global events, where the 'war zone' might be in the local suburb or the soldiers had familiar faces. There is little doubt that a similar survey carried out on Baghdad civilians would have contained many more 'global events' as for them, these would have been very localised and personal events. Clearly, people tend to dream most about what immediately and personally concerns them, rather than about events which occur outside of their immediate environment. For the population in this study, personal events were more important to them and most likely were perceived as having an impact on their wellbeing. This was reflected in the fact that participants who were not greatly distressed by personal events had less distress in relation to nightmares. They were also less likely to be influenced by their concerns about global issues, in terms of their general wellbeing, than people who experienced more initial distress as a result of a personally difficult life event. It would therefore seem that, regardless of the level of preoccupation with personal difficulties, individuals' wellbeing tended to be influenced little by events occurring in their external environment. This was reflected in their dream content, which tended to contain personal rather than global themes, generally, and more particularly when personal events were causing distress.

CONCLUSION

The participants in this study had concerns in their waking lives that were personal and others that related to their wider environment. However, their dreams tended to reflect their personal concerns much more. This was particularly the case when the personal concerns were causing them a lot of distress. It seems very likely that their dreams were providing a fairly accurate barometer of their level of waking preoccupation with personal concerns. It also seems likely that, although most participants expressed concern about global and environmental events, they were not *preoccupied* with these events in the way that they might be preoccupied with personal events, even temporarily. These findings may be considered instructive when attempting to understand why concerns about environmental or global issues does not translate often into behavioural change. People often simply don't see environmental or global issues as having, or likely to have, an immediate, *personal* impact on them. If they did, they would probably dream about things like climate change and food shortages to the same degree that they dream about personal issues, such as family conflict and exam pressures.

REFERENCES

Amato, P. R. (2000). The consequences of divorce for adults and children. *Journal of Marriage and Family, 62* (4), 1269-1287.

Andrew, F. M., and Robinson, J. P. (1991). Measures of subjective well being. In J. P. Robinson, P. R. Shaver, and L. S. Wrightsman (Eds.), *Measures of personality and social psychological attitudes* (pp. 61-69). San Diego: Academic Press.

Antrobus, J. (1993). Dreaming: Could we do without it? In A. Moffitt, M. Kramer, and R. Hoffmann (Eds.), *The functions of dreaming* (pp. 549-558). New York: State University of New York Press.

Barrett, D. (1993). The "committee of sleep": A study of dream incubation for problem solving. *Dreaming, 3*, 115-122.

Barron, F. (1953). An ego-strength scale which predicts response to psychotherapy. *Journal of Consulting Psychology*, *17*, 327-333.

Belicki, K. (1992a). The relationship of nightmare frequency to nightmare suffering with implications for treatment and research. *Dreaming*, *2* (3), 143-148.

Belicki, K. (1992b). Nightmare frequency versus nightmare distress: relations to psychopathology and cognitive style. *Journal of Abnormal Psychology, 101* (3), 592-597.

Belicki, K., Chambers, E., and Ogilvie, R. (1997). Sleep quality and nightmares. *Sleep Research, 26*, 637.

Belicki, K., and Cuddy, M. A. (1991). Nightmares: facts, fictions and future directions. In J. Gackenbach and A. Sheikh (Eds.), *Dream images: a call to mental arms* (pp. 99-113). Farmingdale, NJ: Baywood.

Berquier, A., and Ashton, R. (1992). Characteristics of the frequent nightmare sufferer. *Journal of Abnormal Psychology, 101* (2), 246-250.

Blagrove, M. (1992). Dreams as a reflection of our waking concerns and abilities: A critique of the problem-solving paradigm in dream research. *Dreaming, 2*, 205-220.

Blagrove, M. (1996). Problems with the cognitive psychological modeling of dreaming. *The Journal of Mind and Behaviour, 17* (2), 99-134.

Blagrove, M. (2000). Dreams have meaning but no function. *Behavioural and Brain Sciences, 23* (6), 910-911.

Blagrove, M., and Akehurst, L. (2000). Personality and dream recall frequency: further negative findings. *Dreaming, 10*, 139-148.

Booth, A., and Amato, P. (1992). Divorce, residential change and stress. *Journal of Divorce and Remarriage, 18*, 205-213.

Breger, L. (1967). Function of dreams. *Journal of Abnormal Psychology Monograph, 72*(5), 1-28.

Breger, L., Hunter, I., and Lane, R. W. (1971). *The effect of stress on dreams.* New York: International Universities Press.

Brunkow, K. A. (1996). Working with dreams of survivors of violence: Facilitating crisis intervention with a psychoanalytic approach. In J. Edward and J. B. Sanville (Eds.), *Fostering health and growth: A psychoanalytic social work approach* (pp. 212-225). Northvale, NJ: Jason Aronson.

Butcher, J. N., Graham, J. R., Ben-Porath, Y. S., Tellegen, A., Dahlstrom, W. G., and Kaemmer, B. (2001). *MMPI-2 (Minnesota Multiphasic Personality Inventory-2) Manual for administration, scoring, and interpretation* (Revised Edition). Minneapolis: University of Minnesota Press.

Cartwright, R. D. (1991). Dreams that work: The relation of dream incorporation to adaptation to stressful events. *Dreaming, 1*(1), 3-9.

Cohen, O., and Dekel, R. (2000). Sense of coherence, ways of coping, and well being of married and divorced mothers. *Contemporary Family Therapy, 22* (4), 467-486.

De Koninck, J. (1996). Vertical inversion of the visual field and REM sleep mediation. *Journal of Sleep Research, 5* (1), 16-20.

De Koninck, J. (2000). Waking experiences and dreaming. In M. H. Kryger, T. Roth and W. C. Dement (Eds.). *Principles and practice of sleep medicine* (pp. 502-511). Philadelphia, PA: W. B. Saunders Company.

Delorme, M-A., Lortie-Lussier, M., and De Koninck, J. (2002). Stress and coping in the waking and dreaming states during an examination period. *Dreaming, 12* (4), 171-183.

Demo, D. H., and Acock, A. C. (1996). Motherhood, marriage, and remarriage: The effects of family structure and family relationships on mothers' well-being. *Journal of Family Issues, 17*, 388-407.

Diener, E., Emmons, R. A., Larsen, R. J., and Griffin, S. (1985). The Satisfaction With Life Scale. *Journal of Personality Assessment, 49*(1), 71-76.

Domhoff, G. W. (1993). The repetition of dreams and dream elements: A possible clue to a function of dreams. In A. Moffitt and M. Kramer and R. Hoffmann (Eds.), *The functions of dreaming* (pp. 293-320). New York: State University of New York Press.

Domhoff, G. W. (1996). *Finding meaning in dreams: A quantitative approach.* New York: Plenum Press.

Domhoff, G. W. (2001a). A new neurocognitive theory of dreams. *Dreaming, 11*, 13-29.

Domhoff, G. W. (2003). *The scientific study of dreams: Neural networks, cognitive development, and content analysis.* Washington, DC: American Psychological Association.

Esposito, K., Benitez, A., Barza, L., and Mellman, T. (1999). Evaluation of dream content in combat-related PTSD. *Journal of Traumatic Stress, 12* (4), 681-687.

Flanagan, O. (2000). *Dreaming souls: Sleep, dreams, and the evolution of the conscious mind.* New York: Oxford University Press.

French, T., and Fromm, E. (1964). *Dream interpretation.* New York: Basic Books.

Funkhouser, A. T., Würmle, O., Cornu, C. M., and Bahro, M. (2001). Boundary questionnaire results in the mentally healthy elderly. *Dreaming, 11* (2), 83-88.

Garvin, V., Kalter, N., and Hansell, J. (1993). Divorced women: Individual differences in stressors, mediating factors, and adjustment outcomes. *American Journal of Orthopsychiatry, 63*, 232-240.

Germain, A., Krakow, B., Faucher, B., Zadra, A., Nielsen, T., Hollifield, M., Warner, T. D., and Koss, M. (2004). Increased mastery elements associated with imagery rehearsal treatment for nightmares in sexual assault survivors with PTSD. *Dreaming, 14* (4), 195-206.

Germain, A., and Nielsen, T. A. (2003). Sleep pathophysiology in posttraumatic stress disorder and idiopathic nightmare sufferers. *Biological Psychiatry, 54*, 1092-1098.

Gray, J. D., and Silver, R. C. (1990). Opposite sides of the same coin: Former spouses' divergent perspectives in coping with their divorce. *Journal of Personality and Social Psychology, 59*, 1180-1191.

Greenberg, R., Pearlman, C., Fingar, R., Kantrowitz, J., and Kawliche, S. (1970). The effects of dream deprivation: Implications for a theory of the psychological function of dreaming. *British Journal of Medical Psychology, 43*(1), 1-11.

Hall, C. and Nordby, V. (1972). *The individual and his dreams.* New York: New American Library.

Hartmann, E. (1990). Thin and thick boundaries: Personality, dreams and imagination. In R. G. Kunzendorf (Ed.), *Mental imagery* (pp. 71-78). New York: Plenum Press.

Hartmann, E. (1995). Making connections in a safe place: Is dreaming psychotherapy? *Dreaming, 5*, 213-228.

Hartmann, E. (1996). Outline for a theory on the nature and functions of dreaming. *Dreaming, 6*, 147-170.

Hartmann, E. (1998a). *Dreams and nightmares: The new theory on the origin and meaning of dreams.* New York: Plenum Trade.

Hartmann, E. (1998b). Nightmare after trauma as paradigm for all dreams: A new approach to the nature and functions of dreaming. *Psychiatry, 61*, 223-238.

Hartmann, E., and Basile, R. (2003). Dream imagery becomes more intense after 9/11/01. *Dreaming, 13* (2), 61-66.

Hartmann, E., Rand, W., and Rosen, R. (1998). Personality and dreaming: Boundary structure and dream content. *Dreaming, 8 (1),* 31 – 39.

Hobson, J. A., Stickgold, R., and Pace-Schott, E. F. (1998). The neuropsychology of REM sleep dreaming. *Neuroreport, 9*, R1-R14.

Horowitz, M., Wilner, N., and Alvarez, W. (1979). Impact of Event Scale: A measure of subjective stress. *Psychosomatic Medicine, 41*(3), 209-218.

Jones, R. M. (1962). *Ego synthesis in dreams*. Cambridge, Mass.: Schenkman.

Kaminer, H., and Lavie, P. (1991). Sleep and dreaming in Holocaust survivors: Dramatic decrease in dream recall in well-adjusted survivors. *Journal of Nervous and Mental Disease, 179*, 664-669.

Kaminer, H., and Lavie, P. (1993). Sleep and dreams in well-adjusted and less adjusted Holocaust survivors. In M. S. Stroebe, W. Stroebe, and R. O. Hansson (Eds.), *Handbook of bereavement: Theory, research, and intervention* (pp. 331-345). Cambridge: Cambridge University Press.

Koulack, D. (1993). Dreams and adaptation to contemporary stress. In A. Moffitt, M. Kramer, and R. Hoffman (Eds.), *The functions of dreaming* (pp. 321-340). New York: State University of New York Press.

Koulack, D., Prevost, F., and De Koninck, J. (1985). Sleep, dreaming, and adaptation to a stressful intellectual activity. *Sleep, 8*, 244-253.

Kramer, M. (1991). The nightmare: A failure in dream function. *Dreaming, 1*, 277-285.

Kramer, M. (1993). The selective mood regulatory function of dreaming: An update and revision. In A. Moffitt, M. Kramer, and R. Hoffman (Eds.), *The functions of dreaming* (pp. 139-195). New York: State University of New York Press.

Kramer, M. (2007). *The dream experience: a systematic exploration*. New York: Routledge.

Kramer, M., Schoen, L. S., and Kinney, L. (1984). The dream experience in dream-disturbed Vietnam veterans. In B. A. van der Kolk (Ed.), *Post-traumatic stress disorder: Psychological and biological sequelae* (pp. 81-95). Washington, D.C: American Psychiatric Press.

Kroth, J., Garcia, M., Hallgren, M., LeGrue, E., Ross, M., and Scalise, J. (2004). Perinatal loss, trauma, and dream reports. *Psychological Reports, 94* (3), 877-882.

Lavie, P., and Kaminer, H. (1991). Dreams that poison sleep: dreaming in Holocaust survivors. *Dreaming, 1*, 11-21.

Lavie, P., and Kaminer, H. (1996). Sleep, dreaming, and coping style in Holocaust survivors. In D. Barrett (Ed.), *Trauma and dreams* (pp. 114-124). Cambridge, MA: Harvard University Press.

Levin, R. (1989). Relations among nightmare frequency and ego strength, death anxiety, and sex of college students. *Perceptual and Motor Skills, 69 (3, Pt 2),* 1107 – 1113).

Levin, R., Fireman, G., and Rackley, C. (2003). Personality and dream recall frequency: Still further negative findings. *Dreaming, 13* (3), 155-162.

Lovibond, S.H., and Lovibond, P.F. (1996). *Manual for the Depression Anxiety Stress Scales* (2nd ed.). Sydney: Psychology Foundation.

Mack, J. (1992). Toward a theory of nightmares. In M. R. Lansky (Ed.), *Essential papers on dreams* (pp. 343-375). New York: New York University Press.

Marshall, R. D. and Gallea, S. (2004). Science for the community: Assessing mental health after 9/11. *The Journal of Clinical Psychiatry, 65*, 37-43.

Mellman, T. A. (2000). Sleep and the pathogenesis of PTSD. In A. Y. Shalev and R. Yehuda and A. C. McFarlane (Eds.), *International handbook of human response to trauma* (pp. 299-306). New York: Kluwer Academic / Plenum Publishers.

Mellman, T. A., Nolan, B., Hebding, J., Kulick-Bell, R., and Dominguez, R. (1997). A polysomnographic comparison of veterans with combat-related PTSD, depressed men, and non-ill controls. *Sleep*, *20*, 46-51.

Mellman, T. A., Ramsay, R. E., and Fitzgerald, S. G. (1991). Divergence of PTSD and narcolepsy associated with military trauma. *Journal of Anxiety Disorders, 5*, 267-272.

Muller, K. H. (1996). Jasmine: Dreams in the psychotherapy of a rape survivor. In D. Barrett (Ed.), *Trauma and dreams* (pp. 148-158). Cambridge, Massachusetts: Harvard University Press.

Najam, N., and Malik, I. (2003). Psychological problems and dream content of nightmare sufferers in Pakistan. *Dreaming, 13* (3), 135-141.

Nielsen, T. A., Kuiken, D., Alain, G., Stenstrom, P., and Powell, R. A. (2004). Immediate and delayed incorporations of events into dreams: further replication and implications for dream function. *Journal of Sleep Research, 13*, 327-336.

Pearlin, L. I., and Schooler, C. (1978). The structure of coping. *Journal of Health and Social Behaviour, 19*, 2-21.

Pietrowsky, R., and Köthe, M. (2003). Personal boundaries and nightmare consequences in frequent nightmare sufferers. *Dreaming*, *13* (4), 245-254.

Propper, R. E., Stickgold, R., Keeley, R., and Christman, S. D. (2007). Is television traumatic? Dreams, stress and media exposure in the aftermath of September 11, 2001. *Psychological Science, 18* (4), 334-340.

Punamäki, R.-L. (1997). Determinants and mental health effects of dream recall among children living in traumatic conditions. *Dreaming, 7* (4), 235-263.

Punamäki, R.-L. (1998). The role of dreams in protecting psychological well-being in traumatic conditions. *International Journal of Behavioral Development, 22*(3), 559-588.

Raymond, I., Nielsen, T. A., Lavigne, G., and Choiniere, M. (2002). Incorporation of pain in dreams of hospitalized burn victims. *Sleep, 25* (7), 765-770.

Revonsuo, A. (2000). The reinterpretation of dreams: An evolutionary hypothesis of the function of dreaming. *Behavioral and Brain Sciences*, *23*, 877-903.

Ross, R. J., Ball, W. A., Dinges, D. F., Kribbs, N. B., Morrison, A. R., Silver, S. M., and Mulvaney, F. D. (1994). Rapid eye movement sleep disturbance in posttraumatic stress disorder. *Biological Psychiatry, 35*, 195-202.

Schlenger, W. E. (2002). Psychological reactions to terrorist attacks: Findings from the National Study of Americans' Reactions to September 11. *Journal of the American Medical Association, 288*, 581-588.

Schredl, M. (1999). Dream recall: research, clinical implications, and future directions. *Sleep and Hypnosis, 1*, 72-81, A2-A4.

Schredl, M. (2003). Continuity between waking and dreaming: A proposal for a mathematical model. *Sleep and Hypnosis, 5*, 38-52.

Schredl, M., Ciric, P., Gotz, S., and Wittmann, L. (2003). Dream recall frequency, attitude towards dreams and openness to experience. *Dreaming, 13* (3), 145-153.

Schredl, M., Landgraf, C., and Zeiler, O. (2003). Nightmare frequency, nightmare distress and neuroticism. *North American Journal of Psychology, 5* (3), 345-350.

Schredl, M., and Piel, E. (2006). War-related dream themes in Germany from 1956 to 2000. *Political Psychology, 27* (2), 299-307.

Schredl, M., Schäfer, G., Hofmann, F., and Jacob, S. (1999). Dream content and personality: Thick vs. thin boundaries. *Dreaming, 9*, 257-263.

Schuster, M. A, Stein, B.D., Jaycox, L. H., Collins, R. L., Marshall, G. N., Elliott, M. N., Zhou, A. J., Kanouse, D. E., Morrison, J. L., and Berry, S. H. (2001). A national survey of stress reactions after the September 11, 2001 terrorist attacks. *The New England Journal of Medicine, 345*, 1507-1512.

Siegel, A. (1996). Dreams of firestorm survivors. In D. Barrett (Ed.), *Trauma and dreams* (pp. 159-176). Cambridge, Massachusetts: Harvard University Press.

Stoddard, F. J., Chedekel, D. S., and Shakun, L. (1996). Dreams and nightmares of burned children. In D. Barrett (Ed.), *Trauma and dreams* (pp. 25-45). Cambridge, MA: Harvard University Press.

Tonay, V. (1993). Personality correlates of dream recall: Who remembers? *Dreaming, 3* (1), 1-8.

Tonay, V. K. (1995). Creativity and dreaming. *Dissertation Abstracts International: Section B: The Sciences and Engineering, 55* (7-B), 3053.

Wang, H., and Amato, P. R. (2000). Predictors of divorce adjustment: Stressors, resources, and definitions. *Journal of Marriage and the Family, 62*, 655-668.

Watson, D. (2001). Dissociations of the night: Individual differences in sleep-related experiences and their relation to dissociation and schizotypy. *Journal of Abnormal Psychology, 110 (4),* 526 – 535.

Weiss, D. S., and Marmar, C. R. (1997). The Impact of Event Scale - Revised. In J. P. Wilson and T. M. Keane (Eds.), *Assessing psychological trauma and PTSD* (pp. 399-411). New York: The Guilford Press.

Wood, J., and Bootzin, R. (1990). The prevalence of nightmares and their independence from anxiety. *Journal of Abnormal Psychology, 99*, 64-68.

Wood, J. M., Bootzin, R. R., Rosenhan, D., Nolen-Hoeksema, S., and Jourden, F. (1992). Effects of the 1989 San Francisco earthquake on frequency and content of nightmares. *Journal of Abnormal Psychology, 101*, 219-224.

Wright, J., and Koulack, D. (1987). Dreams and contemporary stress: a disruption-avoidance-adaptation model. *Sleep, 10*, 172-179.

Zadra, A. L., O'Brien, S. A., and Donderi, D. C. (1997-1998). Dream content, dream recurrence, and well-being: A replication with a younger sample. *Imagination, Cognition, and Personality, 17* (4), 293-311.

Contributed by Tim McCrorey, Toowoomba.

Photo: Wintry Doom. (after S. Weart's essay of same name).

In: Meltdown: Climate Change, Natural Disasters…
Editor: Kathryn Gow
ISBN 978-1-60876-153-1

Chapter 6

ECOLOGICAL CRISES – A PSYCHOLOGICAL PERSPECTIVE AND THE PATH TO TRANSFORMATION

Ann Moir-Bussy
Hong Kong Shue Yan University, Hong Kong

ABSTRACT

Ecological crises and environmental threats are constantly in the news and occur all around us. This chapter will take a stance from transpersonal psychology to view the part that humans play in ecological crises. In addition to this, the general notion of duality - that humans are separate from nature - is challenged. It will be argued that lack of awareness of one's relationship with nature has an impact on the environment. This disconnection can lead to environmental destruction. The author maintains that there is a need for consciousness raising on both the individual and the collective level, if constructive change is to take place and if further degradation of the environment is to be avoided.

Keywords: ecological crises; transpersonal psychology, ecopsychology, interbeing, connectedness

INTRODUCTION

In exploring the psychological effects on humans of the ecological crises around the world, there is a key concept to address which underlies all our discussions. It is a simple concept, but one which is rarely grasped - *The Universe and I are one*. The universe and I are part of the same system, interdependent and interconnected. Cybernetics and general systems theory offer a paradigm with which to explore the influence of climate and ecology on humans and the influence of human behaviour and consciousness on climate and the ecology.

We will also see how humanistic psychology points to the need for us to reconnect to our own guilt in personal, collective and ecological relationships (Yunt, 2001). Jung's psychological and philosophical insights bring further perspectives to understanding the place humans have played in the ecological crises. Psychology has a term for this interconnection – 'ecopsychology'. This form of psychology was born out of the awareness that life on earth is under multiple threats. Ecopsychology is: (1) The study of the human psyche's place within the larger context of the earth as a living system; (2) The examination of the interconnectedness of nature and psyche; and (3) The search for an ecologically-based standard of health (John Mack Institute).

The universe is part of a living system and humans are an integral part of this system, so the health of the universal system 'relies on the development of sustainable and mutually nurturing relationships among the human and non-human aspects of the world'.[1] When these relationships break down, not only are humans affected, but so is the environment. In this chapter, it is proposed that the ecological crisis and human functioning are inter-being and interconnected, that is, that there is a dual perspective on this matter. It will be argued from a psychological philosophical perspective that human functioning and humankind's lack of awareness affects that ecological crisis, and some suggestions for possible transformation will be provided.

The Effect of Ecological Crises on Health

The World Health Organization (WHO) addresses the interrelated effect of human societies changing the ecosystems on a global scale and the consequent instability in the 'biosphere's ecological, physical and socioeconomic systems' (WHO, 2003). It recognizes that human activities are putting increased pressure on the earth's climate systems and that these changes now have an impact on the health of humans. Bernard et al. (2001) note that climate change has increased exposure to air pollutants causing serious respiratory and cardiovascular diseases. According to Connor (2005), the World Health Organization estimated that climate change is responsible for at least 150,000 extra deaths a year and "is causing five million extra cases of severe illness a year" (p.1).

In late 2008, the author watched a TV documentary which showed that when companies built chemical industrial plants in Wuli Village in China and polluted the Hangzhou Qiantang River, 70 out of the 200 villagers developed cancer and died. A diary was kept detailing these cases. In Xiping Village, the fishing livelihood of families was destroyed because of the water pollution and chemical levels. In the case of catastrophes such as chemical and radioactive leaks and industrial accidents, Davis (2008) (http://www.johnvdavis.com/ep/envact.htm) shows that toxic exposure has both neuropsychological effects and stress effects; and, that in the case of long-term environmental degradation, one can hypothesize effects such as psychic numbing, anxiety, helplessness, depression, alienation and the sense of not-belonging, as well as negative effects on child development.

Natural disasters, technological catastrophes, and long-term environmental degradation are categories of ecological crises that are increasingly overlapping (Davis, 2008). The boundaries are blurred. Davis continues that while some disasters are predictable or expected,

[1] http://www.johnemackinstitute.org/projects/eco/what_is_it.asp.

we know they will occur at some stage (forest fires, typhoons, tsunamis), and thus to some degree, we can be better prepared for them in the future. However, technological disasters, such as chemical and radioactive leaks or industrial accidents releasing major toxic pressure, are not supposed to occur because of existing safety legislation, so people are not psychologically prepared for them. The result can be psychological numbing, denial, anxiety, helplessness, depression, alienation, as well as having negative effects on child development, bringing about "generational amnesia" (Davis, 2008.). If, as was noted above, humans are interrelated with, and a natural part of, the whole universe system, the process of ecological crises must be two-way. We must look at the part that human behaviour plays in causing these disasters. The author now turns to the effect that human behaviour and human lack of awareness has on the universe.

The Effect of Humans on the Universe

Too often humans look at the world in a dualistic fashion, yet nature and psyche both flow as expressions of the same absolute source. There is a fundamental non-duality. Interdependence or connection sustains the balance of life on the planet. From the view point of Indigenous peoples, 'everything in the environment is a living being whose awareness process is similar to that of humans' (Mindell, 2000, p. 568). This notion is challenging for those who behave in the world, as though they are the owners of the world and the controllers of the environment. Fisher (1995) questions why there is so much talk about healing the planet and so little talk about the psychological dynamics of wounding the planet in the first place. Physicist John Wheeler (in Mindell, 2000) says:

> The universe does not exist "out there" independent of us. We are inescapably involved in bringing about that which appears to be happening. We are not only observers. We are participators. In some strange sense this is a participatory Universe (p. 399).

Jeremy Yunt (2001) argues that Jung's depth psychology is a tool which can help us to understand and deal with 'the psychic roots' of some of the ecological problems. While the term 'ecopsychology' was not labelled as such in Jung's time, his insights into the effects of humans on nature are very clear. Jung says, "It is becoming ever more obvious that it is not famine, not earthquakes, not microbes, not cancer but man himself who is man's greatest danger to man" (Jung, 197, par.1358. in Yunt, 2001). From the perspective of ecopsychology, Yunt (2001) argued that "these problems are rooted fundamentally in the distorted understandings, repressed contents, conscious denials, and unconscious projections that exist in both the personal and societal dimensions of the human psyche" (p. 98).

Interbeing

Taking a holistic approach, 'environmental problems are caused by our failure to see the whole… the whole is more than the sum of its parts" (Du Nann Winter and Koger, 2004, p. 186). Echoing this, the Buddhist teacher Thich Nhat Hanh (1987) refers to the mutual connectedness between people and all things as 'interbeing' and also speaks of *tiep hien* –

two Vietnamese words of Chinese origin. *Tiep* means ‘to be in touch’ with what is going on in your body, your mind and your feelings’ [and in the environment]; *hien* means ‘in the present time’, because the present moment is the moment in which we can be fully alive. *Tiep hien* is *interbeing.* One can only achieve interbeing, if one is aware or conscious of one’s relationship to all other beings. Awareness or consciousness is about relationship and relationship is about consciousness (De Quincey, 2005). This kind of consciousness was very much part of Indigenous wisdom, Shamanism and the ancient Taoist view of the world. In the Tao Te Ching, Lao Tzu (1997) says: “Knowing others is wisdom; Knowing the self is enlightenment; Mastering others requires force; Mastering the self needs strength”. In another section, he says, “Do you think you can take over the universe and improve it? I do not believe it can be done. The universe is sacred… if you try to change it you will ruin it, if you try to hold it, you will lose it” (Lao Tzu, 1997).

Today quantum physicists are beginning to describe the world in the same manner as the ancient Indigenous people – a world of connectedness and relationships with humans and nature – all part of the one Universe. As Mindell (2000) notes ‘you are not emotionally or physically separable from anything you observe, be it a particle, a tree, or another person’ (p. 568). In other words, there is an energy connecting everything and everyone in the universe. Indigenous attitudes to the land were different from the modern Western view: “They frequently spoke of the land as their parent or as a sacred being, on whom they were dependent and to whom they owed loyalty and service” (Gilman, 1984, p. 6). Some Indigenous peoples referred to the land as “mother’ and ancient religious practices were integrated into the ordinary rhythms of life. Such an example is evidenced at the important Neolithic site at Catal Huyuk, in Anatolia, where the goddess was worshipped as mother of regeneration of the earth and its peoples (Gadon, 1989).

Canty (2006) believes that this connectedness was lost when the Western paradigm, based on mechanistic science and rational linear thinking, began to operate from a psychology of scarcity and so natural resources became objects to fight over, to bring under human control and to serve only the needs of the human inhabitants of the world. People developed a “poverty consciousness”, creating a focus on getting more (greed) thus causing depletion. Boston (1996) acknowledges that there are some ecopsychologists who believe that the dominant systems of society lead people to believe that consumer goods are not needs, but rights meant to relieve the ‘empty soul’. This lack of *interbeing* has created numerous global crises, proliferation of nuclear weapons, and destruction of the environment - all aggravated by human materialism and greed. To heal and preserve the planet, there is an urgent need for humans to become more aware, more conscious and increasingly able to experience *interbeing,* interconnectedness all living things – in order to realize that they are being observed by nature as well as observing nature (Mindell, 2000).

Ecological Crises – Crises of Consciousness

Chalquist (2007) cites the novel *Frankenstein*, written by Mary Shelley as a warning ‘against the consequences of technological idealism lacking sufficient understanding of the psychological shadow or underside (p. 2)’. In other words, humans desperately need to understand how ‘we’ve made things worse, even with good intentions, and that means we need to bring the psychological dimension into our discussions about the environmental crises (p. 2)'. Chalquist states strongly that '[t]he environmental crisis is a crisis of consciousness'. Fisher (1995) also cites Glendinning (1994) asking why "the screaming link between pervasive personal dysfunction and the ecological crisis" (p. 20) is not recognized, *especially* by psychologists. Lack of awareness can often be seen in the cognitive beliefs that many people hold: "There's nothing I can do, or it's not my responsibility. My beliefs/views/ideology are the truth and the only truth. It's their fault not ours, that they are hungry, that there is an arms race, and so on. You can't trust them. Worst of all. They're not really human."(Walsh, 1996, p. 396). Walsh continues that repression and denial become the human defence mechanisms to deal with a pervasive fear, and that 'fear, greed, aversion, ignorance, unwillingness to delay gratification, defensiveness, and unconsciousness are marks of psychological immaturity' (p. 400). What is more disturbing is the reality that this dysfunctional way of being and psychopathology is so widespread, but that few recognize it. This is what was described earlier as 'psychic numbing'. Shields (1994) comments on the many ways that our culture discourages us from knowing what we really feel... 'the tranquilization by the trivial', similar to psychic numbing, and that we can make a difference if we wake up to who we really are. Shields challenges people to 'discover [the] capacity to bring about outer, and inner, change', entailing a purposeful transformation of 'habitual responses to life' (p. 13). She cites Kevin McVeigh: "The denial of power is as endemic to our culture as the denial of feelings" (p. 12). William James (1900) also recognizes a 'certain blindness in human beings' which comes from their separation from the universe and living in duality. Ryland (2000) cites ecopsychologist Laura Sewell (1995, p. 202) who has this to say about this state of denial:

> Our "collective myopia" is one manifestation of psychic numbing – a psychological defence against witnessing the world's pain. It is a form of denial that shields us from fully experiencing the latest reports on ozone depletion, increasing pollution, toxicity, poverty, illness, and the death of the species. Full awareness hurts... We become numbed to our feelings, to what we might hear and see... In the midst of collective denial, we further perpetuate the destruction of the biosphere. Our collective myopia thus becomes both cause and effect of the environmental crises (in Ryland, 2000, p. 387).

Denial that we have any part in the ecological crises is psychologically a form of immaturity. Chalquist (2007) says that 'denial, a useful defence in childhood against overwhelming pain and terror, becomes fatal in adulthood when it strangles the will to change. It is denial that pretends there is no crisis, that the polar icecaps are not melting, that we are not in the midst of mass extinction that threatens half the plant and animal life on the planet, that social justice and environmental justice can somehow be meaningfully separated' (p. 3). Denial leads to psychic numbing which is a lack of consciousness and this means a lack of relatedness. Walsh (2002) cites the psychologist Charles Tart (1986, p. 5) who suggests that people live in a "consensus trance", which is "a much more pervasive, powerful, and artificial state than ordinary hypnosis, and it is all too trance like" (p. 14). All are saying a similar thing – the world is in ecological crisis because humans are 'insane' (Walsh, 2002, p. 14). The 'cure' for this pathology is to draw on the understandings from the fields of psychology, ecopsychology, transpersonal psychology, systems theory and many contemplative or spiritual disciplines.

The next section will examine some of the ways that can be used to raise awareness individually and collectively. It is not until individuals realize their interconnectedness with the universe that a societal response to the ecological crises can be addressed.

Path towards Transformation

Many people are writing about sustainability and supporting action plans to redress some of the degradation. This is not enough; change must come from within humans first. Ecopsychology, transpersonal psychology and the spiritual traditions, all recognize that there is a psycho spiritual dimension to the universe and our place in it. "Ecological psychology shares with some indigenous Native American cultures a mode of thought in which interconnection and its psychological counterpart, caring, are predominant" (Pillusk, 2001, p. 25). Fisher (1995) recognizes that 'whatever else may be necessary to reverse the ecocidal behaviour of our society, such a reversal is unlikely to occur unless we find ways to address the deep fear, anger, grief, despair, hopelessness, denial and so forth which presently block our finding creative responses to our present dilemma' (p. 24). This requires a psychological support of 'inner work', which will change our daily functioning.

Ecopsychologists include practices such as 'wilderness practices, vision quests, shamanism, ecological restoration work, bioregional practice, experiential deep ecology and "eco-counselling"'(p. 24). However, without awareness, these practices may not achieve anything. Mindell (2007) argues strongly that awareness is basic to everything we know and that 'our sense of awareness is connected to noticing, watching, knowing, mindfulness, realizing, wondering, and consciousness itself' (p. 12). Mindell calls this awareness "sentient awareness – a kind of intelligence, a knowingness, a quantum mind" (2007, p. 16). Sentient awareness is non-dualistic. In the Chinese tradition of Taoism, the Tao is the 'way' or 'path', yet more than that it is a power, or source or energy that governs the universe and creates order in the universe. The Chinese character for Tao is 道 and incorporates stillness, head, steps, path and eyes – this is awareness of one's interbeing.

Du Nann Winter and Koger (2004) explore holistic approaches such as gestalt and ecopsychology as ways of addressing the renewing of relationship with all parts of the universe. 'The whole is more than the sum of its parts. From this perspective, environmental problems are best approached by considering the system in which they are created' (Du Nann Winter and Koger, 2004, p. 186). Gestalt is a German word and roughly means 'whole' or 'pattern' to describe a process of sensation, awareness, mobilizing, full contact, withdrawal, assimilation and then beginning the cycle again (Murdock, 2009). Because Gestalt emphasizes an 'holistic nature of human functioning', it allows for recognition of the polarities (p. 209). The process of Gestalt enables a person to gain insight and to come back into balance and harmony with their environment: its major goal is awareness or consciousness. Du Nann Winter and Koger (2004) belive that we can know ourselves through holistic experiences, and so lower our defences. Gestalt has been utitlised to deal with our ecological problems by Abram (1987, 1997), Cohen (1997), and in particular, Sewell (1995).

Cohen argues that we must "sentiently" reconnect with nature in order to achieve wholeness again. Abram advocates ways to expand the senses and pays tribute to those who honour the spiritual practice of wonder in nature. Along with others, Sewell argues for the practising of mindful attention, saying:

> The earth calls continually. She [sic] calls us with beauty, sometimes truly breathtaking, sometimes heart wrenching, and always provocative and visceral. We are embedded in a multidimensional web of beauty. It is where we are, now... The moment calls for the reperceiving of our Earth, for perceiving the myriad and magical relations that may inform an ecological ethic. If we are receptive to the ways in which the landscape speaks to us, or the ways in which perception serves as a channel for communication, we may reawaken and preserve a sense of human integrity within the family of all relations. (Sewell, 1995, p. 215)

Boston (1996) calls this mindful awareness 'ecological groundedness...a dynamic state of the person that includes the sense of confidence, pleasure and wonder resulting from progressively deepening contact with the wild and domesticated natural community of the person's neighbourhood and larger land region, with unpaved ground, soil, or landscape, with weather and the diversity of native plants and animals; as well as with family, friends and local cultural activities' (p. 3).

Yunt (2001) explores how Carl Jung's concepts of archetypes, the collective unconsciousness and individuation also allow us 'to envision the interpenetration of psyche, nature, and spirit'. He cites Jung:

> As scientific understanding has grown, so our world has become dehumanized. Man feels himself isolated in the cosmos, because he is no longer involved in nature... His contact with nature is gone, and with it has gone the profound emotional energy that this symbolic connection supplied. (Jung, 1964, p. 95 in Yunt, 2001, p. 97).

Exploring Jung's ideas of archetypal energy, Yunt concludes that implicitly Jung has shown "the possibility of developing a nondualistic worldview that sublimates the inextricable interconnectedness our psyches, bodies and nature... a worldview characterized by ecocentricity, rather than egocentricity" (p. 119). Jung called this process one of individuation. "This provides the way for the individual psyche to connect with its shared, collective human heritage and to heal its isolation, helplessness and loneliness" (Ryland, 2000, p. 393).

Consciousness is both cognitive and feeling-based, and also motivational. We are sentient beings. This means we need to cognitively know, and be feelingly aware of, what is happening for us biologically, mentally and spiritually within our interbeing. Remember that interbeing 'refers to the mutual connectedness of all things, human and non-human, living and nonliving. Knowing, then is awareness of interbeing even as one *is* interbeing' (Puhakka, 2000, p. 18).

Becoming aware is not particularly easy, especially as the cognitive domain tends to control our subjective activity. Puhakka (2000) says 'the existence of a subjective self is... precariously dependent on the constructive activity of thinking' (p. 19). Awareness leads to knowing not only cognitively, but also bodily, psychically and emotionally. Puhakka recalls how Foucault (1980) saw the danger of authentic knowing posing a threat 'to the power elite in the dynamics of social oppression' (1980, p. 19). So challenging people to become knowledgeable about the ecological crises, and then becoming conscious and aware of their part in it, poses a threat to those who want to continue in the cycle of greed and ignore the degradation it causes in the interrelated domains of humankind and ecology. 'Consciousness is a cognitive-emotional-somatic-spiritual-whole knowing' (Cortwright, 1997, p. 54). One has to move beyond thinking to awareness. However, when dialectic reason is pitted against dialectic experience, 'the "feeling" component of the other's knowledge can rarely withstand the encounter' (De Quincey, 2005, p. 29). De Quincey goes on to say that 'reason doesn't have to decimate feeling – it does so only when unplugged from its roots in the deep wisdom of the body' (p. 33). In other words, there is a disconnection from our *interbeing*, when we are out of touch with the truth of our connection with all living things and our interconnectedness with the Universe. We are ecologically unconscious.

Teaching a conscious awareness of one's *interbeing* with nature can begin in early childhood development. Boston (1996) notes that "wilderness settings can be particularly influential on a child's development" (p. 4). Indigenous children used to grow up with this respect and awareness, but they too are losing this awareness in many ways, particularly in many places in Australia where so many Aboriginal children were "stolen" and brought up in institutions away from their mother earth (*Bringing Them Home Report*, 1997). Taking some children on wilderness camps is one way to help them to reconnect with the power of nature which they come to understand is closely related to their own power.

Arrien (1993) tells us that the word *ecology* comes from the Greek word *oikos*, meaning *house*, and that our task in today's world is to again care for both our inner and outer houses; that is, our inner psychic house and our world house which is where we live our daily lives, including the world of nature. They are inseparable. Looking back at traditional native people's oneness with the earth can bring us guidance for remedying the split that our modern society has created.

Mindfulness

Jon Kabat-Zinn (2005), a Buddhist psychologist, argues that to heal our world we need to learn 'to put our multiple intelligences to work' and that this involves 'coming to our senses and waking up to the possible' (p. 15). One way of learning to tap into our intuition and to learn from our senses in addition to having knowledge and reasoning, is to become mindfully aware or to pay attention to the whole of our being – mind, body, senses and feelings.

> When through the practice of mindfulness, we learn to listen to the body through all its sense doors, as well as to attend to the flow of our thoughts and feelings, we are beginning the process of reestablishing and strengthening connectedness within our own inner landscape.
> (Kabat-Zinn, 2005, p. 123)

When we are in touch with the whole of ourselves, we are more alert as to how we are interconnecting with the universe and what we are doing to create dysfunction and dis-ease within our world which contributes to the crises in the universe. We then begin to act with more awareness and contribute to the healing of the universe. Rinpoche (2007) powerfully argues that we can transform adversity into awakening by mind training. He says: 'The tendency to adopt a defeatist attitude in the face of evil is the biggest obstacle to our everyday lives..." (p. 39). Mind training, for Rinpoche, is the cultivation of an enlightened heart – bodhichitta, and the cultivation of compassion; this enables us to transform our emotional interpretation and personalization of events in our lives into a compassionate response to live in relationship with our universe and to contribute positively towards its healing.

Arrien (1993) links this mindfulness with the ancient shamanic and indigenous traditions, and the importance they attached to balancing their lives on four archetypal levels or roles: leading (being a warrior), healing (being a healer), visioning (becoming a visionary) and teaching (being a teacher). These four levels require a deep awareness and consciousness of one's connection to the whole of the universe. Lao Tsu said: "When men lack a sense of awe there will be disaster" (1997, chapter 72).

Such consciousness is not just personal; it must also reach the highest form of political consciousness. On a global level, transformation is needed in existing social institutions where the mainstream paradigm is still very much one of controlling and developing nature, using it as a resource for humans, materialism, proliferation of nuclear weapons, greed and a rigid adherence to dualistic economic ideologies. Change is possible as each one takes a stand with their whole being. Arrien (1993) argues that the Warrior chooses to stand up and be present. The Warrior shows respect and honour for all living things. Of course, this means taking responsibility for our own actions, as well as challenging the actions of a society which is destructive. The healer pays attention to 'the principle of reciprocity' – which is about giving and receiving and staying in connection with Nature. The Visionary, according to Arrien (1993) needs contemplation and mindfulness, gained by many Indigenous people through vision quests. This quest is an extended time of solitude in nature. Such time alone enables one to awaken again one's relationship and dependence on nature. A vision quest requires a letting go of ego-centric attention and becoming much more aware of eco-centric vibrations. The archetype of Teacher means a development of wisdom and discernment, also gained by sitting in silence and meditation. Within meditation, the practice of mindfulness leads to both observation and letting go – detachment. Detachment respects nature and does not use it for ego purposes. Nature itself is a great teacher to whom Indigenous people have listened and from whom they have learned more about life's unity and wholeness. Grine (2008), a poet from the Lakota Sioux captures this in his prayer of gratitude: "O Great Spirit, Creator of all, allow us to walk as with one, show us the pathway of dignity, teach us to respect your creation, give strength amidst times of need, let it be known you are our guardian" (http://siouxpoet.tripod.com/id2.html).

Choosing money or choosing life as a basic value system is a reality that needs to be faced (Korten, 1999). To Korten, the struggle between life and money is a struggle between good and evil. Each person, who becomes more aware and more conscious of their interbeing, has a cybernetic effect on the whole. By shifting our own consciousness, we have immeasurable power to transform. Frattoroli (2001) says that 'consciousness is also the most powerfully humanizing force in the world' (p. 422)… it is about moving from passion to compassion, being one-with all of nature. Transformation is an ongoing process, not a sudden magical change. As Frattoroli comments, 'you always have within you the resources necessary to solve the problems that confront you' (p. 432). Motivation comes from the words of Jon Kabat-Zinn (2005):

> The challenge is nothing less than a wake-up call to our species. We can seize the opportunity to rotate in consciousness, to undergo a quantum leap through cultivating our capacity for sentience, for mindfulness, for awareness, even though it requires considerable motivation and effort as individual people, as nations, and as planetary citizens with all the

> karma we carry; or suffer the consequences, ever more terrifying, of our heedlessness, our ignoring of what is most important, most fundamental for life on Earth to flourish and for us all to pursue our possibilities to their fullest and their wisest (p. 596).

For example, to treat major river systems such as the Yangtse, Yellow, Mississippi, Missouri, Murray Darling, Nile and the Amazon as canalized waterways, completely under human control, is to forget that these and other river systems are there to organize vital entities which are, so to speak, alive. The diminishment of their natural floodways, wet-lands, the constant enlargement of levees and the construction of dams and weirs act to reconstitute their being as "other", in opposition to humanity's needs, and destroy their vitality and that of the human race: similarly with the construction of chemical plants and the leakage of toxic chemicals into rivers in China (Lorenz, 2005). John Mack (2008) cites Vaclav Havel, the Czechoslovakian president addressing the United States Congress in 1990: "Without a global revolution in the sphere of human consciousness, nothing will change for the better in the sphere of our being as humans, and the catastrophe toward which this world is headed – be it ecological, social, demographic, or a general breakdown of civilization – will be unavoidable". Just before he died, psychologist Carl Jung had a vision of enormous destruction across the earth, caused he said, by 'the psyche of man'. However, it is not too late to heal, if humans develop awareness, mindfulness and consciousness.

CONCLUSION

This chapter began by stating the underlying principle that *The Universe and I are one* – interconnected living organisms that affect each other. The author has taken a radical view, in line with ecopsychology and transpersonal psychology, demonstrating that humans affect the universe and play a part in the ecological crises, through lack of awareness and consciousness. Thich Nhat Hahn (2005) challenges us to stop, look and listen, NOW - in this present moment, and he suggests we take time to digest these words from T.S Eliot's Fourth Quartet (1942):

> Not known, because not looked for,
> But heard, half-heard in the stillness
> Between two waves of the sea.
> Quick, now, here, now always,
> A condition of complete simplicity
> (Costing not less than everything)
> And all shall be well
> All manner of things shall be well.

'Costing not less than everything' requires total commitment on our part and the courage and willingness to engage in transformative work that will heal the universe, a work that is one of being-in-relationship with the world. This work is to be done on both the individual and collective level. To avoid further degradation of our earth, it is imperative that we wake up to our oneness with the universe and, as the Plains Indians suggest: 'to walk in a sacred manner'.

REFERENCES

Abram, D. (1997). *The Spell of the Sensuous: Perception and Language in a More-Than-Human World* (pp. 3-29). New York: First Vintage Book.

Abram, D. (1987). The perceptual implications of Gaia. *Revision,* (2), 7-15.

Arrien, A. (1993).*The Four-Fold Way – Walking the Paths of the Warrior, Teacher, Healer and Visionary.* San Franciso: Harper.

Australian Human Rights and Equal Opportunity Commission. (1997). Bringing them home: Report of the National Inquiry into the separation of Aboriginal and Torres Strait Islander People from their families. Sydney. Retrieved 10.04.09 http://www.hreoc.gov.au/social_justice/bth_report/index.html

Bernard, S., Samet, J., Grambesch, A., Ebi, K. and Romieu, I. (2001). The Potential Impacts of Climate Variability and Change on Air Pollution-Related Health effects in the United States. *Environmental Perspectives, 109*(2), 199-209.

Boston, T. (1996). Ecopsychology: An earth-Psyche Bond, Trumpeter, 13(2). Retrieved: 8/10/08: http://trumpeter.athabascau.ca/index.php/trumpet/rt/printerfriendly/269/402

Canty, J. (2006). Environmental Healing: Shifting from a Poverty Consciousness. Gatherings, Retrieved 15/6/08: http://www.ecopsychology.org.journal/ezine/archive3/Environmental_Healing.pdf.

Chalquist, C. (2007). The Environmental Crisis is a Crisis of Consciousness – Bringing the Psychological Dimension into the Discussion. Speech delivered at "voices for Change", organized by students at SonomaState University, November, 2007. Retrieved 3/3/08: http://www.terrapsych.com/crisis.html

Cohen, M. J. (1995). Ecotherapy: The ecology of Gaia, Sensation and Soul. *Theory and Review in Psychology, Electronic Journal.* Retrieved 9/2/09: http://gemstate.net/susan/EcoPsy.htm

Connor, S. (2005). Climate Change Map Reveals Countries Most Under Threat. The Independent, 17 November, 2005. Retrieved 2/2/09: http://www.countercurrents.org/cc-connor171105.htm

Cortwright, B. (1997). *Psychotherapy and Spirit.* New York: State University of New York Press.

Davis, J. (2008). Ecopsychology, Environmental Problems, and Environmental Action. Retrieved 13/6/08: http:www.johndavis.com/ep/envact.htm

De Quincey, C. (2005). *Radical Knowing – Understanding Consciousness through Relationship.* Rochester, Vermont: Park Street Press.

Du Nann Winter, D. and Koger, S. (2004). *The Psychology of Environmental Problems* (2nd ed.). Mahwah, New Jersey: Lawrence Erlbaum Associated, Inc. Publishers.

Eliot, T.S. (1942). Quartet No. 4: Little Gidding, retrieved 3/3/09: http://www.tristan.icom43.net/quartets/

Fisher, A. (1995). Toward a More Radical Ecopsychology – Therapy for a Dysfunctional Society. *Alternatives Journal*, 22(3) July/August, 10-26.

Foucault, M. (1980). *Power/Knowledge: Selected Interviews and Other Writings 1972-1977,* C. Gordon. (Ed.) London: Harvester.

Frattoroli, E. (2001). *Healing the Soul in the Age of the Brain – Becoming conscious in an unconscious world.* New York: Viking Penguin.

Gadon, E. (1989). *The Once and Future Goddess.* San Francisco: Harper.

Gilman, R. (1984). The Idea of Owning Land. Living With the Land (IC#8), Winter. Retrieved 21/8/08: http://www.context.org/ICLIB/IC08/Gilman1.htm

Glendinning, C. (1994). *My Name is Chellis, and I'm in Recovery from Western Civilization.* Boston: Shambhala.

Hannush, M. J. (2003). Radical Ecopsychology in the Service of Life. *Journal of Phenomenological Psychology.* 34 (1). Expanded Academic ASAP. Gale Dixon Library. Uni of New England, 6 Mar. 2008 Retrieved http://find.galegroup.com.exproxy.une.edu.au/itx/start.do?prodID=EAIM

Hart, T., Nelson, P. and Puhakka, K. (2000). *Transpersonal Knowing – Exploring the Horizon of Consciousness.* New York: State University of New York Press.

James, W. (1900). On a certain blindness in human beings. Retrieved 8/10/08: http://www.des.emory.edu/mfp/jcertain.html

Jung, C.G. (1964). Man and his symbols. New York: Random House.

Kabat-Zinn, J. (2005). *Coming to Our Senses – Healing ourselves and our world through mindfulness.* New York: Hyperion.

Korten, D.C. (1999). *The post-corporate world.* San Francisco: Berrett-Koehler.

Lao Tsu. (1997). Tao Te Ching (Translated by Gia-Fu Feng and Jane English), New York: Vintage books.

Lorenz, A. (2005). The World's Toxic Waste Dump- Choking on Chemicals in China. Spiegel Online International. Retrieved 10/3/09: http://www.spiegel.de/international/spiegel/0,1518,387392,00.html

Mack, J. (2008). Blowing the Western Mind. The John E. Mack Institute. Retrieved http://wwww.johnmackinstitute.org/ejournal/article.asp?id=25

Mindell, A. (2007). *Earth-Based Psychology – Path Awareness.* Chicago: Independent Publishers Group.

Mindell, A. (2000). *Quantum Mind – The Edge between physics and psychology.* Portland, Oregon: Lao Tse Press.

Murdock, N. (2009). *Theories and Techniques of Counselling and Psychotherapy.* Upper Saddle River, New Jersey: Pearson.

Pillusk, M. (2001). Ecological Psychology, Caring, and the Boundaries of the Person. *Journal of Humanistic Psychology,* 41 (20), 25-37.

Puhakka, K. (2000). An Invitation to Authentic Knowing, in T. Hart, P. Nelson and K. Puhakka, (Eds.). *Transpersonal Knowing – Exploring the Horizon of Consciousness* (pp. 11-30). New York: State University of New York Press.

Rinpoche, T. K. (2007). Training the Mind to Transform Adversity into Awakening, in Shambhala Sun, May, 2007, 37-41.

Ryland. E. (2000). Gaia Rising: A Jungian Look at environmental Consciousness and Sustainable Organizations. *Organization Environment*, 13, 381-402.

Sandner, D. and Wong S. (Eds.). (1997). *The Sacred Heritage – The Influence of Shamanism on Analytical Psychology.* New York: Routledge.

Sewell, L. (1995). The Skill of Ecological Perception. In T. Roszak, M. Gomes, and A. Kanner (Eds.), *Ecopsychology: Restoring the Earth Healing the Mind* (pp. xiii-xvi). San Francisco: Sierra Club Books.

Shields, K. (1994). *In the Tiger's Mouth – An Empowerment Guide for Social Action.* Australia: Millenium Books.

Tart, C. (1986). *Waking up: overcoming the obstacles to human potential.* Boston: New Science Library/Shambhala.

Thich Nhat Hahn. (1987). *Being Peace*. Berkeley, C.A: Parallax Press.

Walsh, R. (2002). Terrorism and Other Global terrors: An Integral Analysis. *The Journal of Transpersonal Psychology*, 34 (1), 13-21.

Walsh, R. (1996). Toward a Psychology of Human and Ecological Survival: Psychological Approaches to Contemporary Threats, in B. Scotton, A. Chinen, and J. Battista (Eds.). *Textbook of Transpersonal Psychiatry and Psychology* (pp. 396-405). New York: Basic Books.

World Health Organization. (2003). WHO/ Climate Change and human health – risks and responses. Retrieved 3/6/08: http://www.who.int/globalchange/publications/cchhbook/en/print.html

Yunt, J. (2001). Jung's contribution to an ecological psychology. *Journal of Humanistic Psychology*, 41 (2), 96-121.

Contributed by Tim McCrorey.

Photo: Clarity and Beauty in Nature.

In: Meltdown: Climate Change, Natural Disasters… ISBN 978-1-60876-153-1
Editor: Kathryn Gow

Chapter 7

THE IMPACT OF ENVIRONMENTAL THREATS AND CLIMATE CHANGE ON OUR PHYSICAL AND PSYCHOLOGICAL HEALTH

Grace Wai-man Ip
Hong Kong Shue Yan University, Hong Kong

ABSTRACT

The continued survival and health of human beings depends on a stable environment, and the recent changes in the environment such as global warming, species' extinction, overpopulation, exhaustion of resources and pollution are all part of what constitute environmental threats. This chapter focuses on both the physical and psychological health impacts of such environmental threats as well as some of those resulting from climate change. The first part of the chapter will focus on the various direct and indirect physical impacts of climate, exposure to toxins, chemicals and pollution on our physical health, while the second part will focus on the indirect psychological impacts (through stress and emotional distress) of climate change and environmental threat factors on health and decision making.

Keywords: environmental threats, physical health, climate change, psychological health, pollution

INTRODUCTION

Physical Impacts of Environmental Threats on Our Health

Direct Impact of Climate and Climate Change. Researchers quote astonishing figures on the effect of climate on illness and mortality. For example, Patz, Campbell-Lendrun, Holloway and Foley (2005) stated that over 150,000 people died annually because of anthropogenic climate change; Miller (2007) wrote that 78 million cases of illness, 350,000 hospitalizations and 6,000 deaths per year were due to heatwaves and food borne diseases; the BBC (2008) reported that mortality rates tend to increase by up to 10% when temperatures exceed 38˚C for more than one week and Pret, Lentini and Maugeri (2007) claimed that even the higher suicidal rate could be related to global warming. Without a lot more research into the different areas, we can only use these publications as a guide.

Figure 7.1 summarizes the ideas of McMichael et al. (2003) and the BBC (2008) on the direct effect of climate on health. With regard to temperature and humidity, it was found that during hot and humid summer days, we are more likely to experience sleeplessness, decreased general activity, poorer vigilance, slower reaction times and performance, irritability and lethargy. In addition, the use of air-conditioning in summer also depresses our emotional state, because it depletes the negative ions in the air (BBC, 2008; Howard, 2006). Due to global warming, our summer days are hotter and longer which could result in our spending less time outside the air-conditioned rooms, and thus our bodies will have less exposure to the natural weather conditions and hence be less adaptive to different weather conditions. Schata, Jorde, Elixmann and Linskens (1989) were worried that fungi may be found in air conditioners, while Zaslow and Genter (1993) were concerned that bioaerosols could be present in humidifier use. Both fungi and bioaerosols are allergic agents that can cause allergies to sensitive persons. Looking at the bigger picture and what happens to our mood when the weather changes, Keller et al. (2005) believe that although pleasant weather is related to higher mood states in spring, hotter weather is associated with lower moods in the summer.

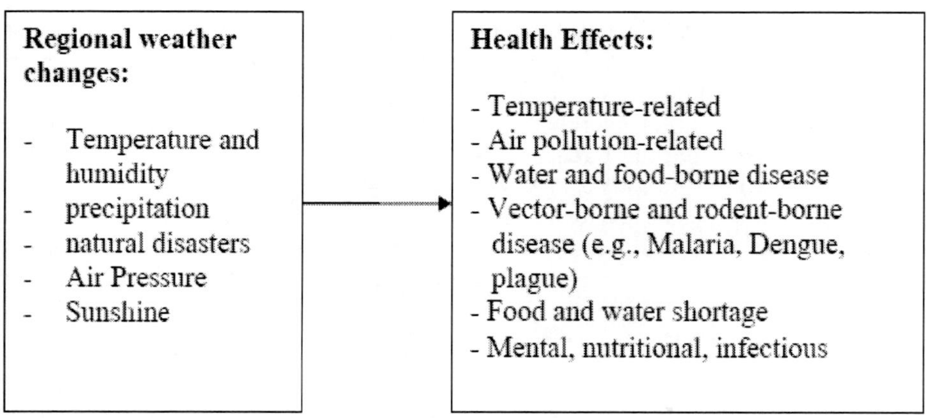

Adapted from McMichael et al. (2003) and the BBC (2008).

Figure 7.1. A simplified model of the consequences of direct effects of weather changes on human health.

Furthermore, the impacts of extreme temperature (such as in heatwaves and very cold spells) on our health are even more serious. For example, heatwaves cause cardiovascular mortality, respiratory illnesses, heatstroke, heart attack and stroke (McMichael et al., 2003; Patz et al., 2005), while in cold winter months, cardiovascular, cerebrovascular, circulatory and respiratory diseases increase and because of lack of sunshine in winter, winter blues may result (Analitis, 2008; McMichael et al., 2003).[1] Since extreme hot summers and cold winters may lead to higher mortality rates, it could be conjectured that severe global weather changes might indirectly cause higher mortality rates because of more extremes of hot and cold periods. The number of deaths in the past decade related to heat waves endorses that concern.

On the positive side, with regard to precipitation, rain is actually good for our health because it brings in negative ions (Howard, 2006). On the negative side, however, more extremes in precipitation, accompanying severe storms, tornadoes, hurricanes and floods can lead to mortality through: physical injury; decreases in nutrition due to lack of food supply (Hashizume, 2007; Wade et al., 2004); increased risk of water-related and infectious diseases including diarrhoea, and the release and dissemination of dangerous chemicals from storage sites and negative mental health issues (Greenough et al., 2001; McMichael et al., 2003). Furthermore, more precipitation and higher temperatures together lead to more waterborne and food borne disease and lower coastal water quality (Curriero, Patz, Rose, Lele, 2001; Gaffield, Goo, Richards, and Jackson, 2003; Rose et al., 2001).

Taken as a whole, local temperature, precipitation, clouds, atmospheric water vapour, all influence atmospheric chemical processes, and hence climate change may affect our exposure to air pollutants by affecting weather, anthropogenic emissions, biogenic emissions and by changing the distribution and types of airborne allergens (Bernard, Samet, Grambsch, Ebi, and Romieu, 2001). The most challenging aspect of climate change is its extreme nature, because our physical body cannot adapt to extreme weather or natural disasters and these situations may result in direct and negative health impacts. Furthermore, air pollution, exacerbated by climate change, also has a significant impact on our respiratory system (Dominici et al., 2006; Schwartz, 1993).

Indirect Impact of Climate and Climate Change

Climate change also results in depletion of the ozone layer, declines in fresh water supply and arable agricultural land, and degradation of land, as well as extinction of species (McMichael, et al. 2003, McMichael, Woodruff, and Hales, 2006; Oskamp, 2000) which produces indirect but synergistic effects on health (Oskamp, 2000). McMichael and colleagues (2003, 2006) believe that climate change can be caused by natural forces and/or by human beings due to increased emissions of carbon dioxide.

Figure 7.2 summarizes the idea of McMichael et al. (2003) on the indirect effect of climate change on our environment. First, depletion of ozone may lead to an increase in ground-level ultraviolet irradiation which could result in more skin cancer. Second, extinction of some species (Thomas et al., 2004; Thuiller et al., 2005) will lead to significant loss of vegetation which will result in loss of nature's medicinal goods. Third, many arable lands and pastures are seriously damaged by erosion and compaction, while many ocean species are becoming extinct due to extra demands on the sources of edible food from fisheries (Commonwealth Scientific and Industrial Research Organisation, 2009). The combined

---

[1] It may be that the cardiovascular and respiratory diseases themselves do not increase in incidence, but that symptoms are exacerbated in extreme temperatures. Also, infectious illnesses like influenza increase during winter months, but this is because the viruses spread more easily amongst people spending more time indoors in groups, not because of the colder temperatures. (S. Sacre, personal communication, 2009).

effect will be a shortage of food supplies which could result in malnutrition and possibly death for many human beings.

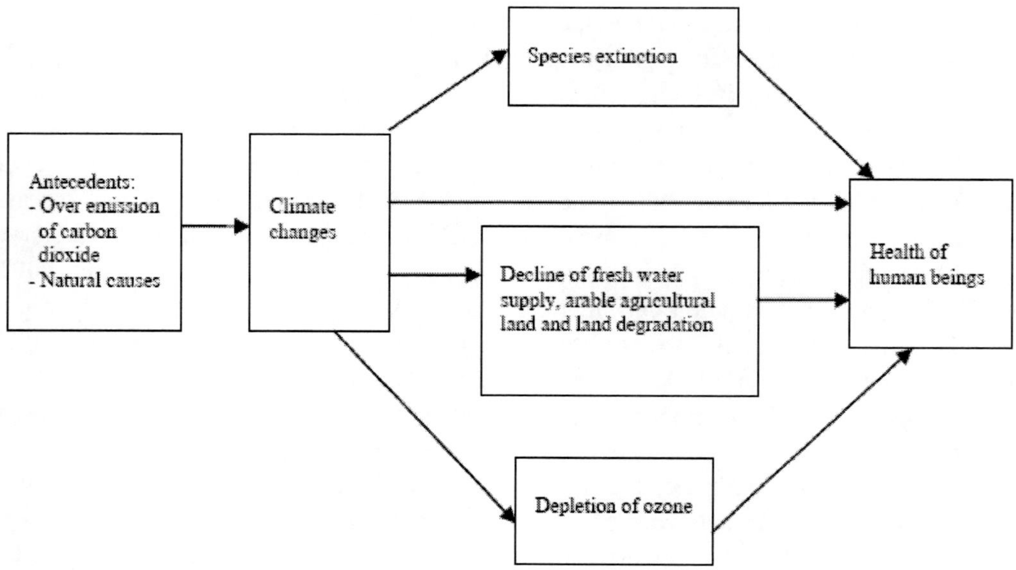

Adapted from McMichael et al. (2003).

Figure 7.2. A simplified model of the antecedents and consequences of climate change.

Further accelerations in climate change may result in more problems and hence might directly or indirectly affect our health and survival.

Toxins and Pollution

Another set of worrying figures and proposals have been promulgated: it has been estimated that two-thirds of people in the USA (i.e., about 200 million people) have an elevated risk of developing cancer due to their exposure to toxic chemicals emitted by vehicles, industrial sources and power plants (Du Nann Winter and Koger, 2004). The World Health Organization (2003) estimated that 3 million deaths annually were due to respiratory and cardiac disease caused by air pollution. Oskamp (2000) commented on greater human exposure to toxic chemicals because of the effect of acid rain and toxic pollution on air and water supplies.

Tauxe (1997) reported in detail on which waterborne pathogens contaminate different kinds of fresh fruit and vegetables. He mentioned that cantaloupe, tomatoes, strawberries, apple cider, scallions, Alfalfa sprouts, orange juice, leaf lettuce, raspberries and apple juices were proven to be contaminated by food borne diseases during food borne outbreaks between 1990-1996. Furthermore, he also mentioned that seafood (e.g., shellfish and fish) could be contaminated by an increase of microbiological agents (e.g., viruses, bacteria, and protozoa) living in contaminated water. Davis (2008) postulated that neuropsychological and stress effects resulting from major toxin exposure (e.g., such as chemical and radioactive leaks, industrial accidents) would be severe, complex, longer-lasting and likely to be chronic, leading to anxiety, psychic numbing, denial, alienation, helplessness and depression. The examples given by Davies (2008), are Chernobyl,[2] Three-Mile Island[3] and Bhopal.[4]

Du Nann et al. (2004) provide details on how common toxin chemicals, including persistent bio-accumulative and toxic pollutants (PBTs) and polychlorinated biphenyls (PCBs), pesticides and heavy mental (such as mercury and lead) affect us. Firstly, the various kinds of chemical toxins can act as neurotoxins, therefore, during prenatal development if the foetus is exposed to such chemical toxins, the babies might potentially have cognitive impairments and developmental disabilities (such as lowered IQ scores, impairments of memory and attention and coordination deficits, attention deficit hyperactivity disorder (ADHD), or mental retardation (Bronstein, et al., 2009; Myers and Davidson, 2000; Weiss, 2000; Weiss and Bellinger, 2006; Weiss and Landrigan, 2000). Weiss (2000) even postulated that prenatal toxin exposure might be the origin of schizophrenia, Alzheimer's and Parkinson diseases. Secondly, toxin chemicals (e.g., PBTs, PCBs) can disrupt hormones potentially leading to deficits in learning and memory (Porterfield, 2000). Garshick et al. (2008) found that among trucking industry workers, exposure to toxic chemicals is related to the incidence of lung cancer. [For a summary on the details of the negative impacts for the exposure to each of the toxin chemicals, see Du Nann et al. (2004)].

In addition, Evans and Campbell (1983) also listed the negative outcomes of air pollution. They included outcomes related to: acute sickness, chronic disease and death; impairments in growth or in physical or cognitive performance; alteration of important physiological functions and some irritating symptoms requiring medical attention; and accumulation of potentially harmful materials in the body which result in discomfort and annoyance.

To sum up this part of the chapter, there is some evidence that the physical impact of ecological crises may result in negative impacts on physical (such as discomfort, illness and mortality) and psychological well-being (stress and mood disturbances), being due either directly or indirectly to weather change, or exposure to toxic chemicals and/or pollution.

Psychological Impacts of Ecological Disturbances on Our Health

Ecological Disturbances as Stressors

Even when we are not aware of all the direct and indirect effects of ecological disturbances on our physical health, they may still affect us and act as stressors. For example, stress may result from being diagnosed with, or suffering from disease, infection, surgery effects, or malnutrition, - all conditions that may have occurred in relation to ecological disturbances. In the most extreme cases, the loss of loved ones, housing or employment, due to natural disasters, may result in more serious forms of stress reactions, such as posttraumatic stress disorder in a small percentage of the affected population.

---

[2] Chernobyl nuclear facility, Ukraine; 1986; steam explosions and resultant nuclear contamination of large areas.

[3] Nuclear accident at Three Mile Island Unit 2 (TMI 2),1979; due to equipment failure and other factors.

[4] Bhopal, India, 1984. Caused by release of methl isocvanate gas from a pesticide plant.

In addition, awareness of the impacts of environmental threats, coupled with helplessness in the face of them, may lead to greater feelings of stress, anxiety and depression; thus, environmental threats can affect us psychologically when experienced as a stressor (Arapis, Goncharova, and Baveye, 2006; Du Nann et al., 2004). Figure 7.3 provides a brief summary on how environmental threats may trigger the psychological mechanism of stress and mood disturbances, and hence affect our health and ability to process information. In the following sections, I will first outline the physiological mechanism and health impact of stress, and then discuss how mood can affect information processing.

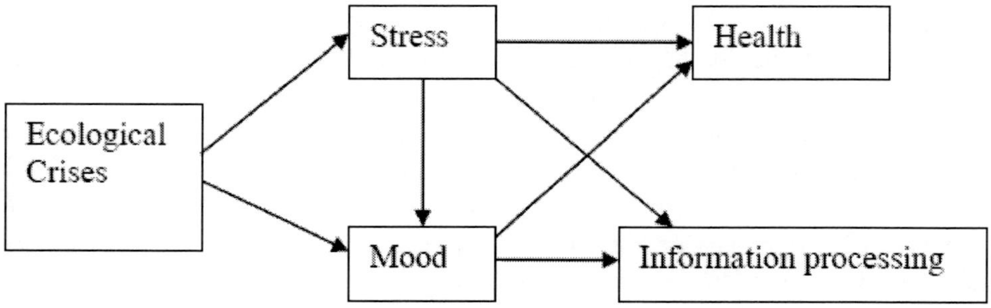

Figure 7.3. A brief summary of the psychological impacts of ecological disturbances.

Physiological Mechanism of Stress

The stress from ongoing and pervasive environmental threats (such as pollution, global warming, and toxic chemicals) is a prolonged, long-term and inescapable one (technically known as chronic stress). This kind of chronic stress follows the full cycle of the generalized adaptation syndrome (Selye, 1979)[5]. In the first stage of stress, known as alarm, the autonomic nervous system is activated, and the body becomes ready for brief emergency activity. Then in the second stage, known as resistance, the activities in the autonomic nervous system decline, but the hormonal system starts to secrete cortisol and other hormones that enable the body to maintain prolonged alertness, fight infections, and heal wounds. However, after intense and prolonged stress, we become tired, inactive and vulnerable as the nervous system and immune system no longer have the energy to sustain their heightened response; this third stage is known as exhaustion (Sapolsky, 1998; Selye, 1979).

Du Nann et al. (2004) mentioned that based on Darwin's evolutionary theory, as our ancestors only needed to deal with more acute kinds of stress (i.e., sudden threatening stimuli), we are more likely to be prepared for different types of fight and flight responses (i.e. intense bursts of energy and nervous system activity). Gualin and McBurney (2001) termed this kind of mismatch between the modern environment and the ancestral environment

[5] Stress activates two bodily systems, the autonomic nervous system and the pituitary/adrenal system. The autonomic nervous system is also known as the sympathetic/adrenal system and it produces brief emergency responses. Once it is activated, it will result in arousal and the fight or flight response. The pituitary/adrenal system, on the contrary, reacts more slowly, but becomes dominant when stressors are prolonged, as it works through many brain areas including the hypothalamus, pituitary gland and adrenal cortex and the slower hormonal system. The activation of the hormonal system increases the release of cortisol which enhances metabolic activity and elevates blood sugar levels and other nutrients (Brannon & Feist, 2007; Carlson, 2006; Franken, 2002; Kalat, 2007).

as mismatch hypothesis and Boaz (2002) had argued that since stressors from environmental threats are ongoing, our automatic stress responses from evolutionary history may not be suitable for an environment which creates even more adverse health effects.

Stress-related Health Risks

When we are under acute stress, that is, in the resistance stage, our immune system and our attention and memory are at its peak. However, under prolonged periods of high stress, that is, in the exhaustion stage, our immune system will shut down and our attention and memory will decline somewhat due to the activity of excessive cortisol.

In dealing with stress, the increase in cortisol level will enhance our immune system activity in the beginning. However since cortisol increases blood sugar metabolism which detracts energy from synthesizing proteins (i.e., fewer proteins are available in the immune system), our immune system is weakened which increases our susceptibility to colds and all kinds of low grade infections. This is known as immune-suppression. Therefore, under prolonged periods of high stress, we will be more likely to have low grade infections and colds. If the cortisol level continues to increase, in some circumstances, it may lead to the complete shutdown of the immune system (Cohen et al., 1998).

Similarly with attention and memory; a moderate level of cortisol would improve attention and memory function when we deal with stress in the early phase. However, during prolonged stress, the cortisol level is too high, leading to a heightened increase in metabolic activity in the hippocampus. Therefore, hippocampal cells become over-stimulated and more vulnerable to toxins and hence more likely than usual to damage or kill neurons in the hippocampus (Sapolsky, 1992); in addition, we cannot produce new hippocampal neurons (Pham, McEwen, LeDoux, and Nader, 2005). The combined effect may result in a decline of memory capability.

Hales (1997) summarized the relationship between stress and disease; he explains that stress can cause minor muscular twitches, aggravate intestinal and bowel syndromes and make us more susceptible to infectious disease. He also mentioned that stress could lead to: headaches and psychological problems (such as anxiety and depression); cardiovascular diseases; skin problems; mouth ulcers; mental disorders in women; and premature ejaculation in men. Thus both physical disease and psychological disorders can be caused by, or exacerbated by stress (Brannon and Feist, 2007; Hales, 1997; Nann, Winter and Koger, 2004).

Stein and Spiegel (2000) have summarized past research findings and concluded that stress can produce stimulant effects through the autonomic system which is a mechanism that possibly results in cardiovascular disease such as essential hypertension, cardiac dysrhythmias, cardiomyopathis and coronary artery disease. Additionally, through affecting the immune system, stress can exacerbate more existing health problems. As reported by Stein and Spiegel (2000), it is well accepted in the medical community that stress exacerbates immune disorders such as rheumatoid arthritis, inflammatory bowel disease and asthma. Also, in both human and animals studies, it has been demonstrated that stress increases our susceptibility to infections such as colds, when our immune system is suppressed (immuneo-suppression) (Du Nann et al., 2004; Stein and Spiegel, 2000). One of the consequences of immuneo-suppression is tumour progression (Du Nann et al., Winter and Koger, 2004; Stein and Spiegel, 2000). As argued by Boaz (2002), stress makes our body more susceptible to viruses and other organisms which in turn may increase our risk of cancer. Furthermore, stress can promote the growth of existing tumours (Stein and Spiegel, 2000; Thaker, Han, Kamat et al., 2006). In addition, experiencing stress may lead us to engage in unhealthy behaviours such as smoking, eating an unhealthy diet or drinking alcohol excessively.

In addition to physical illness, chronic stress may lead to overeating and obesity (Dallmam et al., 2003; Willenbring, Levine and Morley, 1986). Moreover, chronic stress has been found to be associated with addictive behaviours such as alcohol dependence (Heilig and Thorsell, 2002) and smoking (Cohen and Lichtenstein, 1990; Pomerleau and Pomerleu, 2006; Westman, Eden and Shirom, 1985). From a psychological perspective, stress can also be associated with depression (Brannon and Feist, 2007) and anxiety (Brannon and Feist, 2007; Bremner, Krystal, Southwick and Charney, 1998; Maes et al., 2002). In order to deal with life's stressors, people may engage in more "relaxed" behaviours such as smoking, consumption of alcohol, use of illegal drugs, and sitting for long hours in front of the television without any exercise, because they are too tired to do more meaningful, productive and healthy activities (such as exercises and walking or having more time with families and friends); consequently these avoidant behaviours may result in depression or anxiety. The most serious form of stress is posttraumatic stress disorder which sometimes follows serious traumas such as war service, horrific accidents and natural disasters (Du Nann et al., 2004).

Environmental issues can create a cyclical effect in creating stress; for instance: a) stress may lead to eating or addictive disorders which is a type of over consumption and may exert further pressure on the environment; b) under stress, we may consume more fast food or contaminated products which often contain meat, which in turn, if the meats contain growth hormones and/or antibiotics, may increase our risk of absorbing toxins such as PBTs. As these can accumulate in fatty tissue, the eating of such foods can have a negative impact on our health (Motavalli, 2002a; Nann Winter and Koger, 2004). Additionally, environmentalists argue that the amount of natural resources used to produce meat is higher than that of producing vegetables which then adds to the current environmental problems (Motavalli, 2002a; Nann Winter and Koger, 2004).

Mood Changes and Climate

Although researchers have different definitions of mood and emotion, they generally agree that how we respond to specific eliciting stimuli is known as emotion (such as happiness, sadness, fear or anger), while the ways these eliciting stimuli affect our cognitions are known as mood or affective states (Niedenthal, Krauth-Gruber and Ric, 2006). In the first section on the physical effects of weather changes, we explored the direct implications of global warming on our mood states, such as warm winds bringing in positive ions, and air-conditioning depleting negative ions, with the resultant effect being a negative mood state.

Such an indirect effect on our mood may actually affect how people enjoy their holidays and this is an important issue in the economic downturn, as tourism suffers. Often people who travel to some Asian cities complain about the pollution and are concerned about the health risks over the longer term. Indeed some tourists may be concerned about the publicised ultraviolet radiation in some parts of Australia where there is a known ozone hole effect.[6] In

[6] http://www.theozonehole.com/consequences.htm.

order to test how environmental effects (both anticipated and experienced) might influence people's assessment of their holiday destination enjoyment, Böhm (2007) conducted a study with tourists, in which he measured people's subjective feelings in anticipating, as well as experiencing, pollution in one scenario and more ultraviolet radiation in another scenario. Controlling for other factors which might affect their responses, Böhm specifically targeted the environmental threats. He asked the tourists to rate their degree of positive emotions (such as joy, relief or pride), their negative emotions (such as anger, fear or distress) and life quality (such as health, cleanliness or beauty of natural environment), both before and after travel, either to Bangkok (where air pollution is high) or Australia (which has known ozone depletion: Australian Government, Department of the Environment, Water, Heritage and the Arts (2008).

Results indicated that tourists expressed a higher degree of negative over positive emotions for both Bangkok and Australia, before and after their visits, which means that tourists have more negative emotions both in anticipation of visiting a country having environment risks and after they have actually visited those countries. Furthermore, when the intensity of negative emotion was higher, those tourists also perceived their quality of life to be poorer. The implication of Böhm (2007)'s study is that our subjective perception of emotion is affected by both our anticipation and actual experience of pollution and environmental threats.

With natural disasters, naturally people's mood states are greatly affected. Smyth (2002) found that after encountering natural disasters, people usually experience intrusive thoughts about disaster-related events and generally high levels of negative emotions and physical symptoms. Smyth's findings indicated that, in general, people would have higher distress after experiencing natural disasters (Gist and Lubin, 1999). However, if they were vulnerable to depression (such as having the tendency to ruminate on negative thoughts) before they encounter disasters, they would be somewhat more likely to have higher levels of depression and stress symptoms after encountering natural disasters such as earthquakes (Nolen-Hoeksema and Morrow, 1991). As mentioned in the previous section, climate change effects, such as global warming, may cause an increase in certain kinds of natural disasters (although the IPCC is careful not to include all natural disasters in that category) and therefore, it would be likely that more people would suffer from distress, and some may experience depression as a result of the disaster experience.

The consideration of mood state is very important, as it greatly affects our daily functioning in areas such as perceived stress and information processing. While Watson, in 1988, had already concluded that negative affect was highly correlated with perceived stress, Salovey et al. (2002) later determined that if participants know how to alter their bad mood, they tend to perceive the repeated stressors as less threatening, and they are less likely to use active, rather than passive, coping methods to handle stress. Moreover, in a follow up study, they confirmed that mood state repair was associated with less rumination, but more active coping.

Our mood state affects both the type of information we choose to attend to, store and retrieve, and the manner in which we process information. When we are in a happy mood, we will see and remember more happy things, and conversely, while we are in a negative state, we will see and remember more negative events (Niedenthal, Halberstadt and Setterlund, 1997; Niedenthal, Krauth-Gruber and Ric, 2006). Therefore, since climate change and environmental threats may lead to more negative emotions, because of increased heatwaves and drought, we might be more likely to perceive our life circumstances in a negative way, as well as to remember more negative life events, and thus be inclined to live less contentedly.

Furthermore, as indicated above, our mood also affects how we process information, depending on the type of processing goal we have (Niedenthal, Krauth-Gruber and Ric, 2006). If our processing goal is to save mental energy, we will use heuristics – rule of thumb - such as “the experts are correct” and focus more on the attractiveness, likeability, and credibility of the information providers; on the other hand, if our processing goal is accuracy, we will adopt a more deliberate and systematic strategy, such as paying attention to the strength of the arguments provided (Bless, Mackie and Schwarz, 1992; Mackie and Worth, 1991; Niedenthal, Krauth-Gruber and Ric, 2006.)

Isen (1987) argued that people who are in a ‘bad mood’ state will engage in strenuous systematic processing, because a bad mood indicates that there are problems occupying one’s attention that are not yet solved, and so one has to expend more effort in dealing with them. Furthermore, as a bad mood is considered to be an unpleasant state, most of us would prefer to engage in more effective systematic processing, so as to distract ourselves from ruminating on our bad mood (Sternberg, 2009); that is, if we are engaging in systematic information processing, our working memory will have no space for us to ruminate on our current mood state. At the other end of the spectrum, Isen (1987) argues that individuals in a good mood will tend to engage in a simplified heuristic strategy because a good mood indicates that all is fine, and we do not need to put in further effort to carefully evaluate the environmental threat. Moreover, as a good mood is a pleasant state that we would like to maintain, we will avoid tasks that might hamper our good mood, such as conserving energy or recycling waste.

Based on Isen’s (1987) argument, it is proposed that if governments or environmentalists want to persuade people to act on climate change, they should prepare at least two versions of their message, so as to ensure that people in various mood states are more likely to be persuaded by the message. For example, taking any one particular message, a heuristic version could be created by using some attractive and likeable spokesman to deliver some simple advice on saving water that would cater to people experiencing good moods, as they would be more likely be persuaded by this type of advertisement. To cater for the other category, an elaborate version could be created by incorporating stronger evidence and more elaborate information on potable water supplies to cater for those people in bad moods, as they might be more likely be persuaded by this style of advertisement. Obviously a lot more research needs to be done on this area, in terms of climate change messages about changing current ways of conserving the environment.

CONCLUSION

Environmental threats may affect us physically through certain consequences of global warming, pollution and toxic chemicals use, but they can also exert indirect psychological influence on us through acting as stressors. These factors need to be taken into account when designing homes, workplaces and communities in a future of potential marked climate change. These stressors potentially result in negative emotional states which can, in turn, have an impact on our immune system, as well as our information processing system. Government strategists, concerned with climate change policies, and environmentalists need to explore how different kinds of warnings can be designed to match the various mood states that people experience, so that the general public will more likely be persuaded by the message delivered in the advertisements, no matter what mood states they are in at the time of exposure to the messages.

REFERENCES

Arapis, G., Goncharova, N. N., and Baveye, P. (2006). *Ecotoxicology, ecological risk assessment and multiple stressors*. USA: Springer Science and Business.

Analitis, A., Katsouyanni, K., Biggeri, A., Baccini, M., B. Forsberg, B., et al. (2008). Effects of cold weather on mortality: Results from 15 European cities within the PHEWE Project. *American Journal of Epidemiology, 168(12),* 1397-1408.

Australian Government, Department of the Environment, Water, Heritage and the Arts. (2008). Retrieved 5 March 2008 from http://www.environment.gov.au/atmosphere/ozone/publications/ozone-reports-2008.html

BBC. (2008). Weather and Behavior. Retrieved September 5, 2008. http://www.bbc.co.uk/weather/features/health_culture/behaviour.shtml.

Bernard, S. M., Samet, J. M., Grambsch, A., Ebi, K. L., and Romieu, I. (2001). The potential impacts of climate variability and change on air pollution-related health effects in the United States. *Environmental Health Perspectives, 109(2),* 199-209.

Bless, H., Mackie, D. M., and Schwarz, N. (1992). Mood effects on attitude judgements: Independent effects of mood before and after message elaboration. *Journal of Personality and Social Psychology, 63,* 585-595.

Boaz, N. T. (2002). *Evolving health: The origins of illness and how the modern world is making us sick.* New York: Wiley.

Böhm, G. (2007). Anticipated and experienced emotions in environmental risk perception. *Judgment and Decision Making, 2,* 73-86.

Brannon, L., and Feist, J. (2007). *Health Psychology: An introduction to behavior and health.* (6th ed.), Belmont, CA: Cengage - Wadsworth.

Bronstein, J., Carvey, P., Chen, H., Cory-Slechta, D., DiMonte, D., et al. (2009). Meeting Report: Consensus Statement-Parkinson's Disease and the Environment: Collaborative on Health and the Environment and Parkinson's Action Network (CHE PAN) Conference 26–28 June 2007. *Environmental Health Perspectives, 117(1)*, 117- 121.

Carlson, N. R. (2006). *Physiology of behavior*. (9th ed.). Boston: Allyn and Bacon.

Cohen, S., Frank, E., Doyle, W. J., Skoner, D. P., Rabin, B. S., and Gwaltney, J. M., Jr. (1998). Types of stressors that increase susceptibility to the common cold in healthy adults. *Health Psychology, 17,* 214-223.

Cohen, S., and Lichtenstein, E. (1990). Perceived stress, quitting smoking and smoking relapse. *Health Psychology, 9(4),* 466-78.

Commonwealth Scientific and Industrial Research Organisation (CSIRO). (2009). Fisheries in a future ocean: impacts of climate change. Retrieved 9 March, 2009. http://www.csiro.au/resources/ClimateChangeAndFisheries.html.

Curriero, F. C., Patz, J. A., Rose, J. B., and Lele, S. (2001). The association between extreme precipitation and waterborne disease outbreaks in the United States, 1948–1994. *American Journal of Public Health, 91(8)*, 1194-1199.

Davis, J. (2008). Ecopsychology, environmental problems, and environmental action. Retrieved September 5, 2008. http://www.johnvdavis.com/ep/envact.htm.

Dallmam, M. F., Pecoraro, N., Akana, S. F., la Fleur, S. E., et al. (2003). Chronic stress and obesity: A new view of "comfort food". *Proceedings of the National Academy of Sciences of the USA.* Retrieved 9 March, 2009. http://www.pnas.org/content/100/20/11696.full.

Dominici, F., Peng, R.D., Bell, M. L., Pham, L., McDermott, A. et al. (2006). Fine Particulate Air Pollution and Hospital Admission for Cardiovascular and Respiratory Diseases. *The Journal of American Medical Association, 295,* 1127-1134.

Du Nann Winter, D., and Koger, S. M. (2004). *The Psychology of Environmental Problems.* (2nd ed.), London: Lawrence Erlbaum Associates.

Evans, G. W., and Campbell, J. M. (1983). Psychological perspectives on air pollution and health. *Basic and Applied Social Psychology, 4,* 137-169.

Franken, R. E., (2002). *Human Motivation.* (5th ed.). Belmont, CA: Cengage - Wadsworth.

Gaffield, S. J., Goo, R. L., Richards, L. A., and Jackson, R. J. (2003). Public health effects of inadequately managed stormwater runoff. *American Journal of Public Health, 93(9),* 1527 – 1533.

Garshick, E., Laden, F., Hart, J. E., Rosner, B., Davis, M. E., et al. (2008). Lung Cancer and vehicle exhaust in trucking industry workers. *Environmental Health Perspectives, 116*(10), 1327- 1332.

Gaulin, S. J. C. and McBurney, D. H. (2001). *Psychology: An evolutionary approach.* Upper Saddle River, N. J.: Prentice-Hall.

Gist, R. and Lubin, B. (1999). *Response to disaster: Psychosocial, community, and ecological approaches.* (3rd ed.). NY: Psychology Press.

Greenough, G., McGeehin, M., Bernard, S. M., Trtanj, J., Riad, J. and Engelberg, D. (2001). The potential impacts of climate variability and change on health impacts of extreme weather events in the United States. *Environmental Health Perspectives, 109,* 191-198.

Hales, D. (1997). *An invitation to health.* (7th ed.). Pacific Grove, CA: Brooks/Cole.

Hashizume, M., Armstrong, B., Hajat, S., Wagatsuma, Y., Faruque, A. S. et al. (2007) Association between climate variability and hospital visits for non-cholera diarrhoea in Bangladesh: effects and vulnerable groups. *International Journal of Epidemiology, 36(5),* 1030 - 1037.

Heilig, M., and Thorsell, A. (2002). Brain neuropeptide Y (NPY) in stress and alcohol dependence. *Review Neuroscience,* 13(1), 85-94.

Howard, P. J. (2006). *The Owner's Manual for the Brain: Everyday Applications from Mind-brain Research* (3th ed.). USA: Bard Press.

Isen, A. M. (1987). Positive affect, cognitive processes, and social behavior. In L. Berkowitz (Ed.), *Advances in Experimental Social Psychology, Vol. 20,* (pp. 203-253). New York: Academic Press.

Kalat, J. W. (2007). *Biological Psychology.* (9th ed.). Belmont, CA: Cengage - Wadsworth.

Keller, M. C., Fredrickson, B. L., Ybarra, O., Cote, S., Johnson, K., et al. (2005). A warm heart and a clear head. *Psychological Science, 16(9),* 724-731.

Mackie, D. M. and Worth, L. T. (1989). Processing deficit and the mediation of positive affect in persuasion. *Journal of Personality and Social Psychology, 57*, 27-40.

McMichael, A. J., Campbell-Lendrun, D. H., Corvalan, K. L., Ebi, A. Githeko, A., Scheraga, J. D., and Woodward, A. (2003). *Climate change and human health – risks and responses.* World Health Organization. Retrieved June 3, 2008: http://www.who.int/globalchange/publications/cchhbook/en/print.html

Miller, T. W. (2007). Trauma, change and psychological health in the 21st century. *American Psychologist, 62*(8), 889-898.

Motavalli, J. (2002). The case against meat, *The Environmental Magazine, 8(1),* 26-32.

Myers, G. J., and Davidson, P. W. (2000). Does methylmercury have a role in causing developmental disabilities in children? *Environmental Health Perspectives, 108*(3), 413-420.

Niedenthal, P. M., Krauth-Gruber, S. and Ric, F. (2006). *Psychology of emotion: Interpersonal, experimental and cognitive approaches.* New York: Psychology Press.

Niedenthal, P. M., Halberstadt, J. B., and Setterlund, M. B. (1997). Being happy and seeing "happy": Emotional state mediates visual word recognition. *Cognition and Emotion, 11,* 403-432.

Nolen-Hoeksema, S. and Morrow, J. (1991). A prospective study of depression and posttraumatic stress symptoms after a natural disaster. The 1989 Loma Prieta earthquake. *Journal of Personality and Social Psychology, 61,* 115-121.

Oskamp, S. (2000). A sustainable future for humanity? How can psychology help? *American Psychologist, 55,* 496-508.

Patz, J. A., Campbell-Lendrun, D., Holloway, T., and Foley, J. A. (2005). Impact of regional climate change on human health. *Nature, 438*, 310-317.

Pham, K., McEwen, B. S., LeDoux, J. E., and Nader, K. (2005). Fear learning transiently impairs hippocampal cell proliferation. *Nueroscience, 130,* 17-24.

Porterfield, S. (2000). Thyroidal dysfunction and environmental chemicals – Potential impacts on brain development. *Environmental Health Perspectives, 108*(3), 433-438.

Pret, A., Lentini, G., Maugeri, M. (2007). Global warming possibly linked to an enhanced risk of suicide: Data from Italy, 1974-2004. *Journal of Affective Disorders, 102,* 19-25.

Pomerleau, O. F., and Pomerleu, C. S. (2006). Research on stress and smoking: progress and problems. *Addiction, 86*(5), 599-603.

Rose, J. B., Epstein, P, R., Lipp, E. K., Sherman, B. H., Bernard, S. M., and Patz, J. A. (2001). Climate variability and change in the United States: Potential impacts on water - and foodborne diseases caused by microbiologic agents. *Environmental Health Perspectives, 109,* 211-220.

Salovey P., Stroud L.R., Woolery A., and Epel, E.S. (2002). Perceived emotional intelligence, stress reactivity, and symptom reports: Further explorations using the trait meta-mood scale. *Psychology and Health*, 17(5), 611-627.

Sapolsky, R. M. (1992). *Stress, the aging brain, and the mechanisms of neuron death.* Cambridge, MA: MIT Press.

Sapolsky, R. M. (1998). *Why zebras don't get ulcers: An updated guide to stress, stress-related diseases, and coping.* New York: Freeman.

Schwartz, J. (1993). Particulate air pollution and chronic respiratory disease. *Environmental Research, 62(1),* 7-13.

Selye, H. (1979). Stress, cancer, and the mind. In J. Tache, H. Selye, and S. B. Day. (Eds.), *Cancer, stress, and death* (pp.11-27). New York: Plenum Press.

Smyth, J. M., Hockemeyer, J., Anderson, C., Strandberg, K., Koch, M., O'Neill, K., and McCammon, S. (2002). Structured writing about a natural disaster buffers the effect of intrusive thoughts on negative affect and physical symptoms. *The Australasian Journal of Disaster and Trauma Studies, 2002(1).* Retrieved 5 March 2008 from http://www. massey.ac.nz/~trauma/issues/2002-1/smyth.htm

Stein, S., and Spiegel, D. (2000). Psychoneuroimmune and endocrine effects on cancer progression. In K. Goodkin and A. P. Visser (Eds.), *Psychoneuroimmunology: Stress, mental disorders, and health* (pp. 105-151). Washington, DC: American Psychiatric Press.

Sternberg, R. J. (2009). *Cognitive Psychology* (5th ed.), *Belmont, CA: Cengage-Wadsworth.*

Tauxe, R.V. (1997). Emerging foodborne diseases: An evolving public health challenge. *Emerging Infectious Diseases, 3,* 425-434.

Thaker, P. H., Han, L. Y., Kamat, A. A. et al. (2006). Chronic stress promotes tumour growth and angiogenesis in a mouse model of ovarian carcinoma. *Nature Medicine, 12(8)*, 939-944.

Wade, T. J., Sandhu, S. K., Levy, D., Lee, S., LeChevallier, M. W. et al. (2004). Did a Severe Flood in the Midwest Cause an Increase in the Incidence of Gastrointestinal Symptoms? *American Journal of Epidemiology, 159(4),* 398 - 405.

Watson D. (1988). Intraindividual and interindividual analyses of positive and negative affect: their relation to health complaints, perceived stress, and daily activities. *Journal of Personality and Social Psychology, 54*(6), 1020-1030.

Weiss, B. (2000). Vulnerability of children and the developing brain to neurotoxic hazards. *Environmental Health Perspectives, 108*(3), 375-381.

Weiss, B., and Bellinger, D. C. (2006). Social Ecology of Children's Vulnerability to Environmental Pollutants. *Environmental Health Perspectives, 114*(10), 1479- 1486.

Weiss, B. and Landrigan, P. J. (2000). The developing brain and the environment: An introduction. *Environmental Health Perspectives, 108(3)*, 373-374.

Westman, M., Eden, D., and Shirom, A. (1985). Job stress, cigarette smoking and cessation: the conditioning effects of peer support. *Social Science Medicine, 20(6), 637-644.*

Willenbring, M. L., Levine, A. S., and Morley, J. E. (1986). Stress induced eating and food preference in human: A pilot study. I*nternational Journal of Eating Disorders, 5,* 855-864.

World Health Organization. (2003). *WHO/ Climate Change and human health – risks and responses*. Retrieved June 3, 2008. http://www.who.int/globalchange/publications/ cchhbook/en/print.html

Zaslow, S. A., Genter, M. B. (1993). Mold, dust mites, fungi, spores, and pollen: Bioaerosols in the human environment. Retrieved 9 March, 2008. http://www.ces.ncsu.edu/depts/ fcs/mold.html

Contributed by Tim McCrorey, Toowoomba.

Photo: Polluting the Environment.

In: Meltdown: Climate Change, Natural Disasters… ISBN 978-1-60876-153-1
Editor: Kathryn Gow

Chapter 8

SQUEEZING RELATIONSHIPS DRY: THE IMPACT OF DROUGHT ON AUSTRALIAN FARMERS' INTIMATE RELATIONSHIPS

Zoe J. Pearce
Queensland University of Technology, Australia

ABSTRACT

Climate Change and severe drought conditions are having a strong and pervasive effect on Australia's farming community. Australia has long been stereotyped as a nation of people who persevere to overcome hardship, but the psychological consequences of living and working on drought-stricken properties has pushed many farming families to breaking point. This chapter focuses on the negative impact that drought-related stress is having on Australian farmers' intimate relationships. After first considering the nature of the Australian farming community, the chapter identifies key stressors that contribute to relationship distress in this population. The impact of stress on farming couples is discussed and potential risk factors are identified. The chapter also presents some suggested therapeutic goals to improve farming relationships and briefly discusses issues that couples need to be mindful of when navigating through possible separation. The chapter concludes with a collection of web sites and references that list support services for Australian farming families living with the stress of extended drought conditions.

Keywords: relationships, drought, farming, Australia, drought- related stress

INTRODUCTION

The feature of farm life which distinguishes it from other occupations is the intimate connection between the farm as workplace, home, career and family tradition. A threat to one is a threat to all. (Gray and Lawrence, 1996, p. 175)

Climate change is recognised as an important correlate of drought (CSIRO and BOM, 2007). While variable climate conditions and the experience of drought is not uncommon in Australia (Hennessy et al., 2008), climate change reports show Australia's annual rainfall in central and eastern regions as steadily declining, with the occurrence of drought likely to increase in the future (CSIRO and BOM, 2007). In the years between 1970 and 2007, average rainfall for central and eastern Australia dropped by 50mm every 10 years, with predictions that the situation will get worse in coming years (CSIRO and BOM, 2007). Consider that, from the start of this decade, almost two-thirds of Australia's landmass is maintained by farmers for agricultural production and grazing, on over 150,000 separate farms (Bureau of Rural Sciences, 2002). Linkages between the agricultural community and associated industry results in 17% of Australia's workforce relying on Australian farmers for economic support, with approximately 300,000 people directly working in the farming sector. Farming in Australia contributes 90% of domestic food supply and $20 billion in export earnings; Australian farmers are largely self-sufficient and receive only 6% of their income from regular (non-drought) government support (all statistics supplied by National Farmers' Federation, 2008). During the early part of 2008 and for the previous year, 60% of Australia's agricultural land was drought-declared, with 69 areas across Australia receiving 'Exceptional Circumstances' support from the Australian government (Bureau of Rural Sciences, 2008). In sum, climate change, and the prolonged drought conditions that accompany this change, has the potential to have a very real and lasting impact on all Australians, but most notably those at the coal face of primary production - Australian farmers and their families.

The Cost of Drought on Australian Farming Families

Not surprisingly, the biggest impact that drought has on Australian farming families is to produce elevated stress levels. Many farmers struggle to maintain the financial viability of their properties and in some cases will seek additional employment off the land to supplement on-farm income (Cary, Webb, and Barr, 2002). Farmers' partners (where present) also often seek off-farm employment. Data from the Australian Bureau of Statistics suggests that less than half of Australia's farming community have both partners formally engaged in work on the farm, and in 42% of farming families the spouse or partner of the identified farmer works off the land (ABS, 2006). However, it should be noted the ABS data does not identify the *reasons* why partners elect to work off the land and in addition to attempting to ease the financial stress caused by drought, partners may elect to work off-farm for other reasons. For example, off-farm work for a farm wife has the potential to improve self-esteem and bring about personal change (Stehlik, Gray, and Lawrence, 1999), even if the initial reason for seeking this employment was solely to assist with improving family income and reducing stress.

Families who seek employment off-farm due to drought-induced financial strain undergo a unique transition that has both a practical and psychological impact. Farmers and/or their partners who work off-farm still have a farm to maintain and often this maintenance work needs to take place in the evenings and on weekends, making the average working day very long and tiring. In addition to 'typical' household responsibilities, childcare or home schooling, farm maintenance work such as fencing, feeding, crop care, stock control, building repairs, mechanical support, and farm chores need to be conducted; farmers and their partners have little time to de-stress or participate in relaxing family activities such as family outings, play sessions, or spending quality time together as a family. Interestingly, while the number of hours spent off-farm do impact on stress levels, the effect on farmers is not as great as the mere fact that off-farm work needs to be sought in the first place (Gray and Lawrence, 1996), suggesting the psychological impact of seeking work away from the land is a major source of additional stress for many farming families. Adding to the family's burden is the fact that the off-farm worker may live with the children in an urban setting, returning to the farm on the weekend, thus adding to the sense of isolation that the farmer may experience and further reducing the possibility of the family spending quality time together.

Periods of uncertainty and economic hardship such as those experienced during prolonged drought can lead farmers to adopt new approaches to farm management, resource allocation or crop production to remain viable (Gray and Lawrence, 1996). Introducing new practices can have tremendous benefits for farmers giving farmers the chance to improve productivity and efficiency. But change is typically associated with a period of short-term stress, placing further burdens on drought-stricken properties. Thus the thought of having to introduce new procedures simply to survive can be overwhelming for farming families.

The multigenerational nature of many farming families adds further stress to farmers, especially during times of prolonged drought when stress levels are already high and coping resources are stretched. Farms are often handed down through generations with older parents working and living on the farm alongside their adult children and spouses. A multigenerational farmer seeking work off-farm to ease financial stress may feel like he or she is admitting to failure, causing the farmer to feel as though they are letting their family down by breaking with tradition in this manner. Multigenerational farming families also experience role confusion, with a blurring of the boundaries between family and work relationships, and poorly communicated or unrealistic expectations (Weigel and Weigel, 1987). Conflict can be high, and the resulting family friction can be a further source of stress for farmers (Hedlund and Berkowitz, 1979). In addition, there is evidence to suggest that the stress of multigenerational farming is greater for farm wives than it is for farm husbands with daughters-in-law on multigenerational farms reporting that they feel that they have no involvement in decision making, and feeling like they are hired labour instead of part of the farm management (Weigel and Weigel, 1987). Some wives elect not to participate in farm business further increasing the family's stress, as the farming family may question the wife's commitment to the family and the wife may feel isolated as a result.

In sum, Australian farming couples experience tremendous stress in their relationships. The need for off-farm employment while maintaining the farm, lengthy periods of uncertainty and change that pre-empt the introduction of new practices, and the multigenerational aspects of farming life all contribute to the stress that Australian farming couples must cope with. Unfortunately, couples who experience chronic stress often run the risk of harming their relationship (Bodenmann, 2008).

The Impact of Stress on Couples

Intimate relationships often suffer when couples experience prolonged stress (Bodenmann, 2008), with couples having to work harder to maintain a healthy and satisfying relationship. Drought-related stress impacts on couples via both intra-individual and inter-individual mechanisms. The interdependent nature of intimate relationships means that even if only one person in the couple is reacting to a stressor, the resulting emotions, thoughts and behaviours are likely to have an impact on the couple as a whole.

The experience of prolonged stress can be associated with a greater incidence of mental health problems (e.g. Hammen, 2005). Typical methods of coping with stress can include social withdrawal, conflict-laden communication, substance abuse, altered sleep patterns, negative thought patterns and irrational thinking, all of which pose substantial risk for the later onset of depression and anxiety. Indeed, prolonged stress puts people at risk for the onset of depression (Hammen, 2005), especially where themes of loss and threats to self-worth are concerned (Beck, 1983; Paykel, 2003). Hickey et al. (2005) found depressed individuals are more likely to report poorer quality of life, poorer family communication and role confusion, less effective marital problem-solving and less perceived social support compared to non-depressed individuals.

The presence of depressive symptoms in one partner has a substantial impact on the other partner due, in part, to the couple communication that is taking place in the relationship. Couples reporting depressed affect show diminished self-disclosure, while congenial mutual discussion is less prevalent (Kline, Pleasant, Whitton, and Markman, 2006). The degree to which individuals are emotionally engaged during interactions with their spouse impacts on couple functioning (Weiss and Heyman, 1990). In relationships where at least one partner is depressed, it is common to see a higher incidence of negative couple communication and less frequent positive communication. Consequently, relationship communication that is characterised by a high degree of negative affect and less positive affect has a strong association with declining relationship satisfaction (Gottman, 1994; Gottman, Coan, Carrere, and Swanson, 1998).

The prolonged experience of stress also has a direct association with communication even in couples where typical depressive symptoms are absent. Ineffective, physically aggressive communication strategies are more prevalent in men who report higher levels of prolonged stress compared to men with lower stress levels (Frye and Karney, 2006). Reduced levels of effective problem-solving, mutual decision-making and supportive interaction are exacerbated by stress and are more likely to lead to marital distress (e.g. Cohan and Bradbury, 1997; Conger, Reuter, and Elder, 1999).

Communication is a particularly useful index for assessing the impact of stress on couples because it draws on the interdependent nature of partners in an intimate relationship. Communication behaviour acts as a mechanism through which the internal experience of one partner becomes outwardly expressed. In the event that this outward expression is hostile, distant or unsupportive (frequently the case in people experiencing chronic stress), the result is often the gradual undermining of the other partner's relationship satisfaction. Avoidance, withdrawal, and the escalation of negative patterns of interaction are all reliable predictors of distress in couples (Heyman, 2001), and correlates of stress. Thus, one of the greatest risks that couples face as a result of experiencing prolonged stress is the very real prospect that their relationship may not survive.

Farming families are no more or less immune to the impact of stress on their relationships than are other members of the community. Distinguishing this experience in the farming community, however, are certain unique characteristics surrounding farmers' livelihood and life circumstances that act as risk factors for relationship breakdown in this population.

Risk Factors for Farmers' Intimate Relationships

Farming couples can belong to a tight-knit community. This presents both advantages and disadvantages for farmers and their partners. On the one hand, community membership offers support - entire rural communities suffer during drought conditions and a farmer's neighbours can provide an important social network in times of stress. On the other hand, a strong sense of community can also cause some farming families to want to keep problems, especially relationship difficulties, to themselves. Whether attempting to 'hide' relationship problems or perhaps trying to avoid social criticism (Judd et al., 2005), farmers will sometimes not make use of the social community support available to them, an issue that can be compounded by (1) the tyranny of distance, and (2) the typical Australian mentality of the 'Aussie battler', the stoical, resilient farming personality.

Distance and isolation. One of the most obvious risk factors for farming couples undergoing relationship problems is their remote location. The Australian landscape is such that farmers drive for hours to access services that urban dwellers have on their doorstep (Fox, Merwin, and Blank, 1995; Parslow and Jorm, 2000). Even if farmers are willing to travel to seek assistance, there is often a lack of suitable services available for rural and remote communities (Australian Institute of Health and Welfare, 2002; Beyond Blue, 2008). When counselling and support services are available, accessing these services may involve weeks of waiting between appointments. Regular attendance is therefore quite difficult for a substantial portion of farming couples in remote locations.

Isolation and loneliness also pose significant risk factors for farming couples, especially farm wives. Judd et al. (2005) suggest that women who move to their husband's farm can experience a sense of social isolation that is exacerbated by their husbands' heavy workload. Many farm wives spend hours alone in their home or have their children as their only frequent companions. Female participants in Judd et al.'s research reported frequent difficulty making friends in their new community, heightening feelings of exclusion and a lack of acceptance. This places more pressure on the relationship to provide the social support and acceptance that the community may fail to provide. Keating (1987) found a significant predictor of stress in farm wives was their husband's lack of support for the work they do around the farm. Male farmers in Keating's study did not experience the same 'need' for spousal support because "men's identities are so clearly focused on farm work with general community support for their farm role" (p. 360).

The farming personality. Perhaps the single most frequently cited reason for why farming couples may suffer in times of stress is due to the very nature of the stereotypical Australian farming personality. Farmers, especially men, can be seen as stoic and proud people who avoid discussing their problems for fear of being seen as 'whinging' (Judd et al., 2006). This maintenance of a positive attitude in the face of potential threat can act as both a risk and resilience factor. Farmers who avoid discussing problems, when the need is obvious, run the risk of placing added stress on themselves and their relationship. However, farmers whose successful coping skills focus on self-reliance and personal strength benefit from developing a sense of mastery. Developing this mastery can be an important aspect of a farmer's sense of self and can reinforce the farmer's values surrounding success and achievement. So much of the farmer's life is linked to his or her sense of self, from the size of the farm, to the farm's productivity, that when a farmer's sense of mastery is threatened, the psychological threat may be as great as the threat to the farmer's livelihood. In the event that farmers do reach out, they are more inclined to turn to their families to seek help rather than seek assistance from external agencies (Judd et al., 2005). In part, the tendency to seek assistance from within the family appears to be driven by the stigma that is still attached to mental health concerns, but it may also be due to upbringing, community pressure, or a lack of information surrounding what support services are available to the farming community (Judd et al., 2005).

Farmers do appear to have a unique personality profile. Comparing farmers to non-farmers, Judd et al. (2006) found that openness to experience and neuroticism were significantly lower in farmers compared to non-farmers. Conscientiousness was significantly higher in farmers than non-farmers and the authors suggested that farmers who show such a careful, meticulous, devoted and dedicated manner may persevere with their efforts in the face of adversity. In the short-term, this aspect of the farming personality is a definite strength. On the other hand, constantly battling opposing forces may result in poor adjustment and irrational coping behaviour. Judd et al. (2006) also suggested that farmers' lower levels of openness to experience may explain their discomfort with undergoing the necessary changes in farming practice often required in times of prolonged drought.

The typical coping strategies therefore used by many farmers tend to revolve around maintaining an outward appearance of 'all is well' while, behind the scenes, relationships may be strained. Conscientiousness and perseverance are resilience factors and can buffer against stress, but an unfortunate consequence of this type of stoical attitude is reliance on other methods of coping that may seem effective in the short-term but may pose significant long-term risk. For example, Judd et al.'s (2005) interview data suggest many farmers increase their alcohol consumption during periods of stress, and negative depressed mood can predict problem drinking (Leonard and Homish, 2008). Alcohol and other drug abuse pose a significant risk factor for relationships, with excessive consumption contributing towards ineffective control of angry outbursts (Foran and O'Leary, 2008). It is worth noting that here there is little evidence to suggest that self-reported levels of alcohol or other drug use are significantly different between rural and urban populations (Judd et al., 2006), and rural drinking establishments may provide farmers with a much-needed avenue for seeking social support and staying connected with what is happening in the community. However the risks associated with alcohol and other drug use are just as great, if not greater, for the later development of relationship strain and possible dissolution in rural farming communities.

Perhaps one of the greatest risk factors for farmers' relationships undergoing prolonged stress is the impact of mental health concerns discussed earlier, specifically depression and suicidal ideation. The prevalence of depression and other mood disorders in the farming community does not appear to be higher than in non-farming communities (Eckert, Wilkinson, Taylor, Stewart, and Tucker, 2006; Fraser et al., 2005) although this may be a reporting issue, with farmers more likely to consult their general practitioners over physical concerns rather than psychological ones (Booth, Briscoe, and Powell, 2000). Unfortunately the prevalence of suicide in the farming community is significantly higher than the national average (Booth et al., 2000). Page and Fragar (2002) report that farm suicides occur at the rate of one every four days, a rate similar to non-suicide work-related farming deaths. More recent data (e.g. Caldwell, Jorm, and Dear, 2004; Miller and Burns, 2008) suggests the rate of suicide in Australian farmers remains high. The absence of timely support services or difficulty accessing these services puts farmers at greater risk for suicide than the non-farming population (Page and Fragar, 2002). Additional explanations for this heightened suicide risk in farmers include greater access to lethal means (Malmberg, Simkin, and

Hawton, 1999), for example firearms, trees and hidden sheds to hang ropes from. But closer investigation reveals other critical factors such as isolation, fewer alternative income opportunities, a sense of failure and betrayal of the family inheritance and an increased incidence of severe depression because of the long drought and consequent hardships for the family and farming business.

Farming couples experiencing prolonged periods of drought are coping with elevated stress levels, financial pressure, family strain, communication breakdown, long periods of isolation, limited access to support services, community stigma, resistance to change, the desire to persevere no matter what the cost, and potentially an elevated risk of suicide. It is not altogether unsurprising that a farmer's intimate relationship would suffer when faced with mounting stressors such as these. It is difficult to ascertain exactly how many farming relationships completely break down, as detailed statistics for separation and divorce in farming communities (versus the whole of Australia) are not readily available—in Australia, 33% of all marriages end in divorce (ABS, 2007). While the conscientious nature of Australian farming communities might help farmers build resilience to relationship threats, being confronted with the stress of prolonged drought conditions is straining Australian farming families and relationship distress or breakup may be the consequence.

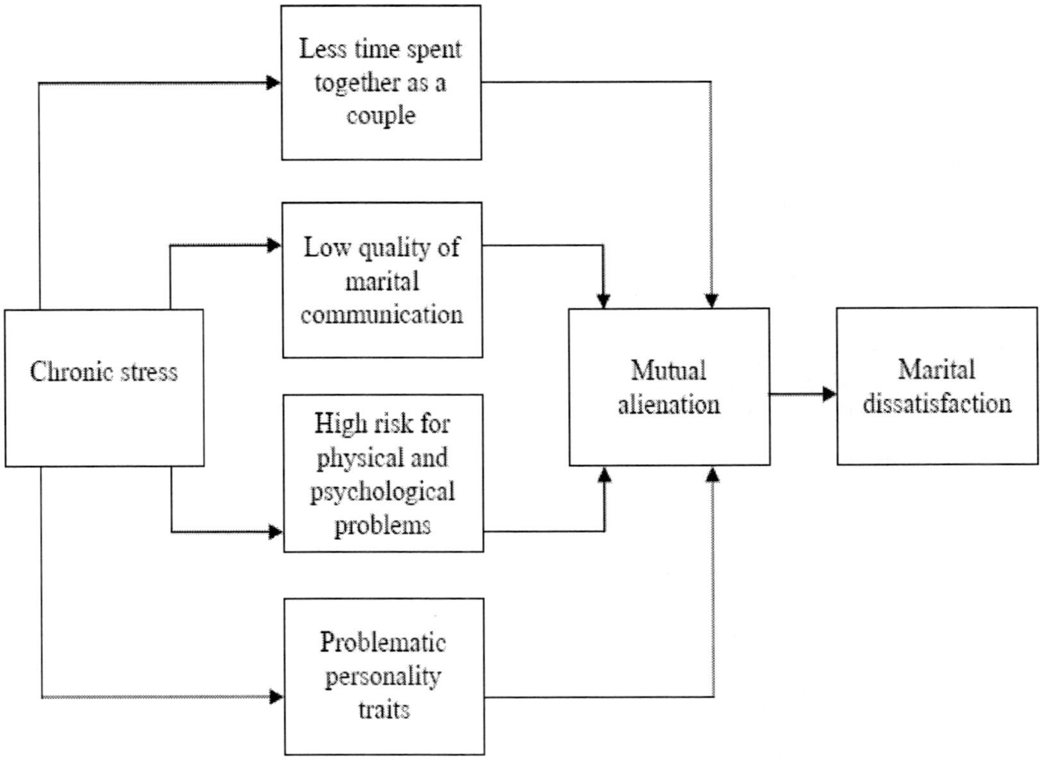

Adapted with permission from Bodenmann, 2008.

Figure 8.1. The impact of stress on relationships.

Improving Farming Relationships

Farming couples who elect to seek outside assistance to build or maintain a healthy relationship stand to gain from their actions. Accessing services may prove difficult, but in the event that a farming couple is willing and able to make contact with welfare or counselling services, there is a good chance their relationship can benefit from the attention the farming couple can give to it.

Communication is one of the key indices for relationship distress and as such many efficacious couples' therapy programs target improvements in communication and successful conflict management in order to bring about improvements in relationship satisfaction (Markman, 1991). Traditional behavioural couple's therapy involves encouraging positive behaviour and affect, improving communication and teaching couples how to be effective problem-solvers (Jacobson and Margolin, 1979). In the short-term, these types of therapies appear to work, but there is some inconsistency regarding long-term sustainable change (Halford, Sanders, and Behrens, 1993). A recent alternative approach to couples therapy, which appears particularly useful for couples experiencing prolonged stress, is Bodenmann's (2008) three-phase Couples Coping Enhancement Training (CCET).

Couples Coping Enhancement Training. What makes CCET particularly applicable for farming couples is the focus this model has on coping with stress. As Bodenmann (2008) explains, and as illustrated in Figure 9.1, stress can reduce relationship satisfaction by impacting on couples' shared time. When couples have fewer activities in common and reduced time together, their sense of what Bodenmann calls 'we-ness' becomes weaker. This is accompanied by reduced rates of affect-laden, intimate self-disclosure and increased rates of superficial daily chat and, in general, declining rates of good quality supportive communication. Depression, poor self-esteem and physical health problems that stem from chronic stress can exacerbate the couple's problems and, as alluded to earlier, personality traits such as being too rigid or intolerant of change, aggressiveness and anxiety also add to the model. The result of all these stress-related pressures on the couple is what Bodenmann refers to as 'mutual alienation', or the slow and continual avoidance by partners to discuss their experiences, needs and interests with one another. Conflicts can escalate and relationship satisfaction can decline. Bodenmann's CCET approach teaches couples how to deal with stress at the same time as incorporating more traditional communication and problem-solving skills training techniques. "Very often stress may be a causal factor for poor communication, thereby indicating the need to improve coping in order to maintain adequate communication levels within the relationship" (Bodenmann, 2008, p. 238). By teaching partners how to explore the emotional aspects of their stress in conjunction with active listening techniques such as paraphrasing, questioning, responding with empathy and reframing, couples learn a dyadic coping technique to buffer against the effects of stress on their relationship.

Unfortunately, not all farming couples will survive the stress that prolonged drought is likely to contribute to their relationship. Whether due to limited access to services, an unwillingness to try, waiting too long to seek help, or to the couple being unable to sustain long-term change, there will be some couples who cannot overcome their difficulties and elect to separate. Bodenmann (2008) argues that couples go through a process of internal evaluation where the advantages and disadvantages of the conditions surrounding the relationship are evaluated. If the facilitating conditions for separation (e.g. high self-esteem, high social competence, financial reward, access to support) are perceived to be greater than inhibiting conditions (e.g. children, religious or social pressure, perceived isolation) the relationship may dissolve. For these couples, the issue now becomes focussed around surviving the separation.

Surviving Separation

It is quite difficult to make general recommendations about how to survive separation and divorce, as each person who is affected by the divorce will likely experience it differently and progress through different recovery stages at different times. The experience of divorce does not exist on a timeframe—as Weingarten (1986, p. 194) writes, "just as marriage comprises more than the details of a wedding ceremony, the experience of divorce cannot be captured by detailing the events involved in the legal dissolution of a marriage". Furthermore, divorce rarely if ever has an impact on just the two partners in the relationship. Children, family members or the community all experience the "divorce" of two people they love and care for.

Amato's (2000) extensive review of divorce research concluded that the key areas in which divorced adults are more at risk compared to married adults were on issues surrounding parenting, navigating the relationship with a former partner, financial strain, and access to support. In families where parenting roles and responsibilities were evenly shared, divorce can present a significant challenge, as two separate parents must now find a way to "maintain roles as co-parents while dissolving co-partner roles" (Coleman, Ganong, and Leon, 2006, p. 163). Clearly the ideal situation for most families is for co-parents to remain in contact with each other and with their children. Positive co-parenting is characterised by mutual respect between parents, constructive communication and problem-solving about the children, the exchange of information, and shared responsibility of the parenting role (Ahrons, 1983). In some cases, regular dialogue between parents is difficult and fraught with conflict. In these situations the child's post-divorce adjustment may be at risk and parents may need to seek assistance in the form of mediation or counselling. Separating parents need to establish boundaries between their parenting relationship and their angry ex-partner relationship to avoid adding to the stress their children are no doubt already experiencing.

The multigenerational nature of farming families means that in the event that a couple chooses to separate, entire generations of families can be torn apart. Grandparents might once have had regular contact with grandchildren, but conflict-laden marital dissolution can limit access to grandchildren, or physical separation can make regular contact quite difficult (Coleman et al., 2006). Successful co-parenting therefore provides extended family members with continuity of contact which, in turn, can act as a source of support for one or both parents.

Seeking social support during separation and divorce is an important, albeit difficult, step for recovery. Whereas in the past, a person's ex-partner may have provided instrumental support and care giving and now this support may need to be sought elsewhere. The provision of support is often an important role that counsellors and community agencies may need to undertake, and family members may also be called upon more regularly. Compounding the problems that a separating couple will face is how to deal with common couple friends and who to approach for support. Research (see Coleman et al., 2006 for a review) tends to suggest that women will perceive their social situation as improving following divorce and, on the whole, women are better able to seek out the support they need.

Women and men tend to experience separation differently. For many men, the experience of divorce can come as a shock; men are rarely the initiators of separation (ABS, 2008). There is some research to support the notion that women are more emotionally aware of the state of a relationship and monitor relationships more closely, acting as emotional barometers for the relationship (Gottman, 1994; Pearce and Halford, 2008), also believing they may lose more by staying in an unhappy relationship, rather than leaving. As initiators, women experience more distress prior to separation whereas men experience more distress after separation (Relationships Australia, 2008). Men appear to cope with separation by diverting their emotions into other avenues such as work or sport and expressing their feelings with humour (Bickerdike, Gee, Ilguaskas, Melvin, and Hearn, 2003), whereas women can experience a more intense emotional response tinged with guilt or blame (Relationships Australia, 2008). In the somewhat stoical atmosphere of Australian farming communities, separation can result in the diversion of emotion into productive work practices which can provide short-term relief from pain, but may not contribute positively to long-term growth.

Conclusion

This chapter has provided an overview of the challenges facing Australia's farming families who are experiencing the devastating effects of severe drought conditions brought on by climate variability. Hundreds of thousands of people stand to be affected by Australia's diminished rainfall with conditions predicted to worsen in the future. The need to seek employment off-farm due to financial strain is a potential stressor for many farming families, along with the possibility of having to be open to trying new farming practices. The multigenerational nature of farming can be both an advantage and a disadvantage and tends to be more stressful for women entering the family farm through marriage. Stress can have a devastating impact on farming relationships, increasing the risk of depression and suicide. Couple communication also suffers when couples face chronic stress, with aggression and conflict-laden communication become more prevalent.

The factors that place farming relationships at risk include the pressure of 'keeping up appearances' in a close-knit community (Judd et al., 2005), confounded by the quintessential farming personality as being someone who rarely complains, persevering in times of hardship (Judd et al., 2006). While a stoic attitude can be a useful coping skill, it can also stop farmers from seeking out valuable support. For many farmers, the active seeking of support in itself adds strain to relationships, as quality services are often hard to come by or require an extra commitment of time to pursue. Isolation and loneliness present another unique risk factor for this community, notably for women entering the farming lifestyle for the first time as newlyweds.

There can be tremendous benefit for farming couples who overcome the tyranny of distance in seeking assistance to build or maintain their relationship. Couples therapy programs that target improving communication can bring about changes in relationship satisfaction in the short-term. For long-term gain, farming couples would be wise to participate in one of the more recent therapeutic interventions designed to teach couples how to communicate the emotional aspects of their stress experience and foster effective dyadic coping techniques. Unfortunately not all farming couples will access or benefit from therapy, and many more will have left it too late to repair the damage that stress has inflicted on their relationship. For these couples, the goal shifts from relationship repair to successful separation. Far from being an oxymoron, successful relationship dissolution is possible, but is made harder when partners need to navigate through issues of changing roles, maintaining respectful co-parenting and resolving any disputes related to financial settlement. Once again, seeking out social support presents challenges for separating farming couples, but remains an important predictor of success in post-separation transitions. Gender differences in the divorce literature suggest the experience of divorce can be unique for men and women, although there is no question the experience is likely to be painful and traumatic for anyone touched by the relationship breaking down.

Severe drought conditions are placing Australian farming families at risk of dissolution. With this risk comes the need to identify innovative strategies that help build resilience in these families. Modern approaches that remove the stigma of mental illness and overcome the problems of residing in a remote location are slowly evolving and may prove attractive to the farming community. For example, the rising popularity of online 'e-counselling' services through interactive websites, live chat rooms and email is providing therapists with access to previously inaccessible clients (Chester and Glass, 2006) and may present the farming community with an alternative avenue of support. Many support agencies (e.g. Lifeline or Kids Help Line, see Appendix) also provide telephone counselling opportunities for couples in distress and their families. Again, this type of service may prove useful in Australia's rural community where accessing typical face-to-face services is limited. Further research is needed to determine the extent to which 'e-therapy' and telephone counselling options are likely to be adopted in a rural community.

The psychosocial effects of worsening drought conditions need to be 'tackled head on' if Australian farmers are going to successfully steer their way through the prolonged interpersonal stress and economic hardship that accompanies this consequence of climate change. While there are many factors that point towards farming relationships suffering during periods of prolonged stress, the 'Aussie battler' personality of Australian farmers may yet prove to push this community into finding methods for protecting their families from the pressures of drought and see them emerge at the other end as an intact family unit.

References

ABS. (2006). *Agriculture in focus: Farming families, Australia, 2006* (No. 7104.0.55. 001). Canberra: Author.

ABS. (2007). *Australian Social Trends, 2007* (No. 4102.0). Canberra: Author.

ABS. (2008). *Divorces, Australia 2007* (No. 3307.0.55.001). Canberra: Author.

Ahrons, C. R. (1983). Predictors of parental involvement postdivorce: Mothers' and fathers' perceptions. *Journal of Divorce, 6*, 55-69.

Amato, P. R. (2000). The consequences of divorce for adults and children. *Journal of Marriage and the Family, 62*, 1269-1287.

Australian Institute of Health and Welfare. (2002). *Mental health services in Australia 1999-2000* (Mental Health Series no. 3). Canberra: Author.

Beck, A. T. (1983). Cognitive therapy of depression: New perspectives. In P. J. Clayton and J. E. Barrett (Eds.), *Treatment of Depression: Old Controversies and New Approaches* (pp. 265-290). New York: Raven.

Beyond Blue. (2008). *Depression in rural and remote communities*. Canberra: Author.

Bickerdike, A., Gee, T., Ilguaskas, R., Melvin, T., and Hearn, J. (2003). Men and separation: Choices in tough times. Retrieved 1 September, 2008, from http://www.relationships.com.au/resources/publications/publications

Bodenmann, G. (2008). Improving dyadic coping in marital distress prevention programs and marital therapy. In L. Vandecreek and J. B. Allen (Eds.), *Innovations in clinical practice* (pp. 235-251). Sarasota: Professional Resource Press.

Booth, N., Briscoe, M., and Powell, R. (2000). Suicide in the farming community: Methods used and contact within health services. *Occupational and Environmental Medicine, 57*, 642-644.

Bureau of Rural Sciences. (2002). *2001/02 Land use of Australia (version 3).*

Bureau of Rural Sciences. (2008). *Exceptional circumstances (EC) and interim assistance (IA) boundaries*. Retrieved 25 August 2008, from http://www.daff.gov.au/__data/assets/pdf_file/0020/153542/national-150808.pdf.

Caldwell, T. M., Jorm, A. F., and Dear, K. B. G. (2004). Suicide and mental health in rural, remote and metropolitan areas in Australia. *Medical Journal of Australia, 181*, S10-S14.

Cary, J. W., Webb, T. J., and Barr, N. F. (2002). *Understanding landholders' capacity to change to sustainable practices: Insights about practice adoption and social capacity for change*. Canberra: Bureau of Rural Sciences.

Chester, A., and Glass, C. A. (2006). Online counselling: A descriptive analysis of therapy services on the Internet. *British Journal of Guidance and Counselling, 34*, 145-160.

Cohan, C. L., and Bradbury, T. N. (1997). Negative life events, marital interaction, and the longitudinal course of newlywed marriage. *Journal of Personality and Social Psychology, 73*, 114-128.

Coleman, M., Ganong, L., and Leon, K. (2006). Divorce and postdivorce relationships. In A. L. Vangelisti and D. Perlman (Eds.), *The Cambridge Handbook of Personal Relationships* (pp. 157-173). New York: Cambridge University Press.

Conger, R. D., Reuter, M. A., and Elder, G. H. (1999). Couple resilience to economic pressure. *Journal of Personality and Social Psychology, 76*, 54-71.

CSIRO, and BOM. (2007). *Climate change in Australia: Technical Report 2007*. Retrieved 25 August 2008, from www.climatechangeinaustralia.gov.au.

Eckert, K. A., Wilkinson, D., Taylor, A. W., Stewart, S., and Tucker, G. R. (2006). A population view of mental illness in South Australia: Broader issues than location [Electronic Version]. *Rural and Remote Health*, 6, 1-13 from http://rrh.deakin.edu.au.

Foran, H. M., and O'Leary, K. D. (2008). Problem drinking, jealousy, and anger control: Variables predicting physical aggression against a partner. *Journal of Family Violence, 23*, 141-148.

Fox, J., Merwin, E., and Blank, M. (1995). De facto mental health services in the rural south. *Journal of Health Care for the Poor and Underserved, 6*, 434-468.

Fraser, C. E., Smith, K. B., Judd, F., Humphreys, J. S., Fragar, L. J., and Henderson, A. (2005). Farming and mental health problems and mental illness. *International Journal of Social Psychiatry, 51*, 340-349.

Frye, N. E., and Karney, B. R. (2006). The context of aggressive behavior in marriage: A longitudinal study of newlyweds. *Journal of Family Psychology, 20*, 12-20.

Gottman, J. M. (1994). *What predicts divorce? The relationship between marital processes and marital outcomes*. Hillsdale, NJ, England: Lawrence Erlbaum Associates.

Gottman, J. M., Coan, J., Carrere, S., and Swanson, C. (1998). Predicting marital happiness and stability from newlywed interactions. *Journal of Marriage and the Family, 60*, 5-22.

Gray, I., and Lawrence, G. (1996). Predictors of stress among Australian farmers. *Australian Journal of Social Issues, 31*, 173-189.

Halford, W. K., Sanders, M. R., and Behrens, B. C. (1993). A comparison of the generalization of behavioral marital therapy and enhanced behavioral marital therapy. *Journal of Consulting and Clinical Psychology, 61*, 51-60.

Hammen, C. (2005). Stress and depression. *Annual Review of Clinical Psychology, 1*, 293-319.

Hedlund, D. E., and Berkowitz, A. D. (1979). The incidence of social psychological stress in farm families. *International Journal of the Sociology of the Family, 9*, 233-243.

Hennessy, K., Fawcett, R., Kirono, D., Mpelasoka, F., Jones, D., Bathols, J., et al. (2008). *An assessment of the impact of climate change on the nature and frequency of exceptional climatic events*. Canberra: CSIRO and BOM.

Heyman, R. E. (2001). Observation of couple conflicts: Clinical assessment applications, stubborn truths, and shaky foundations. *Psychological Assessment, 13*, 5-35.

Hickey, D., Carr, A., Dooley, B., Guerin, S., Butler, E., and Fitzpatrick, L. (2005). Family and marital profiles of couples in which one partner has depression or anxiety. *Journal of Marital and Family Therapy, 31*, 171-182.

Jacobson, N. S., and Margolin, G. (1979). *Marital therapy: Strategies based on social learning and behavior exchange principles*. New York: Guilford Press.

Judd, F., Jackson, H., Fraser, C., Murray, G., Robins, G., and Komiti, A. (2005). *Depression in farmers and farming families: Project report 2005*. Melbourne: Centre for Rural Mental Health.

Judd, F., Jackson, H., Fraser, C., Murray, G., Robins, G., and Komiti, A. (2006). Understanding suicide in Australian farmers. *Social Psychiatry and Psychiatric Epidemiology, 41*, 1-10.

Keating, N. C. (1987). Reducing stress of farm men and women. *Family Relations, 36*, 358-363.

Kline, G. H., Pleasant, N. D., Whitton, S. W., and Markman, H. J. (2006). Understanding couple conflict. In A. L. Vangelisti and D. Perlman (Eds.), *The Cambridge Handbook of Personal Relationships* (pp. 445-462). New York: Cambridge University Press.

Leonard, K. E., and Homish, G. G. (2008). Predictors of heavy drinking and drinking problems over the first 4 years of marriage. *Psychology of Addictive Behaviors, 22*, 25-35.

Malmberg, A., Simkin, S., and Hawton, K. (1999). Suicide in farmers. *British Journal of Psychiatry, 175*, 103-105.

Markman, H. J. (1991). Constructive marital conflict is not an oxymoron. *Behavioral Assessment, 13*, 83-96.

Miller, K., and Burns, C. (2008). Suicides on farms in South Australia, 1997-2001. *Australian Journal of Rural Health, 16*, 327-331.

National Farmers' Federation. (2008). *Farm Facts* Retrieved 25 August, 2008, from http://www.nff.org.au/farm-facts.html

Page, A. N., and Fragar, L. J. (2002). Suicide in Australian farming, 1988-1997. *Australian and New Zealand Journal of Psychiatry, 36*, 81-85.

Parslow, R. A., and Jorm, A. F. (2000). Who uses mental health services in Australia? An analysis of data from the National Survey of Mental Health and Wellbeing. *Australian and New Zealand Journal of Psychiatry, 34*, 997-1008.

Paykel, E. S. (2003). Life events and affective disorders. *Acta Psychiatrica Scandanavia, 108*, 61-66.

Pearce, Z. J., and Halford, W. K. (2008). Do attributions mediate the association between attachment and negative couple communication? *Personal Relationships, 15*, 155-170.

Relationships Australia. (2008). Women and separation: Managing new horizons. Retrieved 1 September, 2008, from http://www.relationships.com.au/resources/publications/publications

Stehlik, D., Gray, I., and Lawrence, G. (1999). *Drought in the 1990s: Australian farm families experiences*. Canberra: Rural Industries Research and Development Corporation.

Weigel, R. R., and Weigel, D. J. (1987). Identifying stressors and coping strategies in two-generation farm families. *Family Relations, 36*, 379-384.

Weingarten, H. R. (1986). Strategic planning for divorce mediation. *Social Work, May-June*, 194-200.

Weiss, R. L., and Heyman, R. E. (1990). Observation of marital interaction. In F. D. Fincham and T. N. Bradbury (Eds.), *The psychology of marriage: Basic issues and applications* (pp. 87-117). New York: Guilford Press.

ACKNOWLEDGMENT

The author wishes to thank Cathryne Lang for her assistance in the early stages of preparing this chapter.

APPENDIX

Further Reading: Resources on the Internet

There are a number of useful services, resources, websites and publications available to assist Australian farming families who are experiencing relationship strain and/or separation. There are also a number of services available for people coping with depression, stress, anxiety and suicidal ideation. A list of the some of these resources is included here for residents of Australia. For other areas and countries, web searches will reveal your nearest helping agencies. Some counselling services are now offered online.

Relationships Australia is one of the most well-known relationships counselling services in Australia. They provide online and face-to-face counselling services for couples who wish to build a stronger relationship or survive the process of separation (www.relationships.com.au or e-counselling at www.relationshiphelponline.com.au).

Family Relationships Online provides a selection of online resources for families, ranging from maintaining healthy relationships through to dispute resolution (www.familyrelationships.gov.au).

Lifeline Australia provides general support and information for anyone in need. Lifeline provides a free 24-hour telephone counselling service across any topic and can direct callers to a wide-range of additional services across Australia (www.lifeline.org.au or telephone 13 11 14 in Australia).

Kids Helpline is a specialised counselling service established specifically with the needs of children and young adults in mind. Like Lifeline, Kids Helpline offers free 24-hour access to telephone support for anyone between the ages of 5 and 25 (www.kidshelpline.com.au or telephone 1800 55 1800 in Australia).

The Australian Government's *Child Support Agency* website provides a lot of support for separating families. In addition to general advice and assistance, this site provides a list of useful publications for couples coping with separation (www.csa.gov.au/services/separation_resources/index.aspx). In particular their publication about coping with co-parenting called "Me, My Kids and My Ex" may be particularly useful for couples negotiating these new boundaries (www.csa.gov.au/publications/pdf/1229.pdf).

Beyond Blue is a national organisation and government initiative designed to increase awareness of mental health problems such as depression and anxiety, and provide support to people living with depression and their families (www.beyondblue.org.au).

The Australian Institute of Family Studies has put together a very comprehensive list of helpful publications on a number of topics surrounding families and parenting (www.aifs.gov.au/afrc/bibs/bib.html).

Mensline Australia is a telephone counselling service dedicated to providing relationship and family support for men (www.menslineaus.org.au; ph: 1300 789 978).

Frontier Services provides health and community services, aged care and pastoral care to rural Australian communities (www.frontierservices.org).

More general climate change and/or drought publications provide Australian farmers with access to much needed information and a further avenue of support. For example, the CSIRO climate change pages detail relevant scientific advances and research partnerships (www.csiro.au/org/ClimateAdaptationFlagshipOverview.html) and the joint CSIRO/Bureau of Meteorology's 'Climate Change in Australia' report discusses international climate change research (www.climatechangeinaustralia.com.au/index.php).

Contributed by Francine Pritchard, Ipswich.

Photo. Farm life in the drought.

Part 2:

What Are Some of the Effects of Climate Change and Natural Disasters?

Contributed by Ray Norton, Victoria.

Photo: Commemorating Forests Struck by Fierce Storms.

In: Meltdown: Climate Change, Natural Disasters… ISBN 978-1-60876-153-1
Editor: Kathryn Gow

Chapter 9

HOW MIGHT THE HEALTH EFFECTS OF AIR POLLUTION CHANGE WHEN THE PLANET GETS WARMER?

***Adrian G. Barnett*[1] *and Craig Hansen*[2]**
[1] Queensland University of Technology, Australia
[2] U.S. Environmental Protection Agency, USA

ABSTRACT

In this chapter, we examine in depth how a warmer planet could affect human health by increasing our exposure to air pollution. The possible mechanisms of this increased exposure are diverse, and include a dryer earth leading to more dust storms, and a hotter earth leading to more forest fires. The consequences of increased exposure are more serious for the most vulnerable people in society; increased pollution has been linked to increased death in the elderly, and to increased visits to hospital and reduced lung function in the young. The pollution sources that cause these health problems are the same sources that have contributed to global warming, namely the combustive pollutants of traffic and industry. Cleaning up these sources will therefore, not only assist in curbing future global warming, but will have an immediate beneficial effect on worldwide health.

Keywords: air pollution, health effects, climate change, pollutants, global warming

INTRODUCTION

The interesting thing is that no one realised at the time that the number of deaths were increasing. There weren't bodies lying around in the street and no one really noticed that more people were dying. One of the first indications was that undertakers were running out of coffins and florists were running out of flowers.

(Dr Robert Waller, working at St Bartholomew's Hospital during the 1952 London smog)

Historical Concerns about Air Pollution

Air pollution is "the presence of substances in air at concentrations, durations and frequencies that adversely affect human health, human welfare or the environment" (Murray and McGranahan, 2003). Air pollution has been a health problem for centuries; in 1230, the air pollution in London was so bad that coal burning was prohibited (Boubel et al., 1994). During the mid-twentieth century a number of extreme pollution episodes caused a noticeable increase in deaths and created worldwide attention. The extreme levels of pollution were caused by a combination of already high levels and unfavorable weather conditions. The first episode occurred during 1930 in the Meuse Valley in Belgium where high levels of sulfur dioxide from industry were trapped for three days by an inversion layer of cold air causing 63 deaths. During 1948, in Donora, Pennsylvania, a similar combination of pollution and the weather over a five day period caused 20 deaths and made approximately14, 000 people ill. The most catastrophic episode happened in London over a 5 day period in December 1952 when 4000 people died due to exposure to air pollution (Brunekreef and Holgate, 2002). A particularly cold period meant that Londoners burned huge amounts of coal to keep themselves warm. The burning released massive amounts of sulfur dioxide and particulates which became trapped in an inversion layer formed by the cold air. The pollution was so thick that it made driving difficult and it also pervaded indoors, so exposing the entire population of London.

The London smog episode resulted in the formation of the Clean Air Act in 1956 (Boubel et al., 1994) and since then there has been extensive research demonstrating the negative health effects of air pollution. From the 1980s until now, much of this research has explored the more subtle health effects associated with lower levels of air pollution (Hertz-Picciotto, 1998). There is now a wealth of literature showing that exposure to low level ambient (or community) air pollution is associated with a multitude of health effects, including mortality, respiratory and cardiovascular hospitalizations, changes in lung function, asthma attacks, days lost from work, and preterm birth and restricted fetal growth. Groups that are most at risk are children, the elderly, and those with asthma or cardiac conditions.

Recent Concerns about Global Warming

Public concerns about air pollution have existed for centuries, but it is only in the last 10 to 20 years that the public have also become concerned about global climate change. In 1979, the World Meteorological Organization (WMO) organized the first conference on world climate and expressed concern that "continued expansion of man's activities on Earth may cause significant extended regional and even global changes of climate" (IPCC, 2004). In 1988, the WMO decided to form the Intergovernmental Panel of Climate Change (IPCC) with the purpose being to provide policymakers with a source of information about the causes of climate change, its potential environmental and socio-economic consequences, and the adaptation and mitigation options to respond to it. The first IPCC assessment report on climate change was published in 1990, the second in 1995, the third in 2001, and the most recent in 2007.

The most recent IPCC report stated that "It is *very likely* that heat waves will be more intense, more frequent and longer lasting in a future warmer climate. Cold episodes are projected to decrease significantly in a future warmer climate. Almost everywhere, daily minimum temperatures are projected to increase faster than daily maximum temperatures, leading to a decrease in diurnal temperature range" (Meehl et al., 2007). In addition, over the past 100 years the average global temperature has increased by about 0.74°C and this increase is greater at higher northern latitudes; also 11 of the last 12 years (1995–2006) have been the warmest since 1850 (IPCC, 2007).

The Combined Effects of Global Warming and Air Pollution

Air pollution is a complex mixture of gases and fine particles that have different interactions with the atmosphere and human health depending on the source of the pollutants and level of human exposure. Air pollution is strongly influenced by temperature (inversions and ozone formation), wind speed and direction, precipitation (see later section on wash-out), water vapor and cloud cover. Hence, the weather has a significant role in the extent to which populations are exposed to air pollution. In a planet affected by global warming, an important question is whether any new patterns in the weather will interact with air pollution and so pose a greater risk to health. In this chapter, we discuss some of the ways in which this interaction might happen. We start with a summary of recent trends in pollution followed by a brief summary of what is currently known about the health effects of air pollution. A previous review addressed the possible interaction between global warming and air pollution (Bernard et al., 2001), although it focused on the implications for the US.

Air Pollution: Sources and Recent Trends

Air pollution has both natural and anthropogenic (manmade) sources. Natural sources include volcanoes (sulfur oxides and particles), forest fires (carbon monoxide and particles), plants (hydrocarbons and pollen), soil (dust, nitrogen oxides, viruses), and sea spray. Common anthropogenic sources of air pollution are burning of biomass, industry, transport and domestic heating.

Whilst a wide variety of pollutants are now being monitored and researched, the main anthropogenic air pollutants investigated are particulate matter (PM), carbon monoxide (CO), nitrogen dioxide (NO_2), ozone (O_3), sulfur dioxide (SO_2) and lead (Pb). These pollutants are collectively referred to as the "criteria" pollutants. Over the past few decades, trends in air pollution worldwide have generally been positive; however, here are still cities that have notably high levels of criteria pollutants (e.g., Mexico City, Sao Paulo and Beijing).

Particle pollution, or particulate matter, is a complex mixture of solid particles and liquid droplets suspended in the air (aerosols). The more coarse particles are generally emitted from industry, constructions sites, and fires, while fine particles are formed when gases react in the atmosphere. These gases can be volatile organic compounds (VOCs), oxides of nitrogen (NO_x) and SO_2, and the main source of these gases are vehicles, industry, and power plants (US Environmental Protection Agency, 2008a).

Since the 1990s, $PM_{2.5}$ (particles less the 2.5 microns in diameter) concentrations in the United States declined by 14% (ibid). There have also been substantial reductions in PM_{10}

levels in Mexico City, which is one of the World's most polluted cities; however, PM_{10} levels in Mexico City are still dangerously high as the daily maximum is around 250 μg/m^3 (Fernandez-Bremauntz, 2008). Similarly, PM_{10} levels in India have reduced by up to 10% over the last 10 years, but levels are still well above air quality standards (Gupta and Kumar, 2006). Whilst there have been reductions in particulate matter (PM) emissions throughout Europe (up to 44%), there has been little change in the observed levels (European Environment Agency, 2007). This is also the case for Australia where PM levels have remained fairly static from 1991 to 2001, although the levels are relatively low compared to industrialized countries (e.g., the average level of PM_{10} in New York was 55% higher than the average level in Sydney in 2000) (Department of the Environment and Heritage, 2004).

CO is a colorless and odorless gas formed by the incomplete combustion of fossil fuels. The main source of CO is vehicles, which contribute up to 95% of anthropogenic CO in some cities. Other sources of CO emissions include industry, domestic wood burning, and forest fires (US Environmental Protection Agency, 2008a). As a result of cleaner cars and industrial processes, CO emissions (and levels) have declined substantially since the 1990s. For example, in the United States, between 1990 and 2006, CO levels reduced by 62% and emissions by 38%. These reductions in emissions and levels occurred, despite a 43% increase in vehicle miles travelled (ibid). Similarly, across Europe, there has been a reduction in CO emissions of up to 50% and the reported levels are no longer considered a threat to human health (European Environment Agency, 2007). There have also been significant reductions in ambient CO levels within Australian cities between 1991 and 2001 (Department of the Environment and Heritage, 2004). For example, in Sydney the average CO level in 1993 was 11.8 parts per million (ppm), but by 2001 this had reduced to 5.1 ppm (a 57% reduction).

Nitric oxide is emitted from high temperature combustion of fossil fuels and then oxidized in the atmosphere to form NO_2, which is a more toxic compound and a precursor to photochemical smog (O_3). The main anthropogenic source of nitric oxides is transport. In the United States, NO_2 levels have reduced by 30% over the past 17 years (US Environmental Protection Agency, 2008a), while throughout Europe there has been a reduction in O_3 precursor emissions of up to 36% (European Environment Agency, 2007). NO_2 levels in Australia have exhibited a slight downward trend between 1991 and 2001 but to a much lesser extent than CO and SO_2 reductions (Department of the Environment and Heritage, 2004); for example, in Sydney, the highest daily maximum 1-hour NO_2 level in 1992 was 0.134 ppm, but by 2001 this had reduced to 0.071 ppm (a 47% reduction).

O_3 is a naturally occurring gas in the stratosphere where it plays an important role in filtering out ultraviolet radiation; however, at ground-level, O_3 is a secondary pollutant and is the prime oxidant in smog, which is highly toxic to humans and vegetation. O_3 is formed by reactions of oxides of nitrogen (NO_x), volatile organic compounds (VOCs) and sunlight, and this reaction is accelerated by high temperatures. The main anthropogenic sources of O_3 precursors are vehicles and power plants (responsible for NO_x), and petrochemical industries (responsible for VOCs).

Between 1990 and 2006, O_3 levels had only reduced by 9% in the United States, which is much less than the reductions in other pollutants (U.S. Environmental Protection Agency, 2008a). In Mexico City, O_3 levels have decreased to less than half of the levels reported in the 1990s; however, despite this decrease, O_3 levels in Mexico City are still a major problem as the daily air quality standard for O_3 is exceeded for more than half of the year (Fernandez-Bremauntz, 2008). Across Europe, the emissions for O_3 precursors reduced by approximately

36% between 1990 and 2004 (European Environment Agency, 2007a). In Australia, annual O_3 levels remained relatively constant from 1991 to 2001 (Department of the Environment and Heritage, 2004); for example, in Sydney the highest daily maximum 4-hour O_3 level in 1992 was 0.118 ppm, but by 2001 this had increased to 0.137 ppm (a 16% increase).

SO_2 is a colorless gas formed by burning fuels that contain sulfur, such as coal, oil, and diesel. The main source of SO_2 is diesel vehicles in urban areas (e.g., buses, trucks) and industry in regional areas (e.g., coal fired power plants, non-ferrous metal smelters). SO_2 levels have reduced significantly in the United States with reductions up to 53% between 1990 and 2006 (U.S. Environmental Protection Agency, 2008a). There has been a similar reduction in SO_2 emissions throughout Europe since the mid-1990s (European Environment Agency, 2007). In Australia, the levels of SO_2 have fallen to very low levels in major cities, most likely due to low sulfur fuels (Department of the Environment and Heritage, 2004); for example, in Sydney, the highest daily maximum 1-hour SO_2 level in 1993 was 0.041 ppm, but by 2001 this had reduced to 0.026 ppm (a 37% reduction).

Despite reductions in many pollutant emissions and levels, it seems that PM and O_3 still continue to present air quality concerns with large populations exposed to "unsafe" levels. For example, approximately 20–30% of the European population lives in cities where the PM, O_3 and NO_2 levels exceeded the European Union air quality standard between 1997 and 2004 (European Environment Agency, 2007). Similarly, in the United States 100 million people live in counties where the national air ambient quality standards for O_3 and $PM_{2.5}$ were exceeded during 2000 (U.S. Environmental Protection Agency, 2008a).

Health Effects of Air Pollution

The effects of air pollutants on human health can seem small when viewed on a day-to-day basis. For example, in urban Australia and New Zealand increases in $PM_{2.5}$, NO_2 and CO of an inter-quartile range (which is the difference between the 75th and 25th percentile of the pollutant level and so is roughly equivalent to the difference between moderately polluted and moderately clean day) increased hospital admissions by between 1–3% in the elderly (Barnett et al., 2006). A meta-analysis of worldwide studies found that a 10 ppb increase in O_3 increased daily deaths by 0.4% (Ito, Leon and Lippmann, 2005). However, the overall impact on public health of these increases is large, because air pollution is such a consistent and ubiquitous exposure (at least in urban environments) (Maynard, 2004).

Many pollutants are known to be detrimental to human health. Specific negative effects have been found with CO, NO_2, O_3, PM, Pb, SO_2 and VOCs. Although these pollutants can occur naturally, they are most often found in the atmosphere due to combustion processes, namely industry and vehicles; for that reason those people living near major roads have often been found to be the most negatively affected by air pollution.

Children are especially vulnerable to the effects of air pollution. A longitudinal study of children in California found that those children living nearer to major roads had significantly reduced lung function compared to children who lived away from major roads (Gauderman et al., 2007). Children in urban Australia and New Zealand were found to be at increased risk of hospitalization for respiratory illness due to exposure to NO_2, PM_{10} and SO_2 (Barnett et al., 2005); children exposed to NO_2 and SO_2 were found to be at increased risk of sudden infant death syndrome (Dales et al., 2004). Even before birth, air pollution can have negative effects on health, because O_3, PM_{10} and SO_2 have been shown to significantly reduce fetus size (Hansen, Barnett and Pritchard, 2008). Also, maternal exposure to ambient air pollution is associated with an increased risk of low birth weight (Parker et al., 2005; Wilhelm and Ritz, 2005) and birth defects (Gilboa et al., 2005; Ritz et al., 2002). The other group most vulnerable to the effects of air pollution is the elderly, whose cardiovascular system is often already overworked and so cannot cope with the extra insult of pollution (Barnett, 2006).

Researchers have been able to estimate the public health impact of air pollution exposure; for example, Kunzli et al. (2000) assessed the effect of outdoor traffic related air pollution in three European countries and estimated that it caused 6% of all deaths. In recent studies focusing on the global burden of disease, outdoor air pollution was found to account for approximately 1.4% of total mortality (Ostro, 2004) and 0.8 million premature deaths and 6.4 million years of life lost, with the greatest burden being noted in developing countries (Cohen, 2005). An Australian report estimated that in the year 2000, motor vehicle pollution accounted for between 900 and 4500 of morbidity cases (cardiovascular disease, respiratory disease and bronchitis) and for between 900 and 2000 of early deaths (Bureau of Transport and Regional Economics, 2005). The economic cost of this premature mortality was between $1.1 billion and $2.6 billion (central estimate $1.8 billion), while the estimated economic cost of morbidity was between $0.4 billion to $1.2 billion (central estimate $0.8 billion) (Bureau of Transport and Regional Economics, 2005).

Climate Change and How It Might Worsen Air Pollution

In this section, we first give a very brief explanation of climate change and how climate change and air pollution might interact. We then list a number of possible ways in which climate change might worsen the health effects of air pollution. These interactions are summarized in Figure 9.1.

A Brief Explanation of Climate Change

The Earth receives radiated solar energy from the Sun in short wavelengths of energy; approximately two-thirds of this energy is absorbed (mostly by the Earth's surface and some by the atmosphere), while the rest is reflected back towards space. Much of this reflected radiation is absorbed by the atmosphere and re-radiated back to Earth. This naturally occurring process is called the greenhouse effect and without it the Earth's temperature would be below the freezing point of water (Le Treut et al., 2007). The gases in the atmosphere responsible for the reflection are called greenhouse gases (GHGs); the naturally occurring GHGs are water vapor, carbon dioxide (CO_2), methane (CH_4), nitrous oxide (N_2O) and O_3. Although many complex interactions take place in the atmosphere, it is estimated that water vapor contributes approximately 60% of the reflection, CO_2 about 26%, and O_3 about 8% (Kiehl and Trenberth, 1997). Other gases such as CO and oxides of nitrogen (NO_x) also play an indirect role in solar radiation absorption by influencing the formation or destruction of GHGs.

An alteration in the energy balance of the Earth-atmosphere system leads to a change in global climate as a result of radiative forcing. Radiative forcing is a change in the radiation balance of the surface-troposphere system which can be due to greenhouse gases. The radiation balance controls how much of the Sun's energy is reflected back into space and how much is absorbed by the Earth. The term "climate change" can be defined as "a change in the state of the climate that can be identified (e.g., using statistical tests) by changes in the mean and/or the variability of its properties, and that persists for an extended period, typically decades or longer" (Hegerl et al., 2007, p. 667).

Although the main GHGs occur naturally, human activities have resulted in increased GHG emissions over the past 250 years, with a sharp increase occurring during the 1970s. Global CO_2 levels have increased by 36% and this increase is largely associated with combustion of fossil fuels and cement production (Forster et al., 2007; Hansen and Sato, 2004). Between 1995 and 2005 the amount of global radiative forcing caused by CO_2 had increased by 20%, which is the largest change for any decade in the last 200 years (Forster et al., 2007). In the United States, fuel combustion accounts for approximately 94% of CO_2 emissions (U.S. Environmental Protection Agency, 2008b).

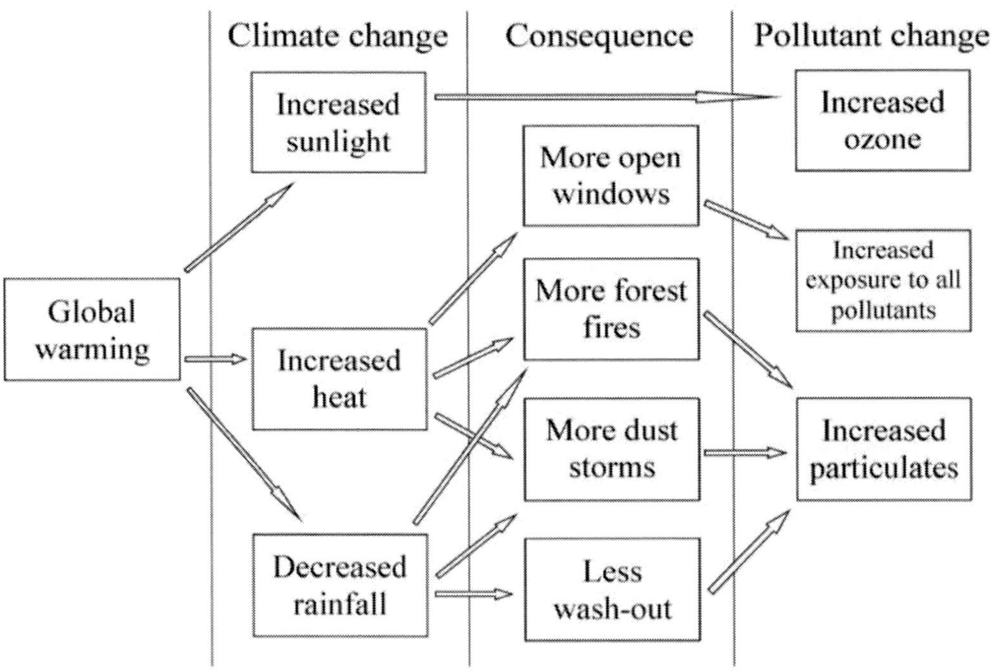

Figure 9.1. The possible pathways between global warming and increased air pollution.

Methane has the next largest radiative forcing and has increased by about 30% over the past 250 years; however, while the overall abundance of CH_4 has increased the overall growth rate has declined from highs during the 1970s and early 1980s to growth rates close to zero during the late 1990s and early 2000s (Forster et al., 2007). In the United States the largest source of anthropogenic CH_4 emissions is enteric fermentation (23%), followed by landfills, natural gas systems and coal mining (U.S. Environmental Protection Agency, 2008b).

Despite low levels in the atmosphere, N_2O is the fourth largest radiative forcing contributor behind water vapor, CO_2 and CH_4. Global N_2O levels have increased by about by

18% over the past 250 years (Forster et al., 2007). The main anthropogenic sources of N_2O in the United States are agriculture fuel combustion in motor vehicles, and nitric acid production (U.S. Environmental Protection Agency, 2008b). Although atmospheric levels and the radiative forcing effect are much lower for N_2O than CO_2, the capacity of N_2O to trap heat is much greater than CO_2.

Interactions between Greenhouse Gases and Air Pollution

There are a number of different chemical interactions between anthropogenic air pollutants and greenhouse gases. Some interactions cause an increase in GHGs, whilst other interactions lead to a decrease (we give some examples below). These divergent effects make it difficult to predict how increasing pollution and global warming will interact in the coming years.

An increase in temperature is associated with an increase in O_3 levels (see next section), which is a major GHG; an increase in oxides of nitrogen (NO_x) can lead to a decrease in CH_4 and in turn reduced radiative forcing. However, an increase in NO_x (a precursor to O_3) also leads to an increase in $O_{3.}$ Also, aerosols (including water vapor) intercept incoming sunlight thereby reducing O_3 production, and reducing the amount of energy reaching the Earth creating a cooling effect known as negative radiative forcing. Additionally, more aerosols also produce more droplets in clouds making them more reflective. While the current day "cooling" effect of aerosols may be protective of climate change, the effect will be short lived. This is primarily due to reductions in aerosol emissions, as the air quality improves and the lifetime of aerosols being much shorter than GHGs (which can survive for centuries), hence this protective "cooling" effect will diminish over time and lead to greater climate change with hotter temperatures (Andreae, Jones, and Cox, 2005).

Hydroxyl (OH) radical, the main oxidant that removes traces gases from the atmosphere, is often referred to as a "pollutant scavenger" or "atmospheric-scrubber" simply because it is involved in the chemical reactions that breakdown pollutants and GHGs. Sunlight breaks down O_3 which further reacts with H_2O to form OH which in turn breaks down various pollutants (e.g., CO, hydrocarbons, CH_4, NO_2, SO_2). However, NO_x is a by-product of the hydroxyl cleaning system and increased NO_x leads to more O_3.

This brief introduction highlights the complexity of the chemical reactions occurring in the atmosphere and how these interactions play a vital role in ambient air pollution levels and climate change. Despite this complexity, it is predicted that climate change will worsen air quality by accelerating the chemical reactions that lead to the formation of pollutants (Giorgi and Meleux, 2007). In the United States, climate change is predicted to increase the severity and duration of summertime air pollution episodes in the Midwest and northeastern parts due to a decrease in the frequency of cyclones tracking across southern Canada, as the cold fronts from these cyclones are the main ventilation for the region (Mickley et al., 2004).

A Sunnier Climate Means More O_3

O_3 is formed by the reaction between sunlight and NO_x and VOCs (pollutants caused by combustion). The need for sunlight means O_3 levels often peak on summer afternoons and thus future levels of O_3 have been predicted to increase substantially over Europe due to global warming (as warmer temperatures speed up the production of O_3) (Meleux, Solmon, and Giorgi, 2007), and because of increases in anthropogenic emissions of O_3 precursors. A study investigating the influence of meteorological factors on O_3 levels over Eastern United States predicted increased O_3 levels and potentially more severe air pollution episodes and this was largely due to an increase in temperature (Dawson, Adams, and Pandis, 2007). Also, in the Northern Hemisphere it is expected that O_3 levels will be higher with increasing latitude as a result of climate change (Hedegaard et al., 2008).

Many air pollution studies have found associations between O_3 and mortality and hospital admissions (Ito, Leon and Lippmann, 2005); however, there has been limited research focusing on the health effects associated with the predicted increase in the levels of O_3 relating to climate change. A recent study investigated how climate change could affect ambient O_3 levels and subsequent mortality in 50 cities in the United States. It was predicted that O_3 levels during the summer would increase by almost 5 parts per billion from the 1990s to the 2050s, which would be accompanied by a significant rise in mortality (total, cardiovascular and respiratory) (Bell et al., 2007). A similar study predicted a median increase of 4.5% in O_3 related mortality across 31 counties in New York due to the effects of climate change (Knowlton et al., 2004).

Although CO_2 is not classified as a pollutant to human health, due to its large role in climate change, it does have an indirect effect on human health; for example, increased CO_2 is associated with increased water vapor and temperatures, both of which may exacerbate O_3 in already polluted areas. Modeled simulations have shown that a one degree Celsius rise in CO_2-induced temperature is associated with an increase of approximately 1000 deaths annually in the United States and about 40% of these additional deaths may be attributed to increased O_3 (Jacobson, 2008).

In the summer of 2003, many parts of Europe experienced the hottest summer in 50 years, which may have been exacerbated by climate change. In France, the heat wave caused 14,947 excess deaths over a two week period (Poumadère et al., 2005), whilst in Italy there was an overall increase in mortality of 3134 (Conti et al., 2005). During this heat wave, levels of O_3 were also extremely high. In a study across nine French cities, the combined effect of O_3 and temperature was associated with an excess risk of death ranging from 10.6% in Le Havre to 174.7% in Paris and the relative contribution of O_3 to this joint effect ranged from 2.5% in Bordeaux to 85.3% in Toulouse (Filleul et al., 2006).

More Forest Fires Means More Pollution

Warmer temperatures increase the risk of forest fires; a hotter, drier Earth means more combustible material and fires can breakout in extreme heat, even in well managed forests. Although forest fires may only last a few days, the plumes of smoke they release can stretch for thousands of kilometers, potentially exposing great numbers of people to increased pollution.

The health effects of forest fires have been noted in many studies; for example, in 2003, a forest fire in California burned more than 3000 km^2 and catapulted the levels of particulate matter up to 10 to 20 times normal levels (Künzli et al., 2006), and children who were exposed to the smoke for prolonged periods showed increased risks of eye and respiratory symptoms, physician visits, and days off school. A forest fire in British Columbia, in 2003, led to elevated levels of particulate matter for a five week period and an increase in physician visits for respiratory diseases of between 46% and 78% (Moore et al., 2006). Also, time-series studies in Australia (Johnston, 2002; Morgan et al., 2006) and Malaysia (Mott et al., 2005) have reported increased hospital admissions for respiratory illnesses associated with particulate matter from biomass burning (e.g., bushfires and back burning).

Based on predictive models of increased temperature due to global warming, an Australian study forecast that the number of "very high" fire danger days will increase by up to 13% by the year 2020 and by up to 23% by the year 2050. The number of "extreme" fire danger days will increase by up to 25% and 50% by the years 2020 and 2050, respectively (Lucas et al., 2007).

A Drier Earth Means More Dust Storms

Dust storms pose a threat to human health because the particles irritate the eyes and the lungs and may trigger allergic reactions or cardiovascular events. Although dust is not a direct pollutant from anthropogenic sources, dust storms are an indirect result of various human activities such as increased deforestation and poor land management. Dust storms are particularly prevalent in Northern China and West Africa with occasional dust storms being experienced in Australia, but more frequently since 1992 when the droughts became more severe and sustained. The main source of dust storms in China has been debated with some suggesting that the dust originates from the deserts, while other suggest the outer areas of the deserts and degraded farming and grasslands (Wang et al. 2004). Dust clouds that originate in Northern China travel long distances to areas of Southern China. A study of dust storms in China during the past 1000 years showed that the frequency of the storms was negatively associated with precipitation and temperature; more dust storms occurred during cold-dry periods than did warm-wet periods (Yang et al. 2007). Similarly, each year in Africa during the dry winter periods large dust clouds originate in the Sahara desert and travel to the west coast of Africa. These dust clouds are then further transported across the Atlantic ocean by the Trade Winds and reach the Caribbean and Americas approximately a week later (Propsero 2006; Sunnu et al. 2008). A recent study reported that African dust clouds are associated with increased pediatric admissions for asthma on the Caribbean island of Trinidad (Gyan et al. 2005).

In 2002, a dust storm extending along a 2400 km front swept across a large part of eastern Australia. This was a result of widespread drought with record maximum temperatures and low rainfall in the previous 12 months. Air quality was dramatically reduced in four cities, with PM_{10} levels reaching measures almost 10 times higher than the Australian national ambient air quality standard (50 $\mu g/m^3$) (Chan et al., 2005).

With hotter and drier periods predicted as a result of climate change (IPCC, 2004), it seems inevitable that there will be an increase in dust storms within the Asian, African and Australian continents. For example, the IPCC report predicts decreased rainfall in Australia (Hennessy, et al., 2007) and Africa (Boko, et al., 2007) causing even more widespread drought.

Less Rain Means Less Wash-Out

Rainfall is a primary removal mechanism for particulate matter (PM pollutants; raindrops collide with particles in the air and some particles may be deposited in the rain and washed away (which is good news for the air, but bad news for waterways). Increased rainfall has been shown to reduce the health effects of PM (Barnett et al., 2006).

A predicted effect of global warming is a reduction in rainfall in: the Mediterranean, Central America, the western U.S., southern Africa, and the eastern Amazon basin (Giorgi and Meleux, 2007); less wash-out in these areas is likely to lead to higher levels of pollution and increased health problems.

Also of concern is the fact that air pollution has been found to delay and suppress rainfall via a number of different mechanisms (Rosenfeld et al., 2008). Polluted aerosol particles that have been sucked into clouds can slow the formation of raindrops and aerosol particles in the air decrease solar radiation, and so reduce the amount of evaporated water for cloud formation. This hints at a disastrous circle of events where more pollution leads to less rainfall and back to more pollution.

A Change in the Weather Means a Change in Behavior

On warmer days, people are more likely to open windows and doors in their homes and workplaces to let cooler air in, and are more likely to go outside. But by mitigating the health risks of heat, people may increase the risks of pollution as an increased exposure to outdoor air often means an increased exposure to pollution. Studies have shown that leaving windows open increases exposure to NO_2 (Kousa et al., 2001) and $PM_{2.5}$ (Koistinen et al., 2001). Going outdoors is likely to increase such exposure for anyone in a densely populated urban environment, particularly those people who live near roads. A time activity study in Australia found that people spent more time outdoors during summer before 9am and between 3pm and 9pm (compared to winter), and more time was spent in cars during summer (Environment Protection and Heritage Council, 2004). As O_3 levels are generally higher during warm periods and the diurnal O_3 pattern usually exhibits peak levels during mid-late afternoon, it would follow that more outdoor activity in summer would mean more O_3 exposure.

Activity patterns may also be influenced by people's perception of the pollution risk and it seems likely that less people would go outside when either the temperature or air pollution levels are extreme. However, a study in the US compared geographical variability in actual ambient air pollution readings with community perception of ambient air pollution and found that the perception of air quality was more influenced by a sense of place, proximity to source and neighborhood setting, than actual air pollution readings (Brody et al., 2004).

Temperature and Pollution Interact at High Levels

High temperatures and high levels of pollution independently increase risk of death, but it is possible that they also interact to pose an even greater threat to health. This interaction has been observed with PM causing more deaths in summer than winter in Belgium (Nawrot et al., 2007) and Brisbane, Australia (Ren, Williams and Tong, 2006). Similarly in the US, an increase in temperature has been shown to pose a greater threat to health when O_3 levels are high (Ren at al., 2008).

The mechanism of this interaction is not yet known. In warmer weather, more people go outdoors and leave their windows open and so the interaction could simply be caused by a greater exposure to pollution. It is also possible that the interaction is biological. High temperatures and air pollution levels both attack the cardiovascular and respiratory systems making it more difficult to breathe and increasing blood pressure and heart rate. A frail person exposed to either air pollution or high temperature may be able to cope with this increased pressure, but when exposed to both, their cardiovascular system may become overloaded and so fails. It is also possible that the interaction occurs at the chemical level, as hotter temperatures make particles more reactive and thus possibly more potent.

Future Sources of Pollution Exposure

The major sources of anthropogenic air pollution in urban environments are traffic, heating and industry. Traffic volume looks likely to increase around the world as the population grows and more people buy cars. In Australia, the number of cars is predicted to grow from 11 million in 2005 to 14 million in 2020 (Commonwealth of Australia, 2008). Production of passenger cars in China has already jumped from 605,000 in 2000 to 2.33 million in 2004, and worldwide the number of motor vehicles was expected to double in the next 15 years to 1.3 billion (Dhal, 2005), prior to the global financial crisis. More traffic means more air pollution and more greenhouse gases, although some of the increase will be offset by improvements in emissions due to cleaner engines.

A reduction in heating is one positive aspect of global warming. This is because heaters use a relatively large amount of electricity and (depending on the country) some, or most, of that electricity is generated by power stations which pump out more pollution and greenhouse gases. At a more individual level, some small heaters emit NO_2 which has been associated with acute and chronic respiratory disease (Franklin, 2007; Barnett et al., 2006). Indoor wood heaters emit SO_2 and PM. The winter increase of air pollution due to heating and subsequent increase in cardio-respiratory admissions to hospital is exemplified by Christchurch, New Zealand (McGowan, Hider, Chacko and Town, 2002) where around 44% of homes are heated using coal or wood fires (Christchurch City Council, 2000).

The beneficial effects of a reduction in heating use in winter might be somewhat offset by an increase in air conditioning use. Air conditioning use reduces the risk of heat-related death (Barnett, 2007) and the risk of hospitalization associated with air pollution (Janssen, Schwartz, Zanobetti and Suh, 2002). However, like heaters air conditioners are heavy consumers of electricity, although unlike heaters the emissions at source are not dangerous.

Researching a Change in the Health Effects of Air Pollution due to Global Warming

Over the last 15 years, there has been a concerted effort to create an agreement about the best study design to use when examining short-term associations between daily changes in air pollution and daily changes in morbidity and mortality. This involved the combined input of health experts, epidemiologists and statisticians in two main groups: APHEA (Air Pollution and Health, a European Approach) (Atkinson et al., 2001) and the Health Effects Institute in the US. The advantage of these groups was a general standardization of study designs worldwide, which made the results from different regions more comparable; such strength in numbers and reputations also made it difficult for the health effects of pollution to be denied.

Thanks to these groups, there is now a clear consensus that increased air pollution causes short-term increases in morbidity and mortality. Creating a consensus concerning any change in this association due to global warming is likely to be far more difficult for a number of reasons. The most important reason is sample size. Studies of the short-term effects of air pollution were often based on years of daily data and so often had sample sizes of over 1000. Many studies were based on more than one location, which increased the sample size and statistical power even further. Future studies that aim to examine any change in the association between air pollution and health due to global warming will have a much smaller power because the change is likely to be gradual and therefore difficult to detect. These studies will probably be based on examining trends in the annual association between pollution and health outcomes; hence the effective sample size is much smaller, because new observations are only gained every year as opposed to every day. A further problem is the lack of variation. Studies of short-term effects could exploit the differences in average pollution levels and temperatures between locations to estimate modifying effects and spatial patterns (Dominici et al., 2002). So relatively clean cities could be compared to relatively dirty cities during the same time period. However, studies of pollution and global warming are concerned with the change in temperature, and this change is likely to be both small and reasonably consistent around the world. These problems mean that it is likely to be some time before we can be certain whether the effects of air pollution are changing due to global warming, because any real change is likely to be gradual.

CONCLUSION

The world is facing its greatest challenge in dealing with the effects of global warming. The environmental scientist James Lovelock has compared the current time to the period just before World War II when it was clear that something terrible was going to happen and humanity prepared for the worst (Aitkenhead, 2008). Global warming is set to confront the world on multiple fronts, but a front that has largely been overlooked is the increased effects of air pollution. Two of the major causes of global warming are coal-fired power stations and vehicles, which are also responsible for large amounts of pollution. So curbing emissions from these sources will have a doubly beneficial effect. At present, we are not well prepared to deal with a warmer and more polluted environment; this lack of preparedness is exemplified by the fact that 76% of Americans see climate change as a serious problem, but view the environment as a low national priority (Leiserowitz, 2006) and according to Searle and Gow (see Chapter 3), the perceptions of climate change for a portion of Australians are similar.

We have outlined a number of ways in which global warming and pollution might combine to threaten health, although we have also outlined difficulties concerning how any increased health effects will be estimated. The empirical evidence we do have comes from previous forest fires and the European heat wave of 2003. Based on these results we can expect a serious increase in the health effects of pollution in the near future, unless strong measures are taken to significantly reduce pollution levels.

ACKNOWLEDGMENTS

Dr. Hansen's participation was supported by his appointment to the Postgraduate Research Program at the National Center for Environmental Assessment administered by Oak Ridge Institute for Science and Education through an interagency agreement between the U.S. Department of Energy and the U.S. Environmental Protection Agency.

REFERENCES

Aitkenhead, D. (2008). Cheerful in the face of Armageddon. *The Guardian Weekly March 23–April 3,* 14.

Andreae, M.O., Jones, C.D., and Cox, P.M. (2005). Strong present-day aerosol cooling implies a hot future. *Nature, 435,* 1187–1190.

Atkinson, R. W., Anderson, H. R., Sunyer, J., Ayres, J., Baccini, M., Vonk, J.M., Boumghar, A., Forastiere, F., Forsberg, B., Touloumi, G., Schwartz, J., and Katsouyanni, K. (2001). Acute Effects of Particulate Air Pollution on Respiratory Admissions. Results from APHEA 2 Project. *American Journal of Respiratory and Critical Care Medicine, 164*(10), 1860–1866.

Barnett, A., Williams, G., Schwartz, J., Neller, A., Best, T., Petroeschevsky, A., and Simpson, R.W. (2005). Air pollution and child respiratory health: a case-crossover study in Australia and New Zealand. *American Journal of Respiratory and Critical Care Medicine, 171*(11), 1272–8.

Barnett, A., Williams, G., Schwartz, J., Neller, A., Best, T., Petroeschevsky, A., and Simpson, R.W. (2006). The Effects of Air Pollution on Hospitalizations for Cardiovascular Disease in Elderly People in Australian and New Zealand Cities. *Environmental Health Perspectives, 114*(7), 1018–23.

Barnett, A. (2007). Temperature and Cardiovascular Deaths in the US Elderly: Changes over time. *Epidemiology, 18*(3), 369–372.

Bell, M.L., Goldberg, R., Hogrefe, C., Kinney, P.L., Knowlton, K., Lynn, B., et al. (2007). Climate change, ambient ozone, and health in 50 US cities. *Climatic Change, 82,* 61–76.

Bernard, S. M., Samet, J. M., Grambsch, A., Ebi, K. L., and Romieu, I. (2001). The potential impacts of climate variability and change on air pollution-related health effects in the United States. *Environ Health Perspect, 109* (suppl 2), 199–209.

Boko, M., Niang, I., Nyong, A., Vogel, C., Githeko, A., Medany, M., Osman-Elasha, B., Tabo, R., and Yanda, P. (2007). Africa. Climate change 2007: Impacts, adaptation and vulnerability. Contribution of Working Group II to the Fourth Assessment Report of the InterGovernmental Panel on Climate Change. [Eds. Parry, M.L., Canziani, O.F.,

Palutikof, J.P., van der Linden, and P.J., Hanson, C.E.] Cambridge University Press, Cambridge, UK, 433-467.

Boubel, R. W., Fox, D. L., Turner, B., and Stern, A. C. (1994). *Fundamentals of air pollution.* Third Edition. Academic Press, Sydney.

Brody, S.D., Peck, B.M., and Highfield, W.E. (2004). Examining localized patterns of air quality perception in Texas: A spatial and statistical analysis. *Risk Analysis, 24*(6), 1561–1574.

Brunekreef, B., and Holgate, S.T. (2002). Air pollution and health. *Lancet,* 360, 1233–42.

Bureau of Transport and Regional Economics. (2005). Health impacts of transport emissions in Australia: Economic costs (Working paper 63). Canberra, Department of Transport and Regional Services.

Chan, Y., McTainsh, G., Leys, J., McGowan, H., and Tews, K. (2005). Influence of the 23 October 2002 dust storm on the air quality of four Australian cities. *Water, Air, and Soil Pollution, 164,* 329–348.

Christchurch City Council (2000, April). Christchurch City Update 2000. Retrieved from http://archived.ccc.govt.nz/update/2000/Part2-AirQuality.pdf.

Cohen, A. J., Anderson, H. R., Ostro, B., Pandey, K. D., Krzyzanowski, M., and Künzli, N. (2005). The Global Burden of Disease Due to Outdoor Air Pollution. *Journal of Toxicology and Environmental Health, Part A, 68* (13 and 14), 1301–1307.

Commonwealth of Australia. (2008, February). Transport Sector Greenhouse Gas Emissions Projections 2007. Retrieved from http://www.climatechange.gov.au/projections/pubs/transport2007.pdf.

Conti, S., Meli, P., Minelli, G., Solimini, R., Toccaceli, V., Vichi, M., Beltrano, C., and Perini, L. (2005). Epidemiologic study of mortality during the Summer 2003 heat wave in Italy. *Environmental Research, 98,* 390–399.

Dales, R., Burnett, R. T., Smith-Doiron, M., Stieb, D. M., and Brook, J. R. (2004). Air pollution and sudden infant death syndrome. *Pediatrics, 113*(6), e628–31.

Dawson, J.P, Adams, P.J., and Pandis, N.S. (2007). Sensitivity of ozone to summertime climate in the Eastern USA: A modeling case study. *Atmospheric Environment, 41*, 1494–1511.

Department of the Environment and Heritage. (2004). State of the air: National ambient air quality status and trends report 1991–2001. Canberra, Department of the Environment and Heritage. Available at: http://www.environment.gov.au/atmosphere/airquality/publications/status/index.html

Dhal, R. (2005). Heavy Traffic Ahead: Car Culture Accelerates. *Environmental Health Perspectives, 113*(4), 239–245.

Dominici F., Daniels, M., Zeger, S. L., and Samet, J. M. (2002). Air Pollution and Mortality: Estimating Regional and National Dose-Response Relationships. *Journal of the American Statistical Association, 97*(457), 100–111.

Environment Protection and Heritage Council. (2004). Time activity study: Summary of findings. Available at: http://www.ephc.gov.au/pdf/EPHC/TASFinalReportSummary.pdf

European Environment Agency. (2007). Air Pollution in Europe 1990-2004. Available at: http://reports.eea.europa.eu/eea_report_2007_2/en

Fernandez-Bremauntz, A. (2008). Air quality management in Mexico. *Journal of Toxicology and Environmental Health, Part A, (71)*, 56–62.

Filleul, L., Cassadou, S., Medina, S., Fabres, P., Lefranc, A., Eilstein, D., Le Tertre, A., Pascal, L., Chardon, B., Blanchard, M., Declercq, C., Jusot, J.F., Prouvost, H., and Ledrans, M. (2006). The relation between temperature, ozone, and mortality in 9 French cities during the heatwave of 2003. *Environmental Health Perspectives, 114*(9), 1344–1347.

Forster, P., Ramaswamy, V., Artaxo, P., Berntsen, T., Betts, R., Fahey, D.W. , Haywood, J., Lean, J., Lowe, D.C., Myhre, G., Nganga, J., Prinn, R., Raga, G., Schulz, M., and Van Dorland, R. (2007). Changes in atmospheric constituents and in radiative forcing. In: Climate Change 2007: The Physical Science Basis. Contribution of Working Group I to the Fourth Assessment Report of the Intergovernmental Panel on Climate Change. [Eds. Solomon, S., Qin, D., Manning, M., Chen, Z., Marquis, M., Averyt, K.B., Tignor, M., and Miller, H.L.] Cambridge University Press, Cambridge, United Kingdom and New York, NY, USA.

Franklin, P. J. (2007). Indoor air quality and respiratory health of children. *Paediatr Respir. Rev, 8*(4), 281–286.

Gauderman ,W. J., Vora, H., McConnell, R., Berhane, K., Gilliland, F., and Thomas, D. (2007). Effect of exposure to traffic on lung development from 10 to 18 years of age: a cohort study. *Lancet, 369* (9561), 571–7.

Gilboa, S.M., Mendola, P., Olshan, A.F., Langlois, P.H., Savitz, D.A., Loomis, D., Herring, A.H., and Fixler, D.E. (2005). Relation between ambient air quality and selected birth defects, seven county study, Texas, 1997–2000. *Am. J. Epidemiol, 162*, 238–52.

Giorgi, F., and Meleux, F. (2007). Modelling the regional effects of climate change on air quality. *Comptes Rendus Geosciences, 339*(11–12), 721–733.

Gupta, I., and Kumar, R. (2006). Trends in particulate matter in four cities in India. *Atmospheric Environment, 40*, 2552–2566.

Gyan, K., Henry, W., Lacaille, S., Laloo, A., Lamsee-Ebanks, C., McKay, S., Antoine, R.M., and Monteil, M.A. (2005). African dust clouds are associated with increased paediatric asthma, accident and emergency admissions on the Caribbean island of Trinidad. *Int. J. Biometeorol, 49,* 371–376.

Hansen, C. A., Barnett, A. G., and Pritchard, G. (2008). The Effect of Ambient Air Pollution During Early Pregnancy on Fetal Ultrasonic Measurements During Mid-Pregnancy. *Environmental Health Perspectives, 116*(3), 362–369.

Hansen, J., and Sato, M. (2004). Greenhouse gas growth rates. *Proc. Natl. Acad. Sci. USA, 101*(46), 16109–16114.

Hedegaard, G.B., Brandt, J., Christensen, J.H., Frohn, L.M., Geels, C., Hansen, K.M., and Stendel, M. (2008). Impacts of climate change on air pollution levels in the Northern Hemisphere with special focus on Europe and the Arctic. *Atmos. Chem. Phys. Discuss, 8,* 1757–1831.

Hegerl, G.C., Zwiers, F.W., Braconnot, P., Gillett, N.P., Luo, Y., Marengo Orsini, J.A., Nicholls, N., Penner, J.E., and Stott, P.A. (2007). Understanding and attributing climate change. In: Climate Change 2007: The Physical Science Basis. Contribution of Working Group I to the Fourth Assessment Report of the Intergovernmental Panel on Climate Change. [Eds. Solomon, S., Qin, D., Manning, M., Chen, Z., Marquis, M., Varyt, K.B., Tignor, M., and Miller, H.L.] Cambridge Univeristy Press, Cambrdige, United Kingdom and New York, NY, USA.

Hennessy, K., Fitzharris, B., Bates, B.C., Harvey, N., Howden, S.M., Hughes, L., Salinger, J., and Warrick, R. (2007). Australia and New Zealand. Climate change 2007: Impacts, adaptation and vulnerability. Contribution of Working Group II to the Fourth Assessment Report of the InterGovernmental Panel on Climate Change. [Eds. Parry, M.L., Caniani, O.F., Palutikof, J.P., van der Linden, P.J., and Hanson, C.E.]. Cambridge University Press, Cambridge, UK, 507-540.

Hertz-Picciotto, I. (1998). Environmental epidemiology. In K. Rothman, and S. Greenland, (Eds.). *Modern epidemiology.* Sydney: Lippincott Williams and Wilkins.

Intergovernmental Panel on Climate Change. (2004). 16 years of scientific assessment in support of the climate convention. Available at: http://www.ipcc.ch/pdf/10th-anniversary /anniversary-brochure.pdf

Intergovernmental Panel on Climate Change. (2007). Intergovernmental Panel on Climate Change Fourth Assessment Report. Climate Change 2007: Synthesis Report. Summary for Policy Makers. Available at: http://www.ipcc.ch/pdf/assessment-report/ar4/syr/ ar4_syr_spm.pdf

Ito, K., De Leon, S. F., and Lippmann, M. (2005). Associations between Ozone and Daily Mortality. *Epidemiology, 16*(4), 446–457.

Janssen, N. A., Schwartz, J., Zanobetti, A., and Suh, H. H. (2002). Air conditioning and source-specific particles as modifiers of the effect of PM (10) on hospital admissions for heart and lung disease. *Environ. Health Perspect, 110* (1), 43–9.

Jacobson, M.Z. (2008). On the causal link between carbon dioxide and air pollution mortality. *Geophysical Research Letters, 35*, L03809.

Johnston, F. H., Kavanagh, A. M., Bowman, D. M., and Scott, R. K. (2002). Exposure to bushfire smoke and asthma: an ecological study. *Med. J. Aust, 176*(11), 535–538.

Kiehl, J. T., and Trenberth, K.E. (1997). Earth's annual global mean energy budget. *Bulletin of the American Meteorological Society, 78*(2), 197–208.

Knowlton, K., Rosenthal, J.E., Hogrefe, C., Lynn, B., Gaffin, S., Goldberg, R., Rosenzweig, C., Civerolo, K., Ku, J.Y., and Kinney, P.L. (2004). Assessing ozone-related health impacts under a changing climate. *Environmental Health Perspectives, 112*(15), 1557–1563.

Koistinen, K. J., Hänninen, O., Rotko, T., Edwards, R. D., Moschandreas, D., and Jantunen, M. J. (2001). Behavioral and environmental determinants of personal exposures to PM2.5 in EXPOLIS-Helsinki, Finland. *Atmospheric Environment, 35*(14), 2473–81.

Kousa, A., Monn, C., Rotko, T., Alm, S., Oglesby, L., and Jantunen, M. J. (2001). Personal exposures to NO_2 in the EXPOLIS-study: relation to residential indoor, outdoor and workplace concentrations in Basel, Helsinki and Prague. *Atmospheric Environment, 35*(20), 3405–3412.

Kunzli, N., Kaiser, R., Medina, S., Studnicka, M., Chanel, O., Filliger, P., Herry, M., Horak, F, Jr., Puybonnieux-Texier, V., Quenel, P., Schneider, J., Seethaler, R., Vergnaud, J.C., and Sommer, H. (2000). Public health impact of outdoor and traffic-related air pollution: a European assessment. *Lancet, 356*, 795–801.

Künzli, N., Avol, E., Wu, J., Gauderman, W. J., Rappaport, E., Millstein, J., Bennion, J., McConnell, R., Gilliland, F.D., Berhane, K., Lurmann, F., Winer, A., and Peters, J.M. (2006). Health effects of the 2003 Southern California wildfires on children. *Am. J. Respir. Crit. Care Med, 174*(11), 1221–8.

Leiserowitz, A. (2006). Climate change risk perception and policy preferences: The role of affect, imagery, and values. *Climate Change, 77,* 45–72.

Le Treut, H., Sommerville, U., Cubasch, U., Ding, Y., Mauritzen, C., Mokssit, A., Peterson, T., and Prather, M. (2007). Historical overview of climate change. In: Climate Change 2007: The Physical Science Basis. Contribution of Working Group I to the Fourth Assessment Report of the Intergovernmental Panel on Climate Change. [Eds. Solomon, S., Qin, D., Manning, M., Chen, Z., Marguis, M., Vaeryt, K.B., Tignor, M., and and Miller, H.L.]. Cambridge University Press, Cambridge, United Kingdom and New York, NY, USA.

Lucas, C., Hennessy, Mills, G., and Bathols, J. (2007). Bushfire weather in Southeast Australia: Recent trends and projected climate change impacts. Consultancy Report prepared for The Climate Institute of Australia. Commonwealth Scientific and Industrial Research Organisation (CSIRO).

Maynard, R. (2004). Key airborne pollutants--the impact on health. *Sci Total Environ, 334-335*, 9–13.

McGowan, J. A., Hider, R. N., Chacko, E., and Town, G. I. (2002). Particulate air pollution and hospital admissions in Christchurch, New Zealand. *Aust. N. Z. J. Public Health, 26*(1), 23–9.

Meehl, G.A., Stocker, T.F., Collins, W.D., Friedlingston, P., Gaye, A.T., Gregory, J.M., Kitoh, A., Knutti, R., Murphy, JM., Noda, A., and Raper, S.C.B. (2007). Global Climate Projections. In: Climate Change 2007: The Physical Science Basis. Contribution of Working Group I to the Fourth Assessment Report of the Intergovernmental Panel on Climate Change. [Eds. Solomon, S., Qin, D., Manning, M., Chen, Z., Marguis, M., Vaeryt, K.B., Tignor, M., and Miller, H.L.]. Cambridge University Press, Cambridge, United Kingdom and New York, NY, USA.

Meleux, F., Solmon, F., and Giorgi, F. (2007). Increase in summer European ozone amounts due to climate change. *Atmospheric Environment, 41*(35), 7577–7587.

Mickley, L.J., Jacob, D.J., Field, B.D., and Rind, D. (2004). Effects of future climate change on regional air pollution episodes in the United States. *Geophysical Research Letters*, *31,* 1–4.

Moore, D., Copes, R., Fisk, R., Joy, R., Chan, K., and Brauer, M. (2006). Population health effects of air quality changes due to forest fires in British Columbia in 2003: Estimates from physician-visit billing data. *Can. J. Public Health, 97*(2), 105–8.

Morgan, G., Sheppeard, V., Khalaj, B., Ayyar, A., Lincoln, D., Lumley, T., Jalaludin, J., Beard, J., and Corbett, S. (2006). The effects of bushfire smoke on daily mortality and hospital admissions in a major city. *Epidemiology, 17*(6), S160.

Mott, J. A., Mannino, D. M., Alverson, C. J., Kiyu, A., Hashim, J., Lee, T., Falter, K., and Redd, S.C. (2005). Cardiorespiratory hospitalizations associated with smoke exposure during the 1997, Southeast Asian forest fires. *Int. J. Hyg. Environ. Health, 208*(1–2), 75–85.

Murray, F., and McGranahan, G. (2003). Air pollution and health in developing countries - the context. *Air pollution and Health in rapidly developing countries.* London: Earthscan publications.

Nawrot, T. S., Torfs, R., Fierens, F., De Henauw, S., Hoet, P. H., Van Kersschaever, G., De Backer, G., and Nemery, B. (2007). Stronger associations between daily mortality and fine particulate air pollution in summer than in winter: Evidence from a heavily polluted region in Western Europe. *Journal of Epidemiology and Community Health, 61*(2), 146–149.

Ostro, B. (2004). Outdoor air pollution: Assessing the environmental burden of disease at national and local levels. Geneva, World Health Organisation (WHO Environmental Burden of Diseases Series, No. 5).

Parker, J.D., Woodruff, T.J., Basu, R., and Schoendorf, K.C. (2005). Air pollution and birth weight among term infants in California. *Pediatrics, 115,* 121–8.

Poumadère, M., Mays, C., Le Mer, S., and Blong, R. (2005). The 2003 heat wave in France: Dangerous climate change here and now. *Risk Analysis, 25*(6), 1483–1494.

Prospero, J.M. (2006). Saharan dust impacts and climate change. *Oceanography, 19*(2), 60–61.

Ren, C., Williams, G. M., Morawska, L., Mengersen, K., and Tong, S. (2008). Ozone modifies associations between temperature and cardiovascular mortality: analysis of the NMMAPS data. *Occup. Environ. Med, 65*(4), 255–60.

Ren, C., Williams, G. M., and Tong, S. (2006). Does particulate matter modify the association between temperature and cardiorespiratory diseases? *Environ. Health Perspect, 114*(11), 1690–1696.

Ritz, B., Yu, F., Fruin, S., Chapa, G., Shaw, G.M., and Harris, J.A. (2002). Ambient air pollution and risk of birth defects in Southern California. *Am J Epidemiol. 155*, 17–25.

Rosenfeld, D., Lohmann, U., Raga, G.B., O'Dowd, C.D., Kulmala, M., Fuzzi, S., Reissell, A., and Andreae, M.O. (2008). Flood or Drought: How Do Aerosols Affect Precipitation? *Science, 321*(5894), 1309–13.

Sunnu, A., Afeti, G., and Resch, F. (2008). A long-term experimental study of the Saharan dust presence in West Africa. *Atmospheric Research, 87,* 13–26.

U.S. Environmental Protection Agency. (2008a). Latest findings on National Air Quality. U.S. Environmental Protection Agency. Office of Air Aquality Planning and Standards, Air Quality Assessment Division, Research Triangle Park, North Carolina. Available at: http://www.epa.gov/air/airtrends/2007/

U.S. Environmental Protection Agency. (2008b). Inventory of U.S. Greenhouse gas emissions and sinks: 1990-2006. Public Review Draft. Available at: http://www.epa.gov/globalwarming/publications/emissions.

Wang, Z., Dong, Z., Zhang, J., and Liu, L. (2004). Modern dust storms in China: An overview. *Journal of Arid Environments, 58,* 559–574.

Wilhelm, M., and Ritz, B. (2005). Local Variations in CO and particulate air pollution and adverse birth outcomes in Los Angeles County, California, USA. *Environmental Health Perspectives, 113,* 1212–21.

Yang, B., Brauning, A., Zhang, Z., Dong, Z., and Esper, J. (2007). Dust storm frequency and its relation to climate changes in Northern China during the past 1000 years. *Atmospheric Environment, 41,* 9288–9299.

Contributed by Barbara Crawford, Ipswich.

Photo: On a clear day, you can see Paradise; but don't swim in the water.

In: Meltdown: Climate Change, Natural Disasters…
Editor: Kathryn Gow
ISBN 978-1-60876-153-1

Chapter 10

CAN WE ANTICIPATE MORE HEATWAVES, WILDFIRES AND DELUGES?

Kathryn Gow
Queensland University of Technology, Australia

ABSTRACT

In this chapter, the author has chosen to highlight some of the natural disaster events that might, or might not, be proven to be linked with climate change. For more than a decade, the media has brought to our attention what appears to be a huge number of calamities including natural disasters that have occurred in different regions of the world; these events do seem to be occurring more often than in the previous 20 years and in some cases more than in the previous 50-60 years. If we read the scientific reports about global warming, the news for the near and not too distant future is nothing short of alarming. With increased publicity about global warming, more often than not the general population attributes these increased natural disasters to climate change. To what extent this is an accurate deduction is not yet clear. What is certain, however, is that we must ensure that preparedness plans are put into place for communities to cope with the different types of disaster events that visit the shores of specific countries or regions.

Keywords: climate change, global warming, heatwaves, wildfires, bushfires, deluges, floods

INTRODUCTION

The wealth of reports now available on the whole gamut of issues linked with climate change and associated issues (not specifically aligned to the climate change debate) that have been of concern for many decades, such as pollution, ozone layer holes, deforestation, droughts, wildfires, cyclones, hurricanes, deluges and floods, etcetera, is both educative and confusing. The confusion occurs because there is considerable disagreement, not about the events and issues, but about the causes of them. There is some consensus about some of the effects of climate change in terms of temperature increases and changes in precipitation and sea level rises. Some of those effects will be witnessed as catastrophic events such as deluges leading to mega floods, and much higher temperatures resulting in more heatwaves, more severe and frequent droughts, and fiercer wildfires.

Without normal precipitation and water availability, there will be shortages in potable water supplies and decreased sanitation effectiveness, decreases in food and stock supplies from the agricultural areas affected, and increases in water borne diseases. There is less consensus about whether or not climate change will cause more natural disasters such as tornadoes, cyclones, earthquakes, volcanic eruptions and minor floods; nor does climate change take the blame for the normal cyclical processes of droughts and floods, wildfires and heatwaves.

While people in the western world may have been able to ignore the perils that have befallen Africa and Asia and the ocean islands, as millions of people have been drowned, made homeless, starved, and been forced to migrate elsewhere, more recently Mother Nature seems to be behaving in a more equalizing manner in Europe, America, Australia, and parts of Asia, as all have been beset by different types of disasters; some wreaked by nature in gargantuan proportions (e.g., tsunamis) and others instituted by mankind's lack of foresight and due diligence (e.g., water shortages).

After a general discussion about what the possibilities are, with respect to what may lie head in terms of natural disasters, the chapter presents firstly the heat issues; there are few who would argue that global warming will not result in more wildfires, more severe droughts (although many farmers may well dispute the greater frequency arguments)[1] and more heatwaves, with resultant deaths in vulnerable members of the community. Then we address the effects of over bountiful precipitation: deluges and floods.

Climate Variability or Climate Change

At the 1992 Rio Earth Summit, climate change was defined as "a change of climate which is attributed directly or indirectly to human activity that alters the composition of the global atmosphere and which is in addition to natural climate variability observed over comparable time periods'." Climate Change has become a major international issue and a serious economic factor. Bearing in mind the droughts, the natural disasters and the global monetary issues, consider the 700 page report by Nicolas Stern who is a senior economist for the British Treasury. Stern said that countries would need to spend 5% to 20% of their GNP to deal with the consequences of Climate Change.

One of the biggest arguments surrounding issues relating to climate change, including carbon emission reductions, is that of causation. In 1980, Ramanathan (in Weart, 2003-2008)[2] had hinted that there might be a host of other greenhouse gases that could be contributing about 40% of the total warming due to CO_2. Then Ramanathan and others, in 1985, isolated 30 gases that "added together could bring as much global warming as CO_2 itself" (p. 5). Each methane molecule had more than 20 times the impact of that of a CO_2 molecule. Ramanathan

[1] The effects of drought are covered in Chapter 16.

[2] Weart, S. 2008. Other greenhouse gasses. DGW 6/08, p. 5. http://www.aip.org/history/climate.

predicted that the warming, unprecedented in human history, would make its presence known before 2010. In Australia, we think that had already happened by 2008.

While the gases debate continued, others pursued the influence of solar activity on climate and eventually they were convinced that "minor solar variations could indeed have been partly responsible for some past fluctuations.....but future warming from the rise in greenhouse gases would far outweigh any solar effects" (Weart, 2008, p. 1). [3]

Are we in fact talking about climate change on an escalating scale, or are we talking about normal climate variability over short cycles (11 years) or long cycles (50-60 years) or very long cycles (centuries)? These are the questions that need to be answered before the general population can even begin to understand the quandary that we may be in.

Most primary producers will argue that they work and live with climate variability, and that the majority of extreme events, such as droughts and floods are due to cyclical variations that can be traced back over time, even across centuries. For them, the debate about climate change lacks adequate analysis of normal cycles. It was J. Murray Mitchell Jr. in 1961 (in Weart, 2008)[4] who declared that the enigma observed with the cooling and warming cycles might be indicative of decades-long rhythms.

Even the World Health Organisation (WHO), in 2003, cautioned about attributing extreme events, such as floods, to climate change. They are not alone. According to Mitchell (2002), "it is difficult to actually quantify the role of climate change in the occurrences of extreme events" (p. 16), because of the lack of climate statistics over the longer term. However, Mitchell says that in recent times, it seems that changes in climate variability might be having adverse effects with respect to drought and flood hazards.

In another essay on "The Public and Climate'[5], Weart (2007) highlighted the scientific findings of the predictions based on the glacial cycles, that CO_2 atmospheric content and temperature covaried. The World Meteorological Organization reported that 2003 and 2004 were the 4th and 5th (respectively) warmest years on record since 1861, with the warmest year remaining as 1998. Pittock (2005) lists the indirect evidence supporting these findings as "tree rings, ice cores, boreholes, and other climate-sensitive indicators" (p. 3). These findings support the conclusion that the last part of the 20th century was the hottest in 1000 years – to substantiate this conclusion, he graphs figures from "thermometers, tree rings, corals, ice cores and historical records." (p. 3).

Whether volcanic eruptions influence neighbouring countries, and even the whole world, to become colder for a short period of time, has been a question asked since observations were made after the 1815 and 1883 eruptions of Krakatoa in the East Indies. Records revealed that the earlier 1815 eruption of Tambora, Indonesia, had affected the whole climate of the world more than had the Krakatoa eruption in 1883. Apparently crops were frozen in New England. Murray in 1969 (in Weart, see footnote 7) maintained that the cooling seen in the northern hemisphere was due to a few volcanic eruptions. In 1991, when Mt Pinatubo erupted in the Philippines, 14-26 million tones of sulfur dioxide led to a global surface cooling of 0.5°C a year after the eruption.[6]

[3] Weart, S. and Feldman, T.S. 2008. Changing Sun, Changing Climate? DGW 6/08, p. 1. http://www.aip.org/history/climate.

[4] Weart, S. June 2008. Aerosols: Volcanoes, dust, clouds and climate. DGW 6/08 Aerosols, p. 2. http://www.aip.org/history/climate.

[5] Weart, S. 2007. The Public and Climate. DGW 7/07, p. 27. http://www.aip.org/history/climate.

[6] http://www.gaspig.com/volcano.htm.

Weart faces the cooling or warming debate objectively and points to the need for more research into how aerosols block sunlight. Back at the research station, Ramanathan's group (2008) was working on the influence of black carbon aerosols, especially their combined effect on global warming with sulfate aerosols. The SO_2 converts to sulphuric acid aerosol in the stratosphere, but the cooling effect can take months to be felt and then the full effect may not occur until up to a year later; then it may take up to seven years before all the cooling effects disappear.

Weart (2008) tracks back the warming and cooling trends, with the populations in the 1930s experiencing a pronounced warming trend which by the 1960s had shifted to cooling. Many Scientists then, he explained, thought that the cooling would continue, regardless of the cause; others thought the effects were temporary and human actions would lead back to warming over time. Edwards and Weart go to lengths to explain that, even with the best climate models, modelers cannot account for the whole picture of what happens in nature and within the atmosphere in particular.[7]

It seems obvious that if the temperature rises, then the ice melts, and thus both the Artic and Antarctic have recorded higher temperatures and melting ice sheets. The IPCC (2007) warns that if the temperature keeps rising as predicted, then ice shelves will melt further and sea levels will rise, leading to the inundation of low lying islands and coastal areas in many countries.

Global Sea level rise was projected, by the IPCC (2007), to be 18-59cm by 2100, and if there were to be further snow melts, then this could increase by 10-20cm. Wang Shourong (2009) informs us that sea level rise would mean that the Bohai Sea would rise by 10-12cm and the Yellow Sea by 13-14cm, thus flowing into the Yantze River and the Pearl River. Floods following cyclones and typhoons are expected to increase by 40%-50%.

Unfortunately, for the east coast of Australia, the sea rise may be greater than the global average, as was its temperature increase over the past century. The Office of Climate change (OCC) (2008) also states that there is high confidence in the rate of sea level rises, based on the period from mid 19th to mid 20th Centuries (p. 20).

Bates and colleagues (IPCC, 2008) noted that changes in radiative forcing (that is, combined anthropogenic, volcanic and solar sources) had a role in trends in precipitation. They also pointed out that there had been an increase in heavy precipitation events, even when total amounts of rain had decreased; this occurred in Australia in the past 10 years. Their final call was that there would be an increase in heavy precipitation events and this trend is more likely than not to continue.

With respect to cyclones, the IPCC team (Bates et al., 2008) declared that "there is no clear trend in the annual number of tropical cyclones (p. 18), but the increase in intense storms seems much larger than anticipated using current models.[8] The Office of Climate change (2008, p. 10) state that, in Australia, the "number of topical cyclones in the Queensland region has declined since the 1970s, largely due to the increased incidence of El Niño, but there has been an increase in the number of severe tropical cyclones (CSIRO et al., 2007)". A 2004 study by Hardy et al., (cited in OCC, 2008) showed that along the east coast of Queensland, cyclones will increase by 10% and cyclones will shift south by 130 klm. The

[7] Weart, S. and Edwards, P. 2008. General Circulation Models of Climate. DGW 6/08, p. 1. http://www.aip.org/history/climate.

[8] "A climate model is a mathematical representation of the Earth's climate system", p. 11. Climate change in Queensland. What the science is telling us. The Office of Climate change (2008). Brisbane: Author/ EPA.

WMO (2006) and the IPCC (2007) reports also indicated that tropical storms and hurricanes would most likely become more intense.[9, 10]

Based on a range of models, they predict that future tropical cyclones will become more intense, with larger peak wind speeds and more heavy precipitation associated with ongoing increase in tropical sea surface temperatures" (OCC, 2008, p. 27).

Weart (2008) documented the milestones in the science of climate change since 1800 to 2007 and notes that in 2005, debates occurred about whether or not Hurricane Katrina and other severe storms had been influenced by global warming.

It is expected that hail storm risk will increase in South East Queensland by up to two days per year by 2030 and by up to four days by 2070; seemingly not a great increase, but important in terms of the widespread damage that hail storms do to vehicles, buildings, and crops, with resultant increased insurance levies for all and major payouts by insurance companies.

The OCC (2008) state that "it is very *likely* that heavy precipitation events will become more frequent" (p. 26). Interestingly, they do affirm that along with the drying, intense precipitation and flooding is likely. The rain will tend to be more concentrated in intense events and there will be high runoffs, but between falls, there will be longer periods of time with lower precipitation occurrences that will lead to droughts. The multi-model climate projections for the 21stC now show "increases in both precipitation intensity and number of consecutive dry days in many regions" (OCC, 2008, p. 26).

Whether or not there is less rainfall, there will be increased evapotranspiration which will increase the aridity, and the severity of droughts, in Australia. For the IPCC to announce to the world that most of the world will be desert by 2100, and then to find that Lovelock really believes the date will be 2040 is indeed frightening. What we need to know is how much of the world is already desert today and how much was desert 30 years ago; otherwise news reports and even government reports are misleading. What tends to happen in the public's mind is that they suspect that they are being duped once again and a conspiracy theory mentality takes hold which prevents true analysis of the facts. Desertification is a most serious scenario. It is defined as: "The processes by which an already arid area becomes even more barren, less capable of retaining vegetation, and progressing towards becoming a desert" (UNDHA, *DM Glossary*, 1992, p. 26).

The EPA (1997) had indicated earlier that there would be an increase in the incidence of cholera and food poisoning from toxic algae following droughts. As indicated in Chapter 16, lack of water is a major factor in climate change scenarios leading to disasters in lack of food, less variety in types of food available, famines, mass migration, poverty and destruction of towns and cities, even in first world countries such as Australia.

Other concerns about the loss of coral reefs, and flooding in the Mekong and Red Rivers, were all warnings for 2050, long after half of the current population would be deceased. However, the glaciers in Latin America with the melting of the Glaciers in the Andes, were to disappear in 15 years, with resultant lack of water for the countries of Peru, Colombia, Ecuador and Bolivia. Additionally, without normal rain, three other countries – Brazil, Chile

---

9 http://www/epa.gov/climatechange/science/futurepsc.html.

10 WMO. 2006. Statement on Tropical Cyclones and Climate Change. http://www.epa.gov/climatechange/downloads/GFDhighlight_Vol1n5.pdf.

and Argentina - would suffer from severe water shortages. Even as early as 2020, up to 77 million people might suffer from lack of water. [11]

The draft UN climate report released in April 2007 indicated that the ecosystems in Africa would experience dramatic changes with possible extinction of some species. In the polar regions, the loss of habitats will result in problems for migratory birds and mammals, and seals and polar bears will be in trouble because of this. According to the IPCC, wildlife will be devastated, because with just a 3˚C rise, up to 30% of all species would die and with a further 5˚C rise, then 70% would die. These are statistics that chill wildlife carers to the bone and haunt the sleep of conservationists.

Lin Erda, Xu Yinlong, Wu Shaohong, Ju Hui and Ma Shiming (2007) warn that warm, dry weather would mean reduced wetland resources, a loss of biodiversity, and more endangered species, while large swamp areas in the Sanjiang Plain of China would turn into grass land. If the sea level rises, then this would degrade and reduce wetland areas and submerge/erode tidal flat land in the Yangtze River Delta.

There will also be more disease in wildlife, and pests that could affect humans such as tick borne encephalitis; already new microbes fungi plants and animals are causing changes to wildlife on the sub-Antarctic islands.[12]

Timothy Ball (2008), a climatologist, is a disbeliever in global warming. He bluntly states that "climate change is not due to human contribution of Carbon Dioxide (CO_2) and they are within natural variability".[13] Apparently the CO_2 levels have already been this high before (in 1936 to 1944) pre and post World War II; before that, high levels were recorded in 1885, 1866, 1857 and 1858.

Some sections of the general community are adamant that "Global warming is a Fraud"[14]. So is it anthropogenic or not? Are these climate changes "resulting from or produced by human activities, in particular, factors that affect the atmosphere due to the burning of fossil fuels, deforestation and other land use change"? (The Office of Climate Change, 2008, p. 30).[15]

That is a good question to end this section on, as the author believes that while the disagreements about what is causing the climate variability is useful and, in terms of mitigation, is essential; not knowing the causes does not 'let us all off the hook'. There are serious problems that need addressing and perhaps bundling them up under that giant umbrella of climate change may be distracting us all from proceeding at a rapid pace to attend to preparedness for, and mitigation and/or adaptation to, each of the natural disasters. Further legislation is necessary to strengthen pollution controls in countries which are still damaging the air in their own countries and that of others, as we all end up breathing the same air. We must take no more excuses about clearing our remaining forests, the plundering of the oceans for marine life, the killing of our native animals and the annihilation of their habitats.

---

[11] The drought and water shortages issues have been taken up in Chapter 16.

[12] Nature's vulnerability has been addressed in chapters 2, 11 and 23, and health issues are covered in chapters 7 and 9.

[13] Timothy Ball. 2007. Global warming: the cold, hard facts? 05.02.07. http://www.canadafreepress.com/2007/global-warming020507.htm.

[14] The New Citizen. (2007). Global warming is a Fraud. 6 (2), p.1. Newspaper of the Citizens Electoral Council of Australia. www.cecaust.com.au.

[15] The Office of Climate change. (2008). Climate change in Queensland. What the science is telling us. Brisbane: Author/EPA.

In the meantime, we need to be able to deal with the heat. There is no doubt that the world has been suffering during a decade of higher temperatures; the next section discusses the factors relating to temperature increases and heatwaves.

HIGHER TEMPERATURES AND HEATWAVES

By 2007, Weart's (2008) timeline demonstrated that global warming was being taken very seriously, as the mean global temperature (taking a five year average) of 14.5˚C was the warmest in hundreds and maybe thousands of year.

As Barnett and Hansen (2009)[16] summarized: "over the past 100 years the average global temperature has increased by about 0.74°C and this increase is greater at higher northern latitudes. Also, 11 of the last 12 years (1995–2006) have been the warmest since 1850 (IPCC, 2007)". The IPCC (2007) has warned that there will be more severe heatwaves with climate change. As indicated in Barnett and Hansen's chapter, the IPCC (2007) has concluded that: "It is very likely that heatwaves will be more intense, more frequent and longer lasting in a future warmer climate. Cold episodes are projected to decrease significantly in a future warmer climate. Almost everywhere, daily minimum temperatures are projected to increase faster than daily maximum temperatures, leading to a decrease in diurnal temperature range".

A heatwave is defined as a "long lasting period with extremely high surface temperature" (EMA Glossary, 1998, p. 95).[17] Distress from heatwaves can be measured via the heat stress index - a four-step index which is "based on the THI (Temperature Humidity Index) or 'humiture' index, expressing the likelihood of heat stroke, sunstroke, or other acute symptoms of bodily stress" (EMA Glossary, 1998, p. 108).[18] However, it depends on where you live in the world, as to whether or not you consider that the hot sweltering weather, when water pours from the body and mineral salts are depleted, is actually a heatwave or just normal weather for that time of year. People who live in the tropics such as Darwin, Cairns, New Guinea, Indonesia and some other Asian countries and Pacific islands expect this type of weather, and it is only after having enjoyed a few years of cooler weather, when the swelter returns, that people start to complain again of heatstroke and heat exhaustion.

As people age, they tend to dislike extremes of weather, both hot and cold, and heatwaves become a fount of conversations. The dangers for young and old lie in their living conditions, and whether or not they take in enough liquid to keep themselves from being dehydrated; the statistics demonstrate that those who are dependent on others to tend to their needs, may dehydrate from lack of fluids, as their carers also struggle to keep operating in the hot and often humid conditions.

This heat stress, however, is a cause for concern with the scenarios about temperatures increasing world wide in the coming years. If in countries considered to be of cooler climes, people are increasingly dying from more frequent heatwaves, then what greater health risks are there for those in countries which already have hot climates and now face greater temperature increases (as proven over the past century) and with greater temperature rises being predicted in the next 1-8 decades?

---

[16] See Chapter 9 by Barnett and Hansen in this text.

[17] EMA Glossary. (1998). Dickson, ACT.: Emergency Management Australia.

[18] EMA Glossary. (1998). Dickson, ACT.: Emergency Management Australia.

Counting heatwaves as a natural disaster, then severe heatwaves tend to cause a lot more deaths than some other natural disasters. However, it seems to be an almost invisible natural disaster, similar in its introverted nature to drought, as there are no dramatic extraverted demonstrations of nature's fury and might, just a wearing down of people's internal physiological resources to cope with the stressors of heat, with resultant dehydration and delirium.

The draft United Nations climate report, released in April 2007, pointed out that amongst the older generations, heatwaves will result in an extra 3200 to 5200 deaths per year by 2050. Surprisingly to those of us who live in the southern hemispheres, hearing about heatwaves in parts of Europe which have much milder climates is confronting.

Disease Increases from Heatwaves and Temperature Increases

It is common to hear about the increase of diseases following the impact of natural disasters on areas around the world, whether it be via floods, earthquakes, wildfires, etcetera, because the infrastructure breaks down and water and sanitation services fail. However, greater heat hastens droughts when rainfall is scarce and leads to a great range of pest invasions, weed infestations, dust storms, and mental health problems, with the heatwaves being experienced as being more overpowering than in non-drought periods. [19]

Thus it is not just the heat and dehydration alone that are going to be problematic, as the world's temperature increases. The World Health Organisation (WHO) concurs that the number of disease carrying pathogens will increase with higher temperatures. The diseases that the WHO (2003) has marked as being of most concern, if the weather heats up in some parts of the world (particularly in tropical and subtropical regions), are the vector borne infections which they foreshadow will increase substantially: Malaria and Dengue fever, along with food-borne infections such as salmonellosis which are always a problem in warmer months of the year.

It has been estimated that by the year 2050, from 0.6 million to 1.4 million people (out of a current total Australian population of 22 million) will be exposed to dengue fever. Such outbreaks are known to occur normally every four years or so in countries like Vietnam, but the presence of dengue in Australia has not been high before in the past.[20] However, in the summer of 2008/2009, dengue fever outbreaks in Queensland caused alarm along the north to mid coast areas; the outbreak had public health officials very concerned, as the mosquito responsible for the illness was found further south and a major public health eradication program was instituted in the affected areas.

In fact, according to their 2002 World Health Report, the WHO had estimated that, in 2000, climate change (unspecified) had led to an increase worldwide in diarrhea of 2.4%, and in malaria of 6% (although they indicated the research focused on middle income countries). The relevance of the statistics being obtained on middle income countries is that the World Health Organisation (2003) predicts that it will be the poorer countries that will suffer significantly more than middle income countries, because of their lack of infrastructure,

[19] See Chapter 16 for issues relating to drought.

[20] Health issues have been taken up in Chapters 7 and 9.

poorer immunity and general health, lack of clean water and sanitation, as well as other factors.

It is chastening to be advised that even small temperature rises can increase the transmission potential of malaria; for example, it would take just a 2%-3% increase in temperature in any part of the world, to increase the disease risk to several hundred million people. Additionally, for those countries which are already besieged by malaria-carrying mosquitoes, their seasonal duration would increase (WHO, 2003, p. 17, Summary).

Hence, it is not just frightening diseases that enter the scene, such as Ebola,[21] that should scare us the most. McMichael et al. (2006) have predicted that climate change will have adverse effects on health leading to thermal stress injury from weather events, infectious/vector borne diseases as the ecosystems change, poor nutrition as a result of impaired agricultural yields following sea-level rises, and poverty and consequent ill health following the displacement of people after environmental degradation.

What is not commonly understood within the general population is that, as our ecosystems are interfered with, the delicate balance in nature between predators, such as mosquitoes and vermin, is weakened and because the ecosystems no longer support the volume of predators as they did in the past, there are not enough control mechanisms in nature to combat the pests, many of which are disease carrying. So yes, this is something that we need to take swift action on, or nature will "sting us back". The World Health Organization (WHO) (2003) carefully outlines what it means by changes in nature's systems and how that can affect our health - see their Table 6.1 on page 17 in the Summary).

Higher temperatures do increase the risk of wildfires (bushfires) in many parts of the world and the next section concentrates on some of the occurrences of bushfires in the past decade.

HIGHER TEMPERATURES AND BUSHFIRES

As indicated in Chapter 13, as the temperature increases and the rain stays away or falls less frequently, and the environment remains affected by drought, it is inevitable that bushfires will increase in intensity and occurrences will become more ferocious. Severe drought, on its own, does not automatically translate into major fires, says Ronald Nielsen, a bio climatologist with the US Forest service.

According to the FAO (Food and Agriculture Organization of the United Nations) (2008), in 1997 and 1998, millions of hectares burned in central America, Mexico, the Amazon basin and Southeast Asia, leading to major smoke hazards,[22] and of great concern was the damage to ecosystems not previously harmed. In 1998-1999, 115,000 fires and in the 1999-2000 fire season, 230,000 fires per year were registered through satellite remote sensing[23] in Australia. In the decade ending 1999, on average 6322 hectares of forest in New Zealand had burned each year.

[21] Scientists hope they have developed a successful vaccine (Clarke and Knight, 2003).

[22] http://www/fao.irg/docrep/004/y1997w/y1997e0d.htm.

[23] Sentinel, is an internet bushfire monitoring system; it obtains data from Nasa Earth Observation Satellites -Terra and Aqua. The information is then processed and thermal images of hotspots are visible on screen.

In 2000, major fires burned in the eastern Mediterranean, western USA and Ethiopia; 200 million hectares burned in Africa in the area south of the equator (FAO, 2008). Twice as many fires burned this decade in the Mediterranean than occurred during the 1970s. But in 1998, when many other countries were experiencing fire hazards, Mexico registered 14,445 wildfires that covered 849,632 hectares, which was the largest area ever recorded in just one fire season. In 2000, drought also exacerbated the conditions in western USA and 2.5 million hectares of forests and grassland burned and only the fall rains quenched these fires (FAO, 2008). In 1997/8, the fires in Southeast Asia destroyed forests and wildlife resources and released greenhouse gases that affected the biodiversity of Borneo and Sumatra (Eaton and Radojevic, 2001).

However, the report does not hint at climate change causes, rather it highlights other factors such as policy, creeping urban fringes, use of fire in land use practices, and lowered economic viability of residents. In relation to analysing what started the fires, 50%-90% of wildfires were attributed to escaped agricultural burns or land clearing (FAO, 2008). In Australia, most bushfires were started by humans deliberately or accidentally, but lightning causes many fires in rural areas. In Africa, lightning was the primary ignition source, but now more recently human-caused fires has increased. The wildfires in Asia in 1997-1998 affected the health of over 100 million people, with the causes there being human induced and prescribed burns (FAO, 2008).

Australia is the most arid country in the world and is bushfire prone. Along with southern California and southern France, Australia is one of the three top fire prone areas in the world (Pritchard and Gow, 2008). As indicated earlier, one of the consequences of drought is that there is little or no available water to fight bushfires with, as occurred in the severe bushfires in Australia during the period 2001-2006. Coupled with high winds, continued absence of rain, and high temperatures, the landscapes exploded and grass fires rolled along for many miles unstopped. Where there should have been creeks, dams and waterways to quench the thirst of the fires, there were no longer any, and additionally when fire-fighters looked for water, there was none available. Water trucks travelled long distances to assist the fire-fighters fighters cope with the worst seasons of forest and grass fires that they had ever endured.

Over the 2002-2003 fire season, 1,456,000 hectares were burned in New South Wales alone. This was a time of serious bushfires (although the 2005/6 summer season proved worse). In that same period, 166,500 plus 244,800 hectares were fuel reduction burns. The amount due to wildfires was 934,600 plus 1,575,000, with 2003 obviously being the worst year on record for New South Wales. Over all, across Australia, the bushfires peaked dramatically between 2002 and 2005. In the 2002/03 bushfire season, across the whole of Australia, more than 67 million hectares of forest and grass burned. In terms of Australian greenhouse emissions, this has added greatly to our carbon emissions debt.

On February 2009, the highest temperatures in living memory hit the forested state of Victoria on the weekend beginning 07.02.09 and wildfires raged for weeks. During the first few days, firestorms cut a path of destruction and desolation through many towns in different parts of Victoria, leaving behind: more than 170 residents dead, over 600,000 hectares burned black, more than 1800 homes reduced to ash, millions of forest trees stripped to black sticks and thousands of farm and domestic animals killed, and tens of thousands of wild animals, birds, monotremes burned or injured and many thousands of fish dead.

This was no ordinary occurrence, and when compared to their two previous worst bushfire disasters, after the shock wore off, questions were asked: how many of the fires were due to arsonists, was there too much emphasis on growing trees which were too close to homes; was there not enough back burning, should the policy of 'to go or stay' change?

Early in the media reports, while little mention was made about the drought conditions, or the dryness, or the nature of the fire and the atmospheric conditions which brought them about, everyone knew about the high temperatures which had left many people in hospital suffering from heat waves effects. The day the fires started (Saturday February 7th 2009), the temperatures reached 48°C in some places- a temperature that had never been recorded in that state before. The fuel load on the ground was massive. The role of the speed and force of the winds was factored in, but nothing before, even in hot and dry Australia, had prepared them for the firestorms and the meltdown.

People could not leave their homes after the fire storms and fire balls were jettisoned across the land. In the post analyses, fuel loads were flagged; high winds were tagged; and survivors told frightening stories that shocked the nation - there was no warning. If they had seen the fire, they had been 20 kilometers away, and even then they had little time to escape, but for most of the victims who survived the inferno, the first they knew of it was the roar or deafening crackling of the fire; decisions had to be instant; get in the car and go now, or find what might be a safe place. Many perished in their cars, on foot, or in their houses and fields. Only a couple of families had bunkers that could withstand the onslaught of hellfire.

Dryness, the new word put into use by Peter Kenny of the 2008 drought report, took on new meaning in this catastrophe; farmers lost their properties, houses, outbuildings, sheds, fences, animals crops and in some cases, their loved ones or their own lives. A few cows and alpacas, by some miracle, made their way to dams and were saved; domestic cats and dogs disappeared or went missing along with other farm animals; while birds, koalas, kangaroos, possum gliders, lizards, echidnas and other creatures were incinerated; the few that made it to safe places were treated, but some were too damaged to be released and were euthanized. Trees exploded such was the heat; the sounds of the oxygen being sucked from the life in the areas was eerie and terrifying. People who had been badly burned were transported to hospital and others with injuries competed for beds with people from other parts of the state who had been admitted to hospital with heat stroke.

Emergency services were shocked, as were all the people who later reviewed the scenes of destruction (see Gow, 2009). Health professionals stated that "if you looked up in the blackened forest areas, as you drove along, it took some moments to realize that there had been rows of houses along the hillside or down in the valleys or near the gullies" and now "there was literally nothing, but ashes; metal had melted, only an occasional brick fire place still stood in the ashes, or the odd pieces of tin from water tanks lay misshapenly over the ashes". The shocked nation mourned and as one rallied in support and poured out their solidarity and gave unstintingly in cash, materials, and physical assistance

If this is the present, then what of the future in terms of bushfires? The draft United Nations climate report released in April 2007, indicated that in Australia, the frequency of days when bushfires would be at their peak would increase to 25% by 2020. The CSIRO (2007) [24] reported that, based on the projected climate change models, the number of days when fire conditions would be extreme could rise by as much as 300%. Perhaps manageable, if the temperature rises by 1°C, but outside most states' fire fighting resources if it rises by

[24] Tanya Nolan. ABC News. 26.09.07. Report links climate change to future bushfire threat. http://www.abc.net.au/news/stories/2007/09/26/2044282.htm.

3˚C, because then the actual fire intensity across Australia would jump by about 20%. Even within the next 12 years, the number of extreme days of fire danger – that is using the highest risk rating - could increase by more than 60%. Indeed within 18 months of publishing that information, Australia did experience these extremes in Victoria and New South Wales. Only the rain and floods saved Queensland from the same fate.

The CSIRO (Hennessy et al, 2005) conducted a review of risk assessment for bushfires in south east Australia, with the specific focus of ascertaining the level of impact of climate change on bushfire frequency and severity. Using the IPCC scenarios of global warming and the several indices of measurement of fire susceptibility such as the Soil dryness index, the Ketch Byram Drought index (BDI), the forest fire danger index (FFDI) and the grassland fire danger index (GFDI), they used two climate change models to foresee what would happen in terms of fire danger in the years 2020 and 2050. Based on data gathered from 1974 to 2003, in New South Wales, the Australian Capital Territory, Victoria and Tasmania, the most likely scenario is that yes there will be an increased fire danger for 2020 and 2050. The average number of days in those years will have FFDI fire ratings at very high or extreme levels; this represents an additional 3 to 5 days of intense fire fighting weather per year.

Water quenches fires, and thus at all times in bushfire seasons, rain is welcomed to lessen the risk of fire and to quench the flames in the fires that are burning in forests and grasslands, near which, and occasionally in which, many people have chosen to live. However, rainfalls differ in intensity and the next section elaborates on deluges.

Climate Change and Deluges

> What's in a Deluge? When the skies open up and delivers a week's rain in half an hour. Like the inevitability of progress, all before is sent marching, swept before an unstoppable current, gouging out previously solid structures and depositing debris from afar. Progress has led to the inevitability of constant confrontation with nature. Winds that sweep the earth bare, fires that blacken the unbreakable, storms that toss supports around like matchsticks, and water that comes like a steamroller, ready to cleanse the earth of the impurities built upon its skin. South East Queensland in the past months has gone face to face with nature and has not been left unscathed. November 2008 saw Brisbane's inner city suburb of The Gap blasted by a sigh from nature, perhaps shaking a little of the confidence of the interloper, and a warning of what to expect with continued population pressures, variable climatic conditions and the surety of being an Australian.
>
> (Francine Pritchard, 31.03.09)

Defined as "an immense rain or fall of matter from the sky", Alfred de Grazia (2008) says that we should preserve the term deluge for a cataclysm or a downfall, while "a flood tide is a body of water in motion". De Grazia points to the likelihood that frogs and rats also fell from the sky at a distance point in the past, but discussion on that point would take us away from the immediate earthly issues at this stage in time. Deluges are micro bursts of precipitation which originate as super-cells within cumulonimbus clouds; in a supercell, there is only one up and one down draught, leaving the cloud space to grow larger (Giles, 2008).[25] They fall as rain drops that sound and feel like hail or strong objects; thus a person can hear

[25] Bill Giles. 14 August, 2008. http://www.bbc.co uk/weather/features/understanding/deluges.shtml.

the rain drops and immediately interpret the sound as hail; but they are not hail stones, just heavy dense raindrops.

Bill Giles (2008) reports that the rainfall can be as much as 8 inches in an hour. Indeed those figures have been matched in deluges in Queensland in 2009. In the past 10 years, we have seen more and more reports of cyclones, hurricanes, tornadoes and severe thunderstorms, which can lead to serious local and generalised flooding. It is such deluges that can lead to flash floods and the longer they continue, and the greater the areas they cover, the more likely they are to lead to mega floods. The drainage systems in most countries have not been constructed to deal with great volumes of water flowing rapidly over short periods of time, hence flash flooding and general flooding occurs; if the deluges keep up their pace for many hours and days, then mega floods can occur as waterways and rivers become swollen and break their banks.

Deluges Occurring World Wide

Apparently one of the highest records of a deluge that could be accorded to a rainfall event occurred on 26.11.70 in Guadeloupe in the Caribbean Islands. In one minute, 38.1 mm of rain fell. Other extremely dense and heavy rain falls have been recorded in many countries. Dublin experienced such a deluge in August 2008,[26] while India was inundated with the more traditional cyclonic rainfall leading to what have become recognized as more common flood events. Alfred de Grazia (2008) warns that such events are not easily forgotten.

In 2002, when deluges occurred in Asia and Europe, 2000 people were killed in the marauding flood waters. In 2005, Hurricane Katrina wiped out a whole area in the USA and New Orleans, where at one point only 181,000 of the original population of 463000 had returned to live, faced a US$23 billion insurance bill. One of the confounding issues here about the role of the actual deluge is the competing role of the high destructive winds.

In October 2003, Seattle - the city of rain - recorded 5 inches of rain in one day- the second rainiest day on record, and Eric Salathe (2007) from the University of Washington's Climate Change Impacts Group indicates that, with changing climate, these types of events will become more common. Not to be outdone, Tonga received 7 inches in 6 hours on the 15th October, 2008.[27] Other areas were inundated with more than 12 inches. La Niña has once again been given the blame for this wet season, whereas La Niña is very much welcomed in Australia, if she stays well mannered. Unlike Cherrapungi, in northeast India, which is one of the wettest places on earth (average rainfall per year is 10,000 mm)[28], Australia is the most arid and suffers from periodic rainfall deficiency and sometimes very severe droughts.

Recent Deluges in Australia

In Australia, the heaviest rainfall events occurred in late January (the river on the border of the town of Charleville was reported by the 'locals' to be more than a kilometer wide in

[26] http://www.foe.ie/news/2008/08/11/leading-climate-scientist-says-expect-more-deluges-like-last-saturdays/.
[27] Bill Giles. World news. 16.08.08. http://www.bbc.co.uk/weather/world/news/09022008news.shtml.
[28] Bill Giles. http://www.bbc.co.uk/weather/features/az/alphabet12.shtml.

some parts) and in early February 2008, Mackay received 625 mm of rain in six hours.[29] Later, in November 2008, across certain suburbs of Brisbane, storms and deluges lead to serious flooding. The Bureau of Meteorology put out warnings about flooding for 12 rivers in the state of Queensland and northern New South Wales. Hundreds of millions of dollars of damage was recorded and many thousands of people were evacuated.

In November and December 2008, some unusual rain events, coupled with high velocity winds occurred. Suburbs in Brisbane City, which is the capital city of Queensland, were accorded natural disaster status, as houses were ripped apart and trees uprooted, snapped and turned into flying missiles, while deluges roared from the heavens.

If we add the compounding problems of hail and high strong winds that accompany the deluges, we can have a natural disaster that partially destroys a suburb such as The Gap (Brisbane) in just minutes [16.11.08] and leaves a large part of the city and country surrounds without power, and fears that the water from the local reservoir had been contaminated. The winds reached 176 km/hr near The Gap at the peak of the storm.[30] Later that week, other suburbs were deluged causing further flood damage; the Brisbane City tunnel flooded, cutting off this important city underpass for days; and road damage took weeks to be repaired. With the force of bad luck, the new deluges once again harassed an already defeated suburb and residents in The Gap wondered what might befall them next.

November 2008 was not a good month for certain areas. Over 24 hours on 20th November 2008, the highest rainfall totals were at Gatton (250 mm) and Tallegalla Alert near the Bremer River where 243 mm were recorded. In the same time period, Rosewood and Brassall, also on the Bremer River, received 204 mm and 194 mm respectively; at Bundamba, 172 mm fell.

Rosewood, about 90 kms west of the Centre of Brisbane city area, received 222 mm in three hours on 19.11.08. On 20.11.08, before The Gap (which had already been severely damaged by the storm described earlier on 16.11.08) was hit by a destructive microburst,[31] storms pelted the city of Ipswich and surrounding areas and rain over several days added to the swollen Bremer river – with flood warnings being issued daily, as the swell rose and widened.[32] Rail lines were cut after landslips occurred following the heavy rain (150 mm) at Booval, and classrooms, businesses, and sales yards were flooded, with flood waters damaging cars, furniture and equipment. The area had only just been besieged by vicious hail stones the size of golf balls, and in some places the size of tennis balls. Not content with the wild wind, storm, hail and rain that had wrecked the houses in The Gap, the next week nature sent more wild storms back for a third battering of exactly the same, now viciously defoliated, areas.

As in 1996 and 1999, the Lockyer River flooded, along with the Bremer River, and at the same time in November 2008, deluges up to 250 mm inundated the Lockyer Valley[33] causing damage to agriculture, infrastructure and housing. Nearby Gatton had received 312 mm of rain in four days and in one of those days received 250 mm. For the first time in many years,

[29] http://www.disaster.qld.gov.au/publications/pdf/Jan-Jun%202008.pdf.

[30] The Gap, 16-20th November 2008. http://www.bom.gov.au/weather/qld/sevwx/ts/20081120/qldts20081120.pdf.

[31] Wikipedia: a microburst is "a very localized column of sinking air, producing damaging divergent and straight-line winds at the surface that are similar to but distinguishable from tornadoes which generally have convergent damage".

[32] The last time the rivers had risen to that height had been 1996 and 1999, although 1974 was the worst last Century.

[33] [Editor's note: This is one of the areas of research interest outlined in Chapter 12 on backbuilding.]

the gates at Somerset Dam were opened to send waters downstream at a rate of 20,000 mm a day.

Further east, Enoggera Creek flooded and Enoggera Dam actually overflowed upstream because of the extreme rain fall. When the storm rains peaked, the Bureau of Meteorology indicated that the dam spillway to Enoggera Dam rose to 60,000 litres a second. Down stream lay The Gap already stripped of her green finery.

The State Premier admitted that Queensland was unprepared for such natural disaster events of this magnitude and effect, and the Ipswich Mayor congratulated the populace for helping out their neighbours and working together as a community. The long drought had lulled residents and organisations into inaction about floods in general and the idea of deluges were far from people's minds, as the dams stayed empty and the rivers looked barren.

The deluges that accompanied the fierce storm seasons in late 2008 in South East Queensland were unexpected and torrential, and could be classified as deluges because the storms precipitated very heavy falls, over a few minutes, at peak times in the rain event. However, the infrastructure, individual householders, communities, and emergency services were not able to cope. Severe weather forecasts were initially received too late and no plans or preparedness for these types of weather events were in place; however, there are such preparations for the rainy season in North Queensland, as this happens regularly almost every year (see Cottrell, 2008).

On the larger screen, Queensland had received torrential rain in January and February 2008; commencing with floods in the southern and south western areas, the rain focus moved to Emerald and Mackay, in the centre of the state, causing widespread damage. The volume of rain was spectacular. Then in early 2009, while bushfires ravaged Victoria, mega floods besieged the northern parts of Queensland and flood waters covered thousands of hectares of farms and towns alike. Later in that first quarter, cyclonic rains brought heavy rains to the eastern coast and south east areas. The Gulf area in North Queensland was beset by a flood which was classified as the second biggest in more than 100 years; the biggest being the 1974 flood.[34,35] Floraville, in the Leichhardt River catchment, received 244 mm in 24 hours on 03.01.09. Falls of 400 to 800 mm were recorded along the north Queensland Coast area near Ingham on 08.02.09. [36]

Queenslanders commented that they had never before seen such heavy and sustained cloud bursts; torrential rain had pelted the state for many weeks and it appeared that the state was now out of drought. However, 41% of Queensland still remained drought declared in May 2009.

While God had promised Noah that never again would the world be destroyed by flood, it might have been an indication that many disasters could destroy the world in the future, and not just floods, as projected by the IPCC. While Noah built an arc, perhaps purchasing a motor boat, with oars on board (in case the fuel supplies become scarce), might be a sensible act of preparedness for future disaster scenarios.

---

[34] ABC 2009a: Queensland Flood Update. An ABC Landline TV program broadcast on Sunday 22/02/2009. (http://www.abc.net.au/landline/content/2008/s2509719.htm).

[35] ABC 2009b: Marooned. An ABC Landline TV program broadcast on Sunday 8/03/2009. (http://www.abc.net.au/ landline/content/2008/s2509719.htm).

[36] Bureau of Meteorology. http://www.bom.gov.au

CONCLUSION

There is no doubt that we have had many different natural disasters and there have been a great many of these in the past decade, resulting in huge losses in lives, property, infrastructure and communities, along with native creatures and wildlife habitats.

To this author, the projected climate changes appear to be occurring now; the past decade has taught us that nature demonstrates her power and fury in different ways. While we have the time left, we need to appreciate the magnificent colourful rainbows, the rising mystic morning mists, the sparking blue oceans, the twinkling stars in clear night skies, the variety of green trees and plants that support our wildlife and our own being. It is not too late to stand up and be counted and get on with the job; if necessary, we must shake ourselves awake, unlike the sleep (turned nightmare) we were in, while the world collapsed financially.

Perhaps it is those people who believe in reincarnation, and others who value the continuity of the human race, and yet others who treasure their grandchildren's lives, together with those who just care about our planet, who must speak out and act wisely and decisively. Then others will follow.

REFERENCES

Bates, B.C., Kundzewicz, Z.W., Wu, S., Palutikof, J.P. (Eds.) (2008): Climate Change and Water. Technical Paper of the Intergovernmental Panel on Climate Change, IPCC Secretariat, Geneva, 210 pp.

Cottrell, A. (2008). Quiet Achievers: Women's resilience to as seasonal hazard. In K. Gow, and D. Paton (Eds.), *The Phoenix of Natural Disasters: Community Resilience.* New York: Nova Science Publications.

de Grazia, A. (1983). *The Lately Tortured Earth*: *Exoterrestrial Forces and Quantavolutions in the Earth Sciences.* US: Metron Publications.

EPA. (1997, October). Climate Change and public health. United Nations Environmental Protection Agency: Office of Policy, Planning and Evaluation. http://www.epa.gov/globalwarming

FAO. (2000). Fires. In *Global Forest Resources assessment 2000.* (Chapter 8). Accessed 24.04.08. http://www.fao.org/docrep/004/y1997e/y1997e0d.htm

FAO. (2007). Fire management global assessment 2006. A thematic study prepared in the framework of the Global Forest Resources Assessment 2005. FAO Forestry Paper 151. Food and Agriculture Organization of the United Nations, Rome.

Garnaut, R. (2008, October). *The Garnaut Climate Change Review.* Final Report. Melbourne: Cambridge University Press.

Gleick, P.H., Cooley, H., Katz, D. Lee, E., and Jason, E.L. (2006). The World's Water 2006-2007. The Biennial Report on Freshwater Resources. Washington, D.C.: Island Press.

Gow, K. (2009). The long road home. ABC News online, 10.02.09. http://www.abc.net.au/news/stories/2009/02/10/2486966.htm

IPCC (Intergovernmental Panel on Climate Change). (2007). Intergovernmental Panel on Climate Change Fourth Assessment Report. Climate Change 2007: Synthesis Report.

Summary for Policy Makers. Available at: http://www.ipcc.ch/pdf/assessment-report/ar4/syr/ar4_syr_spm.pdf

Kenny, P. (2008, September). It's About People: Changing Perspective. A Report to Government by an Expert Social Panel on Dryness. Drought Policy Review Expert Social Panel 2008. Report to the Minister for Agriculture, Fisheries and Forestry. Canberra: Commonwealth of Australia.

Lin Erda , Xu Yinlong, Wu Shaohong, Ju Hui, and Ma Shiming. (2007). China's National Assessment Report on Climate Change (II): Climate change impacts and adaptation. *Adv. Clim. Change Res., 3* (Suppl.), 6-11.

Meehl, G.A., Stocker, T.F., Collins, W.D., Friedlingston, P., Gaye, A.T., Gregory, J.M., Kitoh, A., Knutti, R., Murphy, JM., Noda, A., and Raper, S.C.B. (2007). Global Climate Projections. In: Climate Change 2007: The Physical Science Basis. Contribution of Working Group I to the Fourth Assessment Report of the Intergovernmental Panel on Climate Change. [Eds. Solomon, S., Qin, D., Manning, M., Chen, Z., Marguis, M., Vaeryt, K.B., Tignor, M., and Miller, H.L.]. Cambridge University Press, Cambridge, United Kingdom and New York, NY, USA.

Pittock, A.B. (2005). *Climate Change; turning up the heat*. Collingwood, Victoria: CSIRO publishing.

Pritchard, F. and Gow, K. (2008). Bushfires: Preparation, Resilience and Recovery. In K. Gow and D. Paton (Eds.). *The Phoenix of Natural Disasters: Community Resilience* (pp. 209-233). New York: Nova Science Publications.

Rogers, M. Small towns: Big picture- social sustainability indicators and the Arts. Social Dimensions of the Triple Bottom Line in Rural Australia- Bureau of Rural Sciences (2003). http://www.latrobe.edu.au/smalltowns/social%20dimension%20TBL.htm

Stern, N. (2006, October). Stern Review: The Economics of Climate Change, October 2006. HM Treasury and Cabinet Office.

United Nations Department of Humanitarian Affairs (DHA). (1982). *Glossary: International Agreed Glossary of Basic Terms Related to Disaster Management* (DHA/93/36).

United Nations Department of Humanitarian Affairs (DHA). *Glossary: International Agreed Glossary of Basic Terms Related to Disaster Management* (DHA/93/36). Geneva, Switzerland, UN DHA, December 1992, 81 pgs. At: http://www.em-dat.net/glossary.htm and at: http://www.reliefweb.int/rw/lib.nsf/db900SID/LGEL-5EQNZV/$FILE/dha-glossary-1992.pdf?OpenElement

Wang, Shourong. (2009). Strategies for Climate Change and its impacts on sustainable development in China. National Climate Center. Beijing. http://www.bom.gov.au/climate/cli2000/wShourng.html

Weart, S. (2008). *The Discovery of Global Warming* (2nd ed.). Cambridge, Massachusetts: Harvard University Press.

WHO. (2002). *World Health Report 2002. Reducing risks, promoting healthy life*. Geneva: World Health Organization.

WHO. (2003). *Climate Change and Human Health- Risks and Responses. Summary.* Geneva: World Health Organization.

Contributed by Ray Norton, Victoria.

Photo: The Natural Beauty we stand to Lose in the Future.

In: Meltdown: Climate Change, Natural Disasters… ISBN 978-1-60876-153-1
Editor: Kathryn Gow

Chapter 11

RESPONSES OF FRESHWATER TURTLES TO DROUGHT: THE PAST, PRESENT AND IMPLICATIONS FOR FUTURE CLIMATE CHANGE IN AUSTRALIA

John Roe[1] and Arthur Georges[2]
[1] Indiana-Purdue University, USA
[2] University of Canberra, Australia

ABSTRACT

Australia was not always arid, and the freshwater turtles that have survived to this day are either relicts residing in river systems that have themselves had a history or continuity through the drying of the continent, or they are species that have adapted in some way to meet the challenges of drought. In this chapter, we report on how Australian freshwater turtles cope with periodic loss of habitat through drought at a range of spatial and temporal scales, and consider how they might fare under changing climates predicted to occur through global warming. Global warming is not occurring in isolation of other environmental changes at a landscape scale, and we look at what interactions there might be between climate change and the increasing demands of agriculture, industry and our cities for water, in presenting challenges for our unique freshwater turtle fauna.

Keywords: turtles, drought, climate change, habitat loss, freshwater fauna

INTRODUCTION

Australia has the lowest precipitation of any vegetated continent – only Greenland and Antarctica have less – but few people realize how dry Australia actually is. To put this in perspective, the average annual discharge from all of Australia's rivers is 237,582 gigaliters

(Australian Bureau of Rural Sciences)[1] which is less than half that of the Mississippi River in North America. Consequently, much of the Australian continent is arid or semi-arid, and wetland habitats are small and poorly developed compared to other continents. As air and sea current patterns changed when Australia broke off from Antarctica, major episodes of progressive aridification occurred 30-35 Mya[2], 15 Mya and 2 Mya. The climate became not only drier, but also less predictable, and the biota that persisted had to contend with drought or periodic loss of free-standing water across much of the continent.

Drought is most simply defined as a departure below 'normal' precipitation (Dracup, Lee and Paulson, 1980; Wilhite and Glantz, 1985), although droughts can fall anywhere along a spectrum of climactic variability from long-term aridity at one extreme to annual cycles of wet and dry at the other. Aridity occurs where environmental water inputs are persistently insufficient to balance loss, whereas seasonally variable regions accrue a relatively temporary water deficit. In a sense, aridity can be considered as a long-term drought spanning millennia, whereas seasonally wet-dry climates cycle in and out of drought on a shorter term basis. The distinction between these scenarios is only a temporal one. Drought in the conventional sense falls somewhere in between, defined by water deficits spanning several years or decades compared to expectations derived over a period of historical record keeping, typically spanning decades or centuries.

From the perspective of wildlife responses to drought, this temporal distinction is important to grasp. At either end of the drought continuum, water deficits (whether short or long term) can be a relatively predictable feature of the regional climate. That is not to say that annual fluctuations in the occurrence, amount, timing, and duration of wet and dry periods do not occur – but over the balance of several years the timing and duration of water shortage occurs in a predictable sequence that has persisted for several thousands or millions of years. Such protracted periods of relatively stable climactic patterns can shape biological communities through evolutionary responses so that morphology, physiology and behaviour are suitably matched to the conditions (Ligon and Peterson, 2002; Withers, 1993). However, an abrupt departure from the normal precipitation pattern that spans several years or decades (i.e., drought in the conventional sense) may cause the existing biological communities to struggle to keep pace with the changing climate, even if environmental changes are brought about by natural climactic cycles.

The impact of drought on wildlife should not be discussed without also considering the human dimensions. Human activities may result in the imposition or exacerbation of drought conditions, and our infrastructure in developed landscapes can alter the natural responses of wildlife to drought. For instance, anthropogenic global climate change threatens to increase aridity in several regions of Australia, as well as to increase stochasticity[3] in climate, which can influence the frequency, severity, and duration of extreme weather events such as droughts (Bates et al., 2008; IPCC, 2007). Human water resource development alters water flow and hydrological regimes, in a sense simulating and hastening the onset of drought conditions and attenuating seasonal cycles of wet and dry. Roads, feral predators, and other dimensions of human society may further hamper the ability of wildlife to respond to drought.

---

[1] http://www.daff.gov.au/brs/social-sciences.
[2] Million years ago.
[3] Stochasticity implies irregular and unpredictable variability.

In this chapter, we introduce the Australian freshwater turtle fauna and explore how they cope (or in some cases, do not cope) with periodic loss of aquatic habitat through drought at a range of spatial and temporal scales, and consider how they might fare under changing climates predicted to occur through global warming. Global climate change is not occurring in isolation to other environmental alterations at a landscape level, and thus in assessing some of the challenges for Australia's unique freshwater turtle fauna, we examine what interactions there might be between climate change and the increasing demands of agriculture, industry and cities for water and other natural resources.

THE FRESHWATER TURTLES OF CONTEMPORARY AUSTRALIA

There are no strictly terrestrial turtles or tortoises that inhabit Australia. Instead, the Australian turtle fauna is restricted to freshwaters and the surrounding ocean. In this chapter, we focus only on those that occupy the inland freshwater systems, which are dominated by species belonging to the family Chelidae. This family of side-necked turtles is restricted to the Australasian region and South America - not even fossil forms are found outside this range (Gaffney, 1991; Williams, 1953a, b) - so it is a group with clear Gondwanan origins. Their fossil record extends back to the late Cretaceous (de Broin, 1987) in South America and the Miocene in Australia (Gaffney, Archer and White, 1989). Among extant forms in Australia and New Guinea are the river turtles in the genera *Emydura* (5 species) and *Elseya* (7 species) of northern and eastern Australia and New Guinea, and *Rheodytes* (1 species) and *Elusor* (1 species) from east coastal rivers of Queensland. The critically endangered *Psuedemydura umbrina* resides in seasonally dry wetlands of south Western Australia. Australia is also renowned for its spectacular snake-necked turtles (12 species) that range in habitat from permanent waters (*Chelodina expansa*) to highly ephemeral[4] streams and ponds (*Chelodina steindachneri*). The pig-nosed turtle (*Carettochelys insculpta*) is the only species of Australian freshwater turtle outside of the family Chelidae, belonging to the family Carettochelydidae (the only extant species in this family) that lives in the major rivers and associated wetlands of northern Australia and southern New Guinea (Georges and Rose, 1993).

Not surprisingly, given this diversity, Australia's freshwater turtles occupy a wide range of habitats - from the flooded forests and extensive swamps of the Fly River and Kikori Delta in Papua New Guinea (Georges et al., 2006a, 2008) to the dryland rivers of Australia's arid interior (Georges et al., 2006b); from the tropical north to the snow country around Cooma; from dilute waters of the sandstone plateaus of Arnhem Land and the Kimberleys (Thomson et al., 2000), to the coastal floodplains of northern Australia and the saline or brackish coastal wetlands of the Styx River in Queensland and Lake Alexandria at the mouth of the Murray River in South Australia. However, Australia was not always arid (Quilty, 1994), and the freshwater turtles that have survived to this day are either relicts residing in watersheds that have themselves had a history of continuity through the drying of the continent, or they are species that have adapted in some way to meet the challenges of drought (Georges and Thomson, 2006). Below, we explore several ways that Australian turtles respond naturally to the drying of their aquatic habitat.

[4] Ephemeral bodies of water tend to dry during drought and flood again following precipitation.

NATURAL RESPONSES TO HABITAT DRYING

Along with temperature, precipitation patterns that determine flooding and drying of surface waters are likely to be the most powerful drivers of freshwater turtle distributions and community structure[5] in Australia. Several of Australia's wetlands are under the influence of natural flood-drought cycles, and in these contexts, many turtles have adaptations to drought that allow them to successfully cope with habitat drying by moving to more permanent waters (escape in space) or by remaining in or near the dry waterbody in an inactive state to await the return of flooding (escape in time). However, several species do not regularly contend with the drying of their waterholes and may lack any evolved responses to cope successfully with drying.

Aestivation

In regions where water bodies seasonally or periodically dry, many Australian turtles have evolved specializations to survive extended periods of drought by entering a state of inactivity called aestivation (Burbidge 1981, Kennett and Christian 1994, Roe and Georges, 2007, 2008, Roe et al., 2008). The Western Swamp Turtle, *P. umbrina*, inhabits temporary swamps that flood during winter rains, but by late spring or early summer the swamps have dried and may remain so for 6-9 months (Burbidge, 1981). Several species of turtles in the genus *Chelodina* (*C. canni, C. longicollis, C. rugosa, C. steindachneri*) also inhabit wetlands that dry for several months. In some extreme circumstances, *C. longicollis* can survive aestivation for 13–16 months (Stott 1987; Roe and Georges, 2007), while others (*C. steindachneri*) may occasionally need to survive for longer time periods out of water (Cann, 1998), although in this latter case no empirical work has yet confirmed this.

Aestivation is a strategy to survive the challenges of drought by reducing the body's energy and water demands. This may involve a suite of morphological, physiological, and behavioural traits. When drought occurs, turtles may respond by leaving the wetland and seeking refuge in naturally occurring holes in the ground, or by burrowing several centimeters under soil, leaf litter, woody debris, or other structures (Burbidge 1981; Chessman, 1983; Rees 2008; Roe and Georges, 2007, 2008). Alternatively, some burrow in wetland sediment where they remain entombed until rains return (Cann, 1998; Kennett and Christian 1994). Morphological adaptations for terrestrial aestivation are perhaps best demonstrated in *P. umbrina*, which has an extensively roofed skull and an expanded shell that allows for the complete withdrawal of their head and limbs to reduce water loss. Other turtles employ physiological specializations to further conserve energy and water. For instance, *C. rugosa* reduces metabolism by 72% during dry season aestivation (Kennett and Christian, 1994), while *C. longicollis* is slow to lose water to evaporation and can further conserve water by altering the composition of excreta (Chessman, 1984; Rogers, 1966; Roe et al., 2008). While these behavioral, morphological and physiological specializations highlight adaptations that can greatly extend survival during drought, turtles are nevertheless vulnerable at this time. Predation rates can be high (Fordham et al., 2006b), and if waters do not re-flood before energy stores are depleted, turtles will eventually die of starvation or dehydration (Roe and

[5] Community structure reflects the number and types of species in an area and their relative abundances.

Georges, 2008, Roe et al., 2008). Thus aestivation is only a temporary means of increasing the probability of surviving drought, and the duration that a waterhole remains dry is a critical factor in whether turtles can persist in the area.

Movement to other Waterholes

As an alternative to aestivation, turtles may move to more permanent waterholes when drought strikes. While it is not uncommon to see turtles of several species emigrating from wetlands in Australia (i.e., when they cross roads, golf courses, lawns, or paddocks), this is one of the least well understood aspects of turtle behaviour (Roe and Georges, 2007). We have studied overland movements in *C. longicollis* at Booderee National Park and in the suburbs of Canberra (Graham et al., 1996; Kennett et al. 1990; Rees, 2008; Roe and Georges, 2007). At both sites, drought cues mass emigrations from drying wetlands into nearby permanent waterbodies. These permanent waterbodies offer temporary refuge during drought, where turtles remain (sometimes for years) before staging a return to temporary wetlands following floods. Turtles regularly travel up to three kilometers overland to reach permanent water, but occasional movements of more than five kilometers overland have also been reported in *C. longicollis* (Parmenter, 1976; Roe, 2007). However, just as the temporal duration of dry conditions is a limiting factor for survival through aestivation, an ability to escape to freshwater drought refuges depends upon the availability of waterbodies in the immediate vicinity that are somehow more resilient to drought. Most freshwater turtles could presumably traverse short distances (i.e., a few hundred meters) overland, but when farther distances must be traversed or when the intervening terrain is especially harsh, few would be capable of making such travels. Even the most basic capacities for (and limitations to) overland travel remain largely unknown for the majority of Australia's freshwater turtles.

Life History Specializations to Drying

The onset and duration of habitat drying and the return of floods can vary predictably (i.e., seasonally) or unpredictably (i.e., stochastically). It is in those species, that have to contend with unpredictability in flood-drought cycles, that we see some of the most interesting life history and behavioural "solutions" for survival. *Chelodina rugosa* deals with unpredictability in the recession of wet season flooding by laying its eggs underwater where development is suspended until the ground dries (Kennett et al., 1993a, b). Development is then adjusted so that embryos can compensate for differing durations of inundation to hatch at the optimal time prior to the onset of the ensuing rainy season (Fordham et al., 2006a). *Carettochelys insculpta* overcomes unpredictability in the onset of the wet season for its hatchlings by late-term embryonic aestivation. Eggs develop rapidly in anticipation of the wet-season floods, then enter a state of torpor within the eggshell (Webb et al., 1986). Oxygen deprivation brought about by torrential rain that floods the nest cavity brings the hatchlings out of their slumber and they hatch and make their frenzied way into the rising floodwaters (Doody et al., 2001; Georges, 1992). Hatchlings of other species, like *C. longicollis*, may emerge from eggs but remain within the nest chamber until heavy rain softens the soil and allows their escape. *Chelodina longicollis* adults have a range of adaptations that make them well-suited for both overland travel to refugial waterbodies and for extended bouts of terrestrial aestivation. This flexibility allows them to cope with drought and other unpredictable occurrences of aquatic habitat loss in a variety of ways depending on the context (Roe and Georges, 2008).

Permanent Water Specialists

Permanent water specialists[6] represent the vast majority of turtle genera in Australia, and it is these that are least likely to tolerate the drying of their aquatic habitats for any length of time. Permanent waters include lakes, ponds, rivers, and streams that typically remain inundated or maintain water flow without drying. Permanent water specialists include *Emydura*, *Elseya*, *Rheodytes*, *Elusor*, and *Carettochelys*, but *C. expansa* and *C. oblonga* also rely on habitats that do not typically dry out.

Photo: Queensland EPA Roma. Photo reproduced with permission.

Figure 11.1. When freshwater refuges with a long history of water permanence dry, permanent water specialists typically perish, such as in Lake Numalla, Queensland, where some 10,000 Murray turtles were estimated to have died.

Turtles that are rarely exposed to drying conditions typically lack adaptations that would allow them to successfully contend with habitat drying by evacuating or aestivating (Chessman, 1984; Christiansen and Bickham, 1989; Gibbons et al., 1983). However, permanent water specialists may occasionally enter areas that periodically dry, such as when river flows force them out of channels and into shallow backwaters or downstream waterholes

[6] Habitat specialists tend to occupy a narrow range of environmental conditions, whereas habitat generalists can occupy a wide range of conditions.

that can quickly become isolated when flows subside. In these scenarios, turtles can become unexpectedly stranded by drying floodwaters and many perish as a result (see Figure 11.1).

Nevertheless, some of Australia's permanent water specialists successfully inhabit arid regions, with some of the most widely fluctuating water flows on earth, and where most water can vanish for long periods. The Cooper Creek turtle, *Emydura macquarii emmottii*, has a most interesting life history, as it manages to persist in areas where other turtles cannot. This species occupies the permanent waterholes of the Cooper Creek floodplain in central Australia. Precipitation over much of the Cooper catchment is exceptionally low (approximately 330 mm per year, Australian Bureau of Meteorology,[7] but monsoonal rainfall in its headwaters results in intermittent and widespread flooding of this dryland river.

An exceptionally flat landscape results in an extensive network of distributary channels, peppered with isolated and deeper waterholes. It is in these waterholes that this species resides between episodes of floods, often in exceptionally high population densities (Georges et al., 2006b). Should waterholes dry (which happens during the worst of our droughts) the turtles perish, but when the system floods once again others are able to reinvade from refugial waterholes that did not dry. This dynamic allows this river turtle to persist in Australia's arid centre. The scale of this landscape is immense – waterholes are separated by hundreds of kilometers of arid and semi-arid land, and the temporal scale of flooding is measured in years instead of months, attributes presumably unsuited to the drought survival strategies of any other Australian turtle.

DROUGHT IMPACTS AND THE HUMAN DIMENSION

The human dimensions of drought relevant to freshwater turtles fall into two broad categories. The first is creation of conditions that either cause drought or produce drought-like conditions, and the second includes activities that limit the natural ability of turtles to respond to drought. To assess the impact of drought on turtles, we must understand how these anthropogenic dimensions interact to directly and indirectly influence their populations.

The most visible consequence of drought in freshwater systems occurs when a water body dries completely, and consequences are most severe when water bodies that typically serve as aquatic refugia also dry out. Surface waters can recede and eventually dry from decreased runoff from rainfall, inflow from rivers, or supply by groundwater. Though rainfall ultimately drives all three processes, our demands on water for irrigation, industry and municipal activities hasten the pace of drying and extends the temporal and spatial impact of drought into areas that were once more extensively flooded or served as permanent water refuges (Kingsford and Thomas, 1995; Kingsford, 2000). Reports of mass turtle mortality in drying wetlands of Australia are increasing (see Figures 11.1 and 11.2), though such reports have not yet made their way into the scientific literature, as they have in other countries (Christiansen and Bickham, 1989, Bodie and Semlitsch, 2000). Many turtles die of starvation or dehydration during or following wetland drying, but drought can indirectly cause mortality as well. Turtles emigrating from a drying waterbody are vulnerable to predators including the introduced European Fox (Spencer, 2002), or they may encounter vehicular traffic in developed areas resulting in mass mortality on roads in Australia (Chessman, 1988; Rees,

[7] http://www.adl.brs.gov.au/water2010/pdf/monthly_reports/awap_3_report.pdf.

2008), see Figure 11.3) and elsewhere (Aresco, 2005; Ashley and Robinson, 1996). Exclusion fences can also entrap turtles on land where they eventually die of exposure to predators, high temperatures, or dehydration (Anonymous 1941; JHR and AG pers. obs.).

Feral pigs can even target and depredate turtles aestivating in the mud of dry wetlands, threatening the persistence of turtle populations in northern Australia (Fordham et al., 2006b, 2007). All of the above are examples of how human alterations to natural systems interact with drought to exacerbate impacts upon turtles.

Drought can also result in important changes to habitat heterogeneity with subsequent impacts on turtles. For instance, large river systems naturally include not only the deeper and permanent channel, but during floods the aquatic habitat expands to encompass adjacent floodplain wetlands (e.g., oxbows, billabongs, marshes, and sloughs). Such flooding events resupply these highly productive and important wetlands with water and nutrients and provide a conduit for turtles to travel into them (Doody et al., 2002; Georges et al., 2006). Low rainfall and high use of upstream water during drought reduces both the extent and duration of flooding in these adjacent wetlands such that, in Australia's most developed rivers, these habitats may no longer flood at all (Kingsford, 2000; Roshier et al., 2001).

Photo: John Roe.

Figure 11.2. Receding waters can concentrate turtles, resulting in high densities. Disease and competition for food increase under such conditions, killing many individuals even before the water body dries completely such as in Lake McKenzie, Jervis Bay Territory.

Turtle communities in such systems are often spatially organized according to species habitat preferences (Chessman, 1988), and in Australia several species of turtles including *C. insculpta* (Doody et al., 2002). *C. longicollis* (Chessman, 1988), and *C. expansa* (D. Bower, University of Canberra, unpubl. data) make extensive use of these flooded backwaters. Not only would turtle community diversity decline should river floodplains fail to flood, but our work in Booderee National Park in the Jervis Bay Territory indicates that such drought-induced changes in hydrology can also occur in wetlands isolated from river flows, with subsequent effects on turtle growth, reproduction and regional abundance (Kennett and Georges, 1990, Roe and Georges, 2008).

Drought can also result in important changes to habitat heterogeneity with subsequent impacts on turtles. For instance, large river systems naturally include not only the deeper and permanent channel, but during floods the aquatic habitat expands to encompass adjacent floodplain wetlands (e.g., oxbows, billabongs, marshes, and sloughs). Such flooding events resupply these highly productive and important wetlands with water and nutrients and provide a conduit for turtles to travel into them (Doody et al., 2002; Georges et al., 2006). Low rainfall and high use of upstream water during drought reduces both the extent and duration of flooding in these adjacent wetlands such that, in Australia's most developed rivers, these habitats may no longer flood at all (Kingsford, 2000; Roshier et al., 2001).

Turtle communities in such systems are often spatially organized according to species habitat preferences (Chessman, 1988), and in Australia several species of turtles including *C. insculpta* (Doody et al., 2002). *C. longicollis* (Chessman, 1988), and *C. expansa* (D. Bower, University of Canberra, unpubl. data) make extensive use of these flooded backwaters. Not only would turtle community diversity decline should river floodplains fail to flood, but our work in Booderee National Park in the Jervis Bay Territory indicates that such drought-induced changes in hydrology can also occur in wetlands isolated from river flows, with subsequent effects on turtle growth, reproduction and regional abundance (Kennett and Georges, 1990, Roe and Georges, 2008).

The retention of water in reservoirs is a common way to meet societal water demands in times of drought, but ironically, this too can have negative effects on some turtles. For example, the Fitzroy turtle (*Rheodytes leukops*) inhabits shallow riffle zones (Tucker et al., 2001), a habitat affinity linked to its gill-like cloacal respiration and diet of aquatic invertebrate prey (Gordos, 1999; Priest, 1997). Dams that transform free flowing rivers to lake-like conditions or that otherwise alter natural flow regimes and increase sedimentation rates have degraded much riffle habitat in the Fitzroy River catchment, which has certainly contributed to the listing of *R. leukops* as a Vulnerable species (Georges, 1993). A change in turtle community diversity (typically a reduction in the number of species) often follows the construction of dams, with river (lentic) turtles being replaced by pond (lotic) turtles or species with more general habitat preferences (DonnerWright et al. 1999; Tucker et al., 2001). Our changes to natural hydrology patterns in aquatic systems are not limited to rivers, but also include modifications to still-water wetlands for increased water retention during drought (i.e., farm dams, Brock et al., 1999). In this case, humans' influence on some species may be positive, as farm dams, irrigation channels and reservoirs for water retention in agricultural and pastoral regions now offer drought refuges. The proliferation of these artificial aquatic habitats has perhaps augmented populations of *C. longicollis* in dry regions (Rees, 2008) and may even have facilitated its expansion into areas that were previously too dry (Beck, 1991). However, the persistence of these "unnatural" wetlands and the benefits they offer aquatic turtles into the future is uncertain under changing climate scenarios.

The rising salinity of freshwaters is another major issue with which turtles must contend in Australia. Rising soil salinities brought about by land clearing, intensive irrigation and reduced rates of evapotranspiration[8] have allowed water tables to rise and contaminate freshwater systems (Lemly et al., 2000), but the impact on Australian turtles in these systems has not yet been examined. We know that in some cases increasing salinities have brought saline-tolerant taxa[9] into unwelcome contact with turtles, as in the case of turtle fouling by encrusting estuarine worms in Lake Alexandria, South Australia (see Figure 11.4). Many nominally freshwater turtles may successfully utilize saline environments (Dunson et al. 1986, Georges and Rose, 1993, Moll and Moll, 2004); however, the physiological, behavioral, and population level responses of Australian turtles to rapidly increasing salinities are in immediate need of assessment.

Photo: Damien Michael. Photo reproduced with permission.

Figure 11.3. Turtles forced to emigrate from drying waterbodies frequently cross roads in developed areas, often leading to mass mortality from vehicular collisions. This *Chelodina expansa* was killed on the Riverina Highway adjacent to the Wonga Wetlands and the Murray River.

---

[8] Evapotranspiration is the movement of water from the ground to the atmosphere via evaporation and plant transpiration.

[9] Saline-tolerant taxa are those animals that can successfully inhabit waters with higher concentrations of dissolved salts than in normal freshwater. Freshwater typically has dissolved salt concentrations below 0.5 parts per thousand.

Photo: Sally Grundy. Photo reproduced with permission.

Figure 11.4. Increased salinization of freshwater habitats has led to the encrustation of turtles' shells by estuarine tube worms. In severe cases encrustation can impede turtle movements and limit the ability to retract their head and limbs, such as in this *Emydura* in Lake Alexandrina, South Australia.

THE FUTURE OF AUSTRALIA'S TURTLES IN A CHANGING CLIMATE

Predictions for changes to Australia's climate include a reduction in precipitation across much of the continent, particularly in the southern temperate regions. This will result in expansion of arid and semi-arid zones that are less hospitable to turtles, and an increase in the frequency, duration and intensity of drought. The recent drought in south-eastern Australia, whether or not it is part of a natural cycle, serves as a harbinger of impending climate change and focuses our attention on what might be in store for the future. Turtles in the Murrumbidgee River and elsewhere have perished in waterholes not known to dry since European colonization. The Murray-Darling River now seldom flows into the sea, and the combination reduced water flows with marine incursion and increased salinization is presenting major challenges to the freshwater turtle communities of Australia's largest river system. In the south-west, *P. umbrina*, one of the rarest turtles in the world is already on the brink of extinction, being restricted to a few small ephemeral swamps. Any further reduction in precipitation will require continual human intervention to maintain this species in the wild (Kuchling and Dejose, 1988, Kuchling et al., 1991).

If there is any reason to be optimistic, it is that this unfolding crisis has been experienced to some degree by the Australian biota before, and those species that are with us today are there because they have survived the changes. The challenge will be to determine what humans have done to reduce the capacity of the biota to survive climate change as it has done in the past. In the case of *P. umbrina*, the answer is clear. We have dramatically reduced the extent of available habitat through reclamation of swampland for urbanization and agriculture, rendering this species extremely vulnerable to climate change. In the Murray-Darling drainage and other watersheds that are currently experiencing greatly reduced flow, we need to know where the refugial areas were during past periods of drying so that we can protect their integrity during this drying event. These refugia are areas in which the turtles and other aquatic biota persisted during extended periods of low precipitation, only to expand out to their current distributions when precipitation increased. On a shorter timescale, responses to drought may involve a dynamic of localized extinction and repopulation. What have we done - in terms of fragmenting our freshwater systems through habitat degradation, construction of weirs, dams and roads, and alteration of flow intensity, duration and timing - to disrupt this dynamic? These considerations apply at many geographic scales, from the Cooper Creek watershed in the arid centre to individual wetlands and wetland networks, such as at Booderee National Park in the temperate south-east.

CONCLUSION

Clearly, failure to address several issues, including the maintenance of adequate water during periods of reducing precipitation, the isolation of off-channel waterbodies in the lower Murray to address human demands for water quantity and quality, the construction of more dams and storage facilities without exploring alternatives, and our increasing take from aquifers with little regulation, are all responsible for environmental changes that are likely to have major (and potentially adverse) impacts on freshwater communities, of which turtles are an integral part. Regardless of the ultimate cause of recent droughts in Australia and other parts of the world, the challenge before us is to improve water resource management in a way that allows for more efficient use of water to meet societal demands, while at the same time ensuring that the natural capacity for biota to respond to drought is maintained.

REFERENCES

Anonymous. (1941). Death on the plains-where instinct fails. *Wild life* (Melbourne), 3, 13-14.

Aresco, M.J. (2005). Mitigation measures to reduce highway mortality of turtles and other herpetofauna at a north Florida lake. *Journal of Wildlife Management,* 69, 549-560.

Ashley, E.P., and J.T. Robinson. (1996). Road mortality of amphibians, reptiles, and other wildlife on the long point causeway, Lake Erie, Ontario. *Canadian Field-Naturalist,* 110, 403-412.

Bates, B.C., Z.W. Kundzewicz, S. Wu, and J.P. Palutikof (Eds.). (2008). Climate Change and Water. Technical Paper of the Intergovernmental Panel on Climate Change. IPCC Secretariat, Geneva.

Beck, R.G. (1991). The Common Longnecked Tortoise *Chelodina longicollis* (Shaw 1802) (Testudinata: Chelidae): A comparative study of the morphology and field behaviour of disjunct populations. *South Australian Naturalist,* 66, 4-22.

Bodie, J.R. and Semlitsch, R.D. (2000). Size-specific mortality and natural selection in freshwater turtles. *Copeia 2000,* 732-739.

Brock, M. A., Smith, R.G.B. and Jarman, P.J. (1999). Dam it, drain it: alteration of water regime in shallow wetlands on the New England Tableland of New South Wales, Australia. *Wetlands Ecology and Management*, 7, 37-46.

Burbidge, A.A. (1981). The ecology of the western swamp tortoise *Pseudemydura umbrina* (Testudines: Chelidae). *Australian Wildlife Research,* 8, 203-223.

Cann, J. (1998). *Australian Freshwater Turtles.* Beaumont Publishing, Singapore.

Chessman, B.C. (1983). A note on aestivation in the snake-necked turtle, *Chelodina longicollis* (Shaw) (Testudines: Chelidae). *Herpetofauna,* 14, 96.

Chessman, B.C. (1984). Evaporative water loss from three south-eastern Australian species of freshwater turtle. *Australian Journal of Zoology*, 32, 649-655.

Chessman, B.C. (1988). Habitat preferences of freshwater turtles in the Murray valley, Victoria New South Wales. *Australian Wildlife Research,* 15, 485-491.

Christiansen, J.L., and Bickham, J.W. (1989). Possible historic effect of pond drying and winterkill on the behavior of *Kinosternon flavescens* and *Chrysemys picta*. *Journal of Herpetology,* 23, 91-94.

de Broin, F. (1987). The late Cretaceous fauna of Los Alamitos, Patagonia, Argentina. Part IV. Chelonia. Revista del Museo Argentino de Ciencias Naturales, Bernardino Rivadavia. *Paleontologia*, 3, 131–139.

DonnerWright, D.M., Bozek, M.A., Probst, J.R. and Anderson. E.M. (1999). Responses of turtle assemblages to environmental gradients in the St. Croix River in Minnesota and Wisconsin, U.S.A. *Canadian Journal of Zoology,* 77, 989-1000.

Doody, J.S., Georges, A., Young, J.E., Pauza, M.D., Pepper, A.L., Alderman, R.L and Welsh. M.A. (2001). Embryonic aestivation and emergence behaviour in the pig-nosed turtle, *Carettochelys insculpta*. *Canadian Journal of Zoology,* 79, 1062-1072

Doody, J. S., Young, J. E. and Georges. A. (2002). Sex differences in activity and movements in the pig-nosed turtle, *Carettochelys insculpta*, in the wet-dry tropics of Australia. *Copeia 2002,* 93-103.

Dunson, W.A. (1986). Estuarine populations of the snapping turtle (*Chelydra*) as a model for the evolution of marine adaptations in reptiles. *Copeia 1986,* 741-756.

Dracup, J. A., Lee, K. S. and Paulson, E.G. (1980). On the definition of droughts. *Water Resources Research,* 16, 297-302.

Fordham, D. A., Georges, A. and Corey. B. (2006a). Compensation for inundation-induced embryonic diapause in a freshwater turtle: achieving predictability in the face of environmental stochasticity. *Functional Ecology*, 20, 670-677.

Fordham, D. A., Georges, A. and Corey. B. and Brook, B.W. (2006b). Feral pig predation threatens the indigenous harvest and local persistence of snake-necked turtles in northern Australia. *Biological Conservation*, 133, 379-388.

Fordham, D., Georges, A. and Brook. B.W. (2007). Indigenous harvest, exotic pig predation and local persistence of a long-lived vertebrate: managing a tropical freshwater turtle for sustainability and conservation. *Journal of Applied Ecology,* 45, 52-62.

Gaffney, E.S. (1991). The fossil turtles of Australia. In P.Vickers-Rich, J.M. Monaghan, R.F., Baird, and T.H. Rich (Eds.). *Vertebrate Paleontology of Australasia* (pp. 704-720). Pioneer Design Studio and Monash University Publications Committee, Melbourne.

Gaffney, E.S., Archer, M. and White, A. (1989). Chelid turtles from the Miocene freshwater limestones of Riversleigh Station, Northwestern Queensland, Australia. *American Museum Novelties,* 2959, 1-10.

Georges, A. (1992). Thermal characteristics and sex determination in field nests of the pig-nosed turtle *Carettochelys insculpta* (Chelonia: Carettochelydidae) from northern Australia. *Australian Journal of Zoology,* 40, 511-521.

Georges, A. (1993). Settling conservation priorities for Australian freshwater turtles. In D. Lunney and D. Ayers (Eds.), *Herpetology in Australia, a Diverse Discipline* (pp. 453-476). Royal Zoological Society of New South Wales.

Georges, A., Alacs, E., Pauza, M., Kinginapi, F., Ona, A., and Eisemberg, C. (2008). Freshwater turtles of the Kikori Drainage, Papua New Guinea, with special reference to the pig-nosed turtle, *Carettochelys insculpta. Wildlife Research,* 35,700-711.

Georges, A., Guarino, F., and Bito, B. (2006a). Freshwater turtles of the TransFly region of Papua New Guinea - notes on diversity, distribution, reproduction, harvest and trade. *Wildlife Research,* 33, 373-384.

Georges, A., Guarino, F., and White, M. (2006b). Sex-ratio bias across populations of a freshwater turtle (Testudines: Chelidae) with genotypic sex determination. *Wildlife Research,* 33, 475-480.

Georges, A. and Rose, M. (1993). Conservation biology of the Pig-nosed Turtle, Carretochelys insculpta. *Chelonian Conservation and Biology,* 1, 3-12.

Georges, A. and Thomson, S. (2006). Evolution and zoogeography of Australian freshwater turtles. In *Evolution and Biogeography of Australian Vertebrates* (pp. 281-300). J.R. Merrick, M. Archer, G.M. Hickey, and M.S.Y. Lee (Eds.). Oatlands, NSW: Australian Scientific Publishing.

Gibbons, J.W., J.L. Greene, and Congdon, J.D. (1983). Drought-related responses of aquatic turtle populations. *Journal of Herpetology,* 17, 242-246.

Gordos, M.A. (1999). Physiological ecology of the Fitzroy Turtle, *Rheodytes leukops.* Honours Thesis, University of Queensland.

Graham, T.E., Georges, A., and McElhinney, N. (1996). Terrestrial orientation by the eastern long-necked turtle, *Chelodina longicollis*, from Australia. *Journal of Herpetology*, 30, 467-477.

Intergovernmental Panel on Climate Change. (2007). Synthesis Report. R.K. Pachauri, and A. Reisenger (Eds.). IPCC, Geneva, Switzerland. Available from, http://www.ipcc.ch/ipccreports/ar4-syr.htm (accessed March 2009).

Kennett, R. and Christian, K. (1994). Metabolic depression in estivating long-neck turtles. *Physiological Zoology,* 67,1087-1102.

Kennett, R.M. and A. Georges. (1990). Habitat utilization and its relationship to growth and reproduction of the eastern long-necked turtle, *Chelodina longicollis* (Testudinata: Chelidae), from Australia. *Herpetologica* 46, 22-33.

Kennett, R., Christian, K., and Pritchard, D. (1993a). Underwater nesting by the freshwater turtle *Chelodina rugosa* from tropical northern Australia. *Australian Journal of Zoology,* 41, 47-52.

Kennett, R., Georges, A., and Palmer-Allen, M. (1993b). Early developmental arrest during immersion of eggs of a tropical freshwater turtle, *Chelodina rugosa* (Testudinata: Chelidae) from northern Australia. *Australian Journal of Zoology*, 41, 37-45.

Kingsford, R. T. (2000). Ecological impacts of dams, water diversions and river management on floodplain wetlands in Australia. *Austral Ecology*, 25, 109-127.

Kingsford, R.T., and Thomas, R.F. (1995). The Macquarie Marshes in arid Australia and their waterbirds: A 50 year history of decline. *Environmental Management,* 19, 867-878.

Kuchling, G. and Dejose, J.P. (1988). A captive breeding operation to rescue the critically endangered Western Swamp Tortoise *Pseudemydura umbrina* from extinction. *International Zoo Yearbook*, 31, 103-109.

Kuchling, G, Dejose, J.P. Burbidge, A.A, and Bradshaw, S.D. (1991). Beyond captive breeding: the Western Swamp Tortoise *Pseudemydura umbrina* recovery programme. *International Zoo Yearbook*, 31, 37-41.

Lemly, A. D., Kingsford, R.T., and Thompson, J.R. (2000). Irrigated agriculture and wildlife conservation: conflict on a global scale. *Environmental Management*, 25, 485-512.

Ligon, D.B. and Peterson, C.C. (2002). Physiological and behavioral variation in estivation in mud turtles (*Kinosternon* spp.). *Physiological and Biochemical Zoology*, 75, 283-293.

Moll, D. and Moll, E.O. (2004). *The Ecology, Exploitation, and Conservation of River Turtles.* Oxford University Press, New York.

Parmenter, C.J. (1976). The natural history of the Australian freshwater turtle *Chelodina longicollis* Shaw (Testudinata, Chelidae). PhD Thesis, University of New England.

Priest, T. (1997). Bimodal respiration and dive behaviour of the Fitzroy River Turtle, *Rheodytes leukops*. Honours Thesis, University of Queensland.

Quilty, P.G. (1994.) The background: 144 million years of paleoclimate and paleogeography. In *History of the Australian Vegetation, Cretaceous to Present* (pp. 14-44). R. Hill (Ed.). Cambridge: Cambridge University Press.

Rees, M. (2008). Impacts of urbanization on the eastern long-necked turtle *Chelodina longicollis* (Testudines:Chelidae). Honours thesis, University of Canberra.

Roe, J.H. (2007). The terrestrial ecology of a freshwater turtle, *Chelodina longicollis*, in Booderee National Park, Australia. PhD Thesis, University of Canberra.

Roe, J.H., and Georges, A. (2007). Heterogeneous wetland complexes, buffer zones, and travel corridors: landscape management for freshwater reptiles. *Biological Conservation,* 135, 67-76.

Roe, J.H., and Georges, A. (2008). Maintenance of variable responses for coping with wetland drying in freshwater turtles. *Ecology,* 89, 485-494.

Roe, J.H., Georges, A., and Green, B. (2008). Energy and water flux during terrestrial aestivation and overland movement in a freshwater turtle. *Physiological and Biochemical Zoology*, 81, 570-583.

Rogers, L.J. (1966). The nitrogen excretion of *Chelodina longicollis* under conditions of hydration and dehydration. *Comparative Biochemistry and Physiology,* 18, 249-260.

Roshier, D. A., Whetton, P.H., Allan, R.J., and Robertson, A.I. (2001). Distribution and persistence of temporary wetland habitats in arid Australia in relation to climate. *Austral Ecology,* 26, 371-384.

Spencer, R.J. (2002). Experimentally testing nest site selection: fitness tradeoffs and predation risk in turtles. *Ecology,* 83, 2136-2144.

Stott, P. (1987). Terrestrial movements of the freshwater tortoise *Chelodina longicollis* Shaw as monitored with a spool tracking device. *Australian Wildlife Research,* 14, 59-567.

Thomson, S., Kennett, R., and Georges, A. (2000). A new species of long-necked turtle (Testudines: Chelidae) from the Arnhem Land plateau, Northern Territory, Australia. *Chelonian Conservation and Biology,* 3, 675-685.

Tucker, A. D., Limpus, C. J., Priest, T.E., Cay, J., Glen, C., and. Guarino, E. (2001). Home ranges of Fitzroy River turtles (*Rheodytes leukops*) overlap riffle zones: potential concerns related to river regulation. *Biological Conservation,* 102, 171-181.

Webb, G., D., Choquenot, D., and Whitehead, P. (1986). Nest, eggs and embryonic development of *Carettochelys insculpta* (Chelonia: Carettochelidae) from Northern Australia. *Journal of Zoology, London (B) 1*, 521-550.

Wilhite, D. A. and Glantz, M. H. (1985). Understanding the drought phenomenon: the role of definitions. *Water International,* 10, 111-120.

Williams, E.E. (1953a). Fossils and the distribution of Chelyid turtles (2). Breviora *(Museum of Comparative Zoology),* 13, 1-8.

Williams, E.E. (1953b). Fossils and the distribution of Chelyid turtles (2). Breviora *(Museum of Comparative Zoology)*, 32, 1-6.

Withers, P.C. (1993). Metabolic depression during estivation in the Australian frogs Neobatrachus and Cyclorana. *Australian Journal of Zoology,* 41, 467-473.

In: Meltdown: Climate Change, Natural Disasters... ISBN 978-1-60876-153-1
Editor: Kathryn Gow

Chapter 12

CAN UNDERSTANDING BACKWATER EFFECTS HELP US TO PREPARE FOR AND MANAGE DELUGES?

Francine Pritchard
University of Southern Queensland, Australia

ABSTRACT

One of the accepted scenarios of climate change is that we will experience more deluges, especially in areas that are subject to severe storms. Failure to prepare for, and adequately warn residents about, such rain events was evident in storm deluges in 2008/2009 in South-East Queensland and mega floods in northern Queensland which spread to cover over 70% of the state with continued heavy rainfalls. The lack of adequate infrastructure to mitigate the effects of the precipitation reminds us how important it is to study previous events and to learn from them. One of the neglected areas of flash flooding and mega floods is that of backbuilding. South-East Queensland has experienced flood events of differing magnitudes during many of the summer seasons in the past three decades, with the 1974 flood being the most significant. At the time of writing this chapter, because of the severe drought in Australia in recent years, no further opportunities had arisen to update the information on flooding realities, although SEQWater has comparatively recently completed a $70 million Wivenhoe Dam upgrade project which involved the construction of a 165-meter wide spillway and related works to mitigate a 1 in 100,000-year flood scenario.

Keywords: GIS, deluges, flood modelling, backwater effects, climate change, Queensland

INTRODUCTION

Two flood events of April/May 1996 and February 1999 are focused on in this chapter, as the causes of both were from different rainfall distribution and river system discharge characteristics. These events occurred in the Southern Region of the Esk Shire,[1] Queensland, where the dams for the city of Brisbane, with around three million people, are located. Analysis techniques are reviewed on how flood inundation has been mapped and estimated previously. The line of inquiry taken in this chapter is whether prior attempts of mapping have included the modelling of backwater effects and if not, is it possible to map and model backwater effects when runoff is unable to discharge from the tributaries. The need for an effective model to estimate backwater effects, for use by emergency services personnel, for example, is demonstrated. The projected increases in deluges worldwide adds to the impetus to find solutions to backbuilding.

History of Significant Flooding in Brisbane and Environs

The following history highlights some of the key water movements in three of the major floods in the areas under review. The graph in Figure 12.1 depicts the highest annual flood peaks between 1840 and 2000.

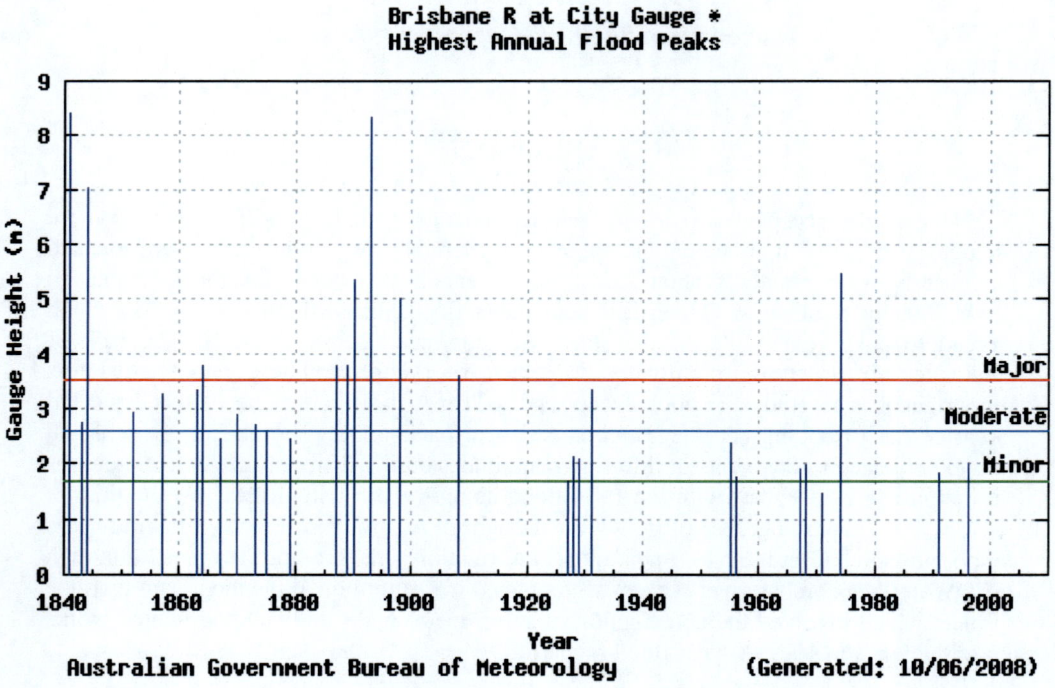

Figure 12.1. Highest annual flood peaks between 1840 and 2000. Reproduced with permission.

An elaborated description of these floods and their effects can be found on the website of the Bureau of Meteorology;[2] the notes date from the 25th January, 1974 through to February 1999.

[1] Now known as the Somerset Region after an electoral redistribution in 2008.

[2] Bureau of Meteorology. http://www.bom.gov.au.

DELUGES, DISPERSAL AND BACKBUILDING

Cyclones, hurricanes, tornadoes and severe thunderstorms are indeed the harbingers of wet times (de Grazia, 2008). One of the more accepted aspects of climate change is that we will be subjected to very heavy rain events that can be classified as deluges (Garnaut, 2008; IPCC, 2007). In order to solve future problems in flooding rains brought about by climate change and normal cyclical weather events, it is useful to analyse previous flood events where back building occurs.

In back building, it is the inability of the river systems to discharge water that causes the back building; apart from the large volume of water and its speed and force of flow, the impeding of its dispersal may be due to blocked gullies, drains, silted up creek beds, etcetera. Generally, there is not enough capacity in rivers to cope with the sudden roar and swell of the wealth of water bestowed in such weather events. Thus backbuilding can be conceptualised as reverse flooding.

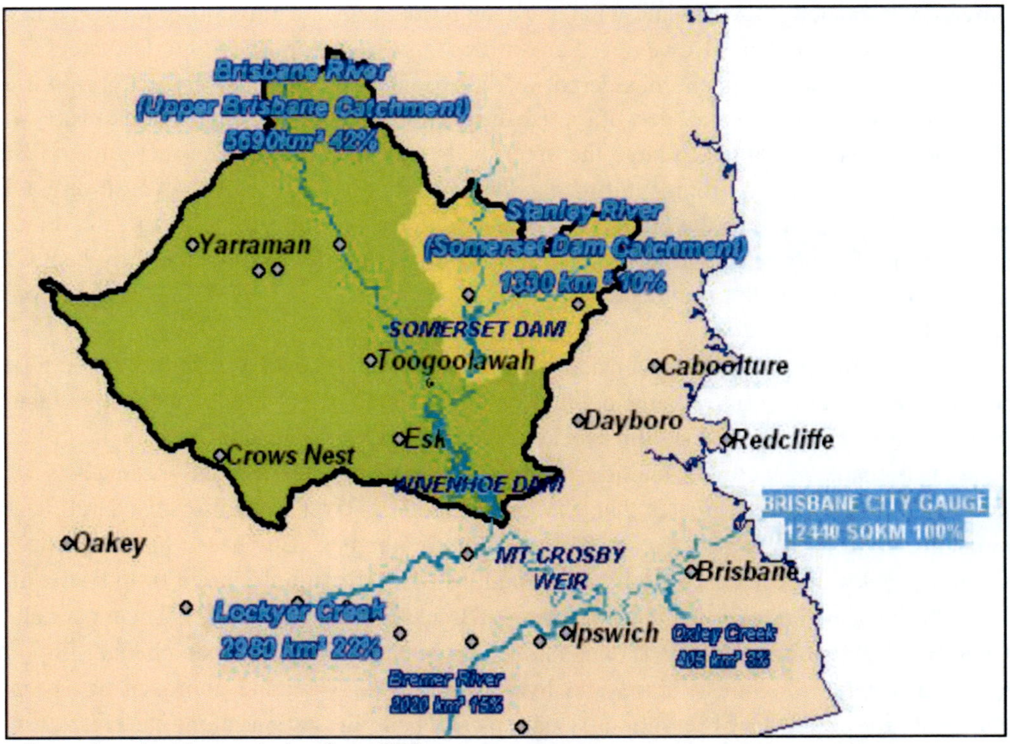

Reproduced with permission from SEQwater.

Figure 12.2. Brisbane River Catchment Areas.

The most recent deluges, in the areas under study in South East Queensland,[3,4] whilst classed as severe weather hazards, were in fact not as severe as those in 1996 and 1999 and

[3] Post script: In the first quarter of 2009, severe flooding rains inundated parts of the north and south coastal areas, flooding some towns and properties. It is noteworthy that Coastal cyclones and storms were a feature of coastal living until the time of the great drought in Australia.

[4] To see map of South East Queensland, go to http://www.mbb.com.au/Maps/seqldmap.htm.

could not match the widespread flooding of 1974, when large tracts of Brisbane were under water.

It appears that little research in the literature has focused on backwater effects, although flood inundation has been mapped and estimated previously and some of that research has estimated siltration and alluvial depositions. Since the catastrophic 1974 flood in South East Queensland (SEQ), several other floods have occurred that have attracted little mitigation attention, although they have occurred in the shire that contains the catchment area for Wivenhoe Dam which provides Brisbane and Ipswich cities with this rich resource. While attention may now be drawn to flood mitigation following the cyclonic floods in June/July 2007 throughout several states of Australia, it has been neglected in South East Queensland due to the ongoing severe drought in Australia, where provision of water has attracted billions of dollars of expenditure in the past three years and will remain the priority resource concern along with energy resources, until sufficient rain (and that means three weeks of flooding rain) has been received in the dam catchment zones (See Figure 12.2).

Fortunately, construction was undertaken on Wivenhoe Dam in the previous three years to upgrade its capacity to withstand what has been termed a 1 in a 100,000 year flood and this has now been completed. However, there are a large number of tributaries, creeks and causeways that have not been considered, to the extent that many are now overgrown and blocked by trees which have been uprooted during long years of drought and severe wind storms. Requests by farmers to have the creek beds excavated have been met with refusals, based on a complex of environmental management protections issues. However, it only takes a lay person to realise that the creeks will not be able to contain the flooding rains that are needed to end the drought, and action on clearing of the creeks, compounded by backwater effects, will come too late to save the low lying farms, animals and residences in many of the rural shires.[5]

Thus it seems appropriate to consider methods to estimate backwater effects due to past flooding events and current extreme conditions (attributable or not to climate change factors) in the southern region of the Esk Shire located in South East Queensland.[6]

What is really needed is a method for detecting when and how floodwater reacts when runoff and stream flow is greater than the capacity of the river system to discharge it. The approaches to this scenario have different scientific themes. Engineers have approached estimating backwater effects by hydraulic equations that estimate discharge from a stream by step back-water methods from a reference point (Brackenridge et al., 1998). The most notable of these is the Manning equation (Featherstone and Nalluri 1995) for open channel flow that allows for surface roughness, slope and hydraulic radius. A second approach has been to gather real time data from observations, locations and gauging stations along the river channel and to compare these with historical floods.

Advances in remote sensing have allowed a variety of methods, independent of weather conditions, to detect flood inundation areas. Integration of remote sensing and GIS (Geographical Information Systems) into a World Wide Web architecture have allowed images to be disseminated real time to interested parties. Combining hydrological studies

[5] This was evidenced when some of the farming areas in the study area were flooded in late 2008 and became lakes until the flood waters receded.

[6] A map of the shire can be seen on http://www.esk.qld.gov.au/the_shire.htm.

with the use of a GIS has provided the final component necessary to create runoff flow models to simulate the path of flood waters.

One of the questions asked of the literature was what methods of mapping and modelling backwater effects had been determined to date. Could, for example, a method be developed to determine floodplain inundation using satellite remote sensing and how this could be integrated within a GIS, because GIS combined with hydrological Modelling software can be used to map and model flood events, predict runoff and discharge. It was anticipated that the findings might indicate a way to simulate the backwater effects associated with discharge from minor tributaries being stymied at confluence points along the main river.

Why Estimate Backwater Effects?

As outlined earlier, with such a serious situation developing in drought stricken Queensland, backwater effects can cause substantial damage to infrastructure, farms, residents, farming animals, wildlife and flora and a range of hazards associated with local tributaries being unable to discharge during heavy rainfall events. These effects include: (1) Closures of roads and bridges; (2) Flooding of houses and properties; (3) Erosion to embankments and surrounding lands due to changing dynamics of water surface and subsurface turbulence; (4) Catchment waters being affected by pollution gain from areas not usually inundated; (5) Exposure to increased salinity; and (6) Chemical and pollutants seeping into the water table.

Since the initial research was undertaken, a population boom has occurred throughout South East Queensland and with it a number of people have taken up the "tree change" phenomenon. With a continuing growth of population and concurrent expansion of industries and businesses, as well as traditional agricultural activities, 'now more than ever' is the time to voraciously test and interrogate methods that will assist residents, business owners, farmers and emergency service workers when the next flood incident, or another flash flood, or backbuilding occurs.

Searching for a Workable System: Application to Water Resources

What is really needed is an outline of the requirements of a system which will simulate the backwater effects showing when causeways and bridges will yield to floodwaters. It is anticipated that this would occur with enough time to allow disaster response efforts, such as those provided by the State Emergency Services, to prepare, warn and evacuate community members if necessary.

A suitable output would be a map showing temporal variations in floodwaters with the associated static layers of roads, bridges and towns for the use of field personnel and those in flood mitigation. The dissemination to the public (Wang et al., 1999), by the use of the World Wide Web would be a contribution to community awareness and increased self efficacy at times when the community has to fend for themselves.

The Bureau of Meteorology in Brisbane issues warnings about river and creek heights, estimated flooding heights and the times that they are expected to occur. SEQWater estimates when to release floodwaters from the spillway gates to reduce the chance of gaining larger flood peaks.

This system would ideally go to a larger spatial scale in order to locate the tributaries, creeks, roads and property boundaries. Also providing maps of potential outbanking and inundation at the local level may help inform or reassure residents that floodwaters are being monitored. As in the case of bushfires, the communities' and individuals' ability to respond is dependent upon receiving adequate warning and accurate information.

Associated Benefits

The flow on effects from the monitoring of floodwaters, outside the immediate disaster mitigating, include the ability to: (i) monitor sedimentation in catchment waters; (ii) estimate erosion; and (iii) calculate silt deposition.

A further benefit of creating a flood monitoring system is the ability to monitor other environmental effects. These effects hamper water quality measures put in place by municipalities and cause degradation to some of the sub-catchments due to erosion. It is not unreasonable to assume that a robust flood monitoring system could be transferred into salinity and water quality monitoring when not being used for flood events.

This would assist in water quality assessment downstream (which is now paramount with the advent of recycled water and management of Moreton Bay), where the sediment is ultimately deposited. Local water catchment groups may be able to assess potential damage to creek and river banks and to devise a response to erosion.

Study Areas: Focus Area and Flood Events

While there were three floods in 1974, 1996 and 1999, it is argued that because the Wivenhoe Dam Construction was completed before the 1996 and the 1999 floods, it should therefore have provided mitigation of these two flood events, unlike that of the 1974 flood. Therefore the conditions surrounding these later floods would have different aspects to be studied than had these floods occurred before Wivenhoe Dam was completed.

However, very little has been published about how to assess backbuilding effects in the smaller tributaries surrounding major catchment areas, such as the Brisbane River which experienced its last major flood (1 in 100 years likelihood) in 1974. This flood was so severe, that it made international media headlines for several days.

The two main flood events that occurred in the Brisbane Valley region of South East Queensland, after the Wivenhoe Dam Construction, could be used as a basis to design or recommend the structure of a system to animate such flood events in the future. Two flood events, together with a description of the typography of the areas in which they occurred, are now outlined to demonstrate the issues involved with backwater effects.

The areas in focus encompass the Brisbane and Lockyer Valleys with some influence being exerted by the Bremer River catchment area. This region was chosen due to the flood events outlined earlier. Historical records from the Bureau of Meteorology have also shown that the district is prone to smaller flash floods and flood events on a regular basis.

South East Queensland lies in a subtropical region that, when not in drought, experiences intense rainfall events at sporadic intervals throughout different areas in the region. Localized flooding, flash flooding and backbuilding of river waters can cause inconvenience to many, but extreme losses to some. The Brisbane Valley region has flood mitigation means in place with the establishment of the Wivenhoe Dam. Currently, flood mitigation is used by SEQWater[7] to assess the most suitable time to release floodwaters from the spillway gates at the Wivenhoe Dam (see Figure 12.3).

Flood Event 1: Brisbane Valley Catchment - April 30/May 1, 1996

This area includes the Brisbane Valley catchment area and the Lockyer Valley catchment area. These two areas merge below the spillway at Wivenhoe Dam. The drainage from these areas is influenced by peak discharges from the Bremer River at the confluence with the Brisbane River.

Rainfall, catchment runoff and discharge characteristics for both these events were of a different nature (flood event two is outlined next). Some of the main locational features of the Southern end of the Esk shire are Lowood, located at 27^0 27' latitude, 152^0 34' longitude, west of the Brisbane Valley highway and on the bend of the Brisbane River. The spillway gates from Wivenhoe Dam are upriver of Lowood at 27^0 33', 152^0 36' and the juncture of the Brisbane and Bremer Rivers is 27^0 34', 152^0 38'. Lockyer Creek joins the Brisbane River north west of Lowood at 24^0 27', 152^0 36'. The Brisbane River snakes its way around and past small towns as do many of its tributaries.

The main creeks contributing to the catchment above the dams are Stanley River, Sheepstation Creek, Gregors Creek, Long Creek, Emu Creek, Linville Creek, Monsilldale Creek and Cooyar Creek. Catchment waters below the spillway gates of Wivenhoe Dam include the runoff from Buaraba Creek and Lockyer Creek, as well as discharges from Wivenhoe Dam.

Table 12.1. April 30/May1, 1996 Rainfall maximums

Location	Actual in mm	Previous Highest mm	Day/Month
Crows Nest	334	227	Month
Esk	375	248	Month
Gatton	573	232	Month
Lowood	468	295	Month
Somerset Dam	303	216	Month
Mount Nebo	946	344	Month
Esk	120.8	114	Day
Lowood	115.4	88.4	Day

Baesed on data from the Bureau of Meteorology, 2001.

Predominately the types of landcover include crops, pastures, water, native forests, woody vegetation, urban areas, and some plantation farms. The terrain consists of small hilly

[7] http://www. Seqwater.com.au.

areas with native forests surrounding much of the Wivenhoe and Somerset Dams.[8] The catchments leading into these dams are larger than 7000KM^2 and the catchments downstream are of a similar scale.

To map the effects of flood waters accurately, a suitable scale needs to be found for the tributaries due to many of the significant features being geographically positioned as a cluster.

During late April 1996 to the 1st May, the topical rain depression covered the area and delivered multiple deluges, thereby bringing on the first flood event. When the stream flow from the Bremer River merged with the Brisbane River, the intensity of the flow from the run off into the Bremer River, resulted in it becoming the primary source of water discharging from the mouth of the Brisbane River. Hence the reaches of the Brisbane River upstream of the confluence of the two rivers back-built and then overflowed into some of the smaller tributaries. Eventually the reaches of the Brisbane River bellow the spillway gates, crossed roads and submerged major bridges on the Brisbane Valley highway and the Gatton Esk Road.

Another factor adding to the intensity of the flood was the presence of a tidal surge and gale force winds at the mouth of the Brisbane River. After the effects from this had contributed to the backwater in the Brisbane River, the Lockyer creek started to discharge the floodwaters near the small town of Lowood. Further roads and bridges were immersed or cut from flood waters in the Lockyer Creek and flooding was tempered by a nil discharge from the spillway gates at Wivenhoe Dam.

Rainfall intensity in the few days leading up to the flood, reached the highest levels since 1974 and the drainage capacity from the smaller tributaries, below the Wivenhoe spillway, became negative, as the backwater effects from the discharge downstream became evident. Table 12.1 contains a summary of the significant falls in the catchment area for Wivenhoe and Somerset Dams.

Localized flooding occurred at all major bridges, causeways, and creeks, isolating many small pockets of homes for differing periods of time. Major highways in the region were cut and movement became dependant on tidal times for the major river crossings. Low tidal changes allowed some runoff to be discharged into the Brisbane River, whilst high tides forced the runoff to back build, re-flooding local tributaries. On April 30, many people commuted long distances to work in the morning, but could not return home that evening, and had to wait until the tide receded on the morning of May 1st, 1996.

It is because of the result of the capacities of the smaller tributaries being hampered by upstream and downstream events, that it is necessary to investigate the backbuilding effects of floodwaters.

Flood Event 2: Upper reaches of the Brisbane River

The second study area is the Bremer and Lower Brisbane catchment which feeds into the Brisbane River east of Ipswich and often affects the runoff and discharge capabilities in the Brisbane Valley. The Bremer River catchment area tributaries consist of the Laidley Creek, Blackfellow Creek, Tenthill Creek, Flagstone Creek and Franklin Vale Creek. This includes the area south to Lake Moogerah ($28^0 43'$, $152^0 32'$), west of Gatton and east to the juncture of the rivers near Karalee ($27^0 34'$, $152^0 51'$).

---

[8] For map of Wivenhoe Dam and Somerset Dam, see http://www.seqwater.com.au/content/standard.asp?name=FloodOperations.

Major features in this area include the Gatton and Laidley shires and the outskirts of the Ipswich city urban area. Landuse catergories include agriculture, open forests, town centres, water, native forests and light industry. These areas are also experiencing rapid population and industry growth.

While this analysis focuses predominantly on the Brisbane and Lockyer catchment areas, it includes the Bremer River because it contributes to flood events in the geographical region. A major location of interest is the merge of the Brisbane and Bremer Rivers. Also of significance is the location of bridges that cross the rivers or the creeks mentioned. Finally the population and property at risk need to be considered, as it is the monitoring of these which has been of concern to the emergency services teams.

The second event took place on February 18th, 1999 and was preceded by high intensity tropical downpours exceeding normal rainfall for the area. The location of the floodwaters was in the catchment areas draining into Somerset Dam, upriver of Wivenhoe Dam (See Figure 12.2). Table 12.2 shows the peak rainfalls at some of the locations discharging into the catchment areas. Major flooding occurred in areas above Somerset dam along the Brisbane River. This region forms part of the upper reaches of the Brisbane River and became inundated, as the runoff from the catchment exceeded the capacity of Somerset Dam. With releases on the 9th February from Wivenhoe Dam causing flooding, the surrounding towns and creeks experienced a backbuilding effect, when the floodwaters began to accumulate in the tributaries without being able to discharge.

As a result of the capacities of the Somerset Dam being reached, discharge was released into Wivenhoe Dam[9]. These waters were in turn released into the Brisbane River with the spillway floodgates (See Figure 12.3) open at maximum discharge. Following this, some roads were flooded and many tributaries were cut off again, as they represented flash flood events. With maximum discharge from Wivenhoe, many residents and tourists were caught unawares. The water plume (see Figure 12.4) was reported to be visible from satellite images for this continuous maximum release. It must be kept in mind that since that date, as indicated earlier, a new spillway, completed in late 2006, was constructed to mitigate the effects of a flood of a magnitude that might occur only once in a 100,000 years.

Table 12.2. Significant rainfall recordings for 9th February 1999

Location	Actual in mm	Previous Highest mm	Day/Month
Somerset Dam	276.2	248.9	Any Day
Crows Nest	211.2	153.4	Day in February
Toogoolawah PO	190.4	144.3	Day in February

Source: Bureau of Meteorology, 2002.

[9] For map of Wivenhoe Dam and Somerset Dam, see. http://www.seqwater.com.au/content/standard.asp?name=FloodOperations.

Reproduced with permission from SEQwater.

Figure 12.3. Release of flood waters through spillway gates.

Reproduced with permission from SEQwater.

Figure 12.4. Example of Water plume at Wivenhoe Dam.

METHODOLOGY

A review of the literature in general and a review of case studies was undertaken in order to assess what methods and techniques were available and had already been utilised for estimation of backwater effects. The intention was to determine what, if anything, gleaned from these published case studies, could assist the Esk Shire in future flood mitigation.

Review of Available Techniques for Backwater Mitigation

As indicated earlier, it is necessary to review the techniques available for floodplain determination, for creation of digital elevation models and for calculation of runoff and discharge of catchment areas. Hopefully then, recommendations can be made as to how this can be applied to the areas under threat of the SEQ flood events outlined in this chapter.

Data sources are the main requirements that need to be determined and the best techniques for the use of this data is to create a floodplain simulation capable of predicting/identifying areas susceptible to backbuilding of floodwaters. Initially, the availability of data sources would need to be confirmed in the desired resolution. Potential data sources are described later in this chapter, along with a discussion about how these could be used for the creation of a backwater flood analysis.

Of the literature reviewed, the common strategies were to: (A) Gain an overall estimation of the land cover types surrounding the study area; (B) Estimate runoff from terrain characteristics; (C) Estimate the discharge from the river system; (D) Compare with remotely sensed images; (E) Validate the accuracy of the estimates achieved; and (F) Create catchment models.

Techniques currently used in determining floodplains and creating discharge models are now examined, along with a brief summary of their relevance to the proposed model. Following this, summaries of three case studies relevant to the issues are hand are analysed and then a synthesis of what contribution the literature and the case studies can make to the problem is laid out.

Floodplain characteristics have been determined through use of remotely sensed images. One case study utilsing GIS is the Amazon Basin, where a significant amount of research has been conducted in mapping the inundation area and the floodplain itself. This has been attempted using SMMR 37 GHz (Sippel et al., 1994; Sippel et al., 1998) to provide the estimation of the surface water present and through SAR (Melack et al., 1994). The difficulties encountered in this example were the effects of mixed pixels and how to assign wet/dry classes to them.

Further studies have indicated that the use of programs such as HEC-RAS (Kraus and Dodson, 1999) and GISStreamPro (Dodson and Li, 1999) may provide a solution to estimating backwater effects of flood events.

In conceptualising this project as a series of stages, the following factors would need to be considered to create a comprehensive system capable of creating catchment models and simulating backwater effects. The final product would include:

1. A user interface on the World Wide Web for members of the public.
2. Localized information for State Emergency Service volunteers.
3. Continuous records of flood events for future analysis.
4. Identification of isolated areas and the pre-planning mitigation of floods.
5. Extra warning systems enhancing longer lead time for residents likely to be isolated.

Review of Case Studies of GIS Usage in Water Resources Management

Reviews were undertaken on examples of GIS usage in water resources management. Each of the three examples covered different aspects from flood inundation mapping, flood risk assessment, effects of floodwaters and representations of catchment hydrology with the aid of a GIS. A summary of findings of their contributions to this problem is given next.

The first case study was situated at the Watershed of Mill Creek located at Lufkin, in Texas. The relevance of this case to the problem scenario in the Esk Shire of modelling backwater effects was in the example of processing smaller reaches of the catchment area in a

GIS and creating a floodplain map. The capabilities to recalculate all the cross section boundaries after redefining the TIN has greater prospects, especially if this occurs with an increase in spatial accuracy and resolution. Recent improvements in Positron Emission Tomography system accuracy will make this even more possible.

The second case study was set around the Ohio River, Cincinnati. Pertinent research from this study included the description of World Wide Web dissemination to the public, overlaying what would be street and property boundaries. However, this study concentrated only on downstream flow and out-banking. Therefore, further examples are needed of the back-building effects of floodwaters. Nevertheless, the scale used is sufficient for the preliminary needs of our problem scenario research interests and it is anticipated that contours of around 3m would be acceptable at the preliminary stage.

The third case study was The River Thames in the United Kingdom. Areas at risk were identified through the use of the UK's unit postcode. These were intersected with the flood depth to give the flood intensity for each location. The interface was developed to allow navigation around the floodplain. Features included the ability to search by geographical locators or grid coordinates, turn themes on and off, enquire into flood depth and geo-demographic information. Data from this project was available on CD-ROM. The collection of SAR data required a data management of resources by the selection of flood prone areas from the study area, as a way to reduce data storage requirements.

Overall, in considering what the reviews of the case studies had to offer towards the resolution of the backwater effects scenarios in South East Queensland, there were several contributions that might be of assistance: (a) TINS; (b) Identify roads, culverts; (c) Method to determine flooded areas (Polygon); (d) River elevators and contours; (e) user interface in ARCIMS inputting river heights for flood area; (f) Choice of web interface; (g) Multi choice for topography; (h) Aerial photos; and (i) Choice of software: Arcview and IMS; 3d Analyst; HEC-RAS.

It thus seems that due to the nature of the recorded events, in the Brisbane valley catchment, a combination of processes would be needed to mitigate the effects of backbuilding and these are outlined in the next section.

DISCUSSION

Proposed Strategy for Developing a Backwater Modelling System for the Esk Shire

One of the most relevant themes in the literature is the collection of Aerial SAR data to obtain the required resolution. In a recommendation made by the first author on Bushfire Detection, the use of airborne remote sensing to collect meaningful data at the local level was suggested. The potential to use any data collected for a variety of monitoring needs, such as flood mapping and land use categorization, is due to the large spatial scale that the data can be acquired in. For the southern regions of the Esk Shire, any intersecting with property boundaries would provide the "at risk" areas.

Investigating the techniques to reduce the size of the data files by targeting the areas at risk of flooding is another of the benefits from following an example of this type.

Furthermore, the extent of the area of interest and the limited amount of data are shown to suit SAR flood risk assessment. Perhaps a more global approach is to map with SAR during real time events, combined with the use of hydrological modelling of specific known areas of risk.

There are six stages in establishing a backwater modelling system. Before mitigation methods can be planned, these six stages have to be in place: (1) collecting information necessary to create the base layers; (2) creation of a catchment model; (3) inputting of historical flood events; (4), testing the accuracy of the elevation models; (5) estimating inundated areas; and (6) obtaining flood location maps. These stages are now articulated.

1. Collect Information

The first stage would be to collect the information necessary to create the base layers. These include details of roads, bridges, towns, infrastructure, and land use classes. Information in these layers does not frequently change and therefore could be used in any of the analysis used for water resources. The layers can be combined in a view suitable for overlaying of flood inundation maps. They can also form part of the runoff and discharge potential and identify hydraulic features in river channels. As software such as ArcIMS allows for the internet display of flood maps with the ability to toggle themes, the availability of static layers would provide a component of the user interface.

2. Create a Catchment Model

After establishing the base layers, the next step would be to create a catchment model by locating streams, creeks and dams within a hydrological model. For this problem scenario, a topographical model has been created by SEQWater, for the Brisbane and Lockyer Valleys. This was created from 50m contour layers available from the Department of Natural Resources. Later work involves the creation of 1m vertical levels for analysis of the surface dynamics of storage waters. Further research and testing of models would need to utilise up-to-date contour models.

The main decision would be to decide on how suitable the scale is for the current resolution of the Digital Elevation Models. For example, would a 1m resolution be the ideal scale for the purpose of estimating backwater effects? How cost effective would it be to create a zone around the tributaries and the main river channel at the 1m resolution scale? In investigating the establishment of a flood model, Townsend and Walsh (1998) found that the resolution from converting topographic maps to DEMs with 30m horizontal and 10m vertical resolution was inadequate for the analysis they were undertaking. The solution they posed was to use the package TOPOGRID and create DEM from contour data that includes contour lines, point elevation and hydrography information. They gained the hydrography data via Digital Line Graphs from the USGS 7.5m quadrangles. The features present included rivers, lakes, streams represented as single line features and all features were coded with the downstream direction with any anomalies being deleted to minimize sinks and internal drainage.

Garbrecht and Martz (1999) indicated that the size of the DEM needs to be relative to the drainage feature and surface features being studied. Discussions have centred on whether to use single flow direction algorithms or a multiple flow approach. Recommendations for large drainage areas are for the use of single flow directions in well developed channels.

3. Input Historical Flood Events

Previous approaches have used satellite imagery to map flood water and inundation. This has been enhanced through change detection using SAR and the input of known flood elevations. The use of Nimbus-7 37Ghz data has been used by Gidding and Choudhury (1989), to map Amazon basins and floodplains by detecting surface water and performing change detection. Initial attempts to represent river elevation through equating pixel value to river elevation at a large size of pixel indicated a way to determine hydrological features. Further research was undertaken by Brakenridge et al. (1998) through the use of SAR, to map river flood and discriminate water/land boundaries.

Other information of a historical nature would enhance preparedness. This includes discharges of known flood events such as 1-2, 1-5 and 1-100 year events, combined with river heights and areas inundated, and would provide data to categorize current flood events.

4. Test accuracy of the Elevation Models

The next step would be to test the accuracy of the elevation models by randomly selecting control points to assess elevation and downstream flow characteristics (Townsend and Walsh, 1998). In Kraus and Dobson (1999), data manipulation was carried out on attribute tables to allow the TIN to match the actual field measurements. The feasibility and costs associated with the accuracy testing (Dobson and Li 1999) may prove to be prohibitive and alternative data integrity checks may be required.

Combining the actual catchment model with flood data would require each set of data to be accurate. Data integrity checks and processing systems compatibility would ultimately ensure accurate reporting of flood maps. The main contributions to accuracy are the terrain model and calculation of discharge outbanking.

5. Estimate Inundated Areas

In the estimation of the area inundated or the area covered with surface water, mapping can take on two forms: remotely sensed inundation area for real time events, or output generated by hydrological modelling. As one of the main needs in this backbuilding problem is the mapping of flood inundation and backwater effects, the type of layout required forms the essential component of the next stage. Of concern to the public, local councils and members of the State Emergency Services is the location of populations at risk, which includes the buildings and property boundaries, as well as roads that will be closed and bridges currently under water. Overlaying the flood data map as a transparency, to show clearly identified roads, bridges and property affected, would be an ideal solution. Obviously it would be necessary to have simulated trial runs to see if it works.

6. Obtain Flood Location Maps

Output required from an estimation of inundated areas are the flood location maps which provide the scale of inundation and a scale representing the likelihood of inundation. A follow on from this is the simulation of the rise and fall of floodwaters as a time series, to show where and when the waters rose and then receded, giving estimates of time elapsed when roads were cut in measures of hours, days, or weeks; for instance, the tropical Cyclone "May" cut off areas around Normanton for several weeks in late February-early March, 1998. [10]

---

[10] In 2009, large areas of North Queensland were cut off for many weeks and roads were impassable.

Maintaining records of prior flood events and the extent of the area inundated would provide the necessary background data for response to flooding and flood mitigation efforts. Accessing the types of events that occurred, be they 1 in 100 year events or 1 in 2 year events, would aid in the types of responses required for a new flood event. The details, such as which roads and bridges were submerged and what areas became isolated, would give a snapshot view of the possible outcomes of the current event.

Data Sources

The Bureau of Meteorology has a comprehensive archive of rainfall data and flood histories. There are also records available indicating the river heights above Wivenhoe and Somerset Dams and below the spillway gates at various points along the Brisbane River. Some of the processing of data has already been performed by government agencies and departments.

The complexity of the catchments of SEQ and the inconsistency in the data available, have shown the need for an integrated approach to floodplain modelling. To carry out the ideal analysis, TINs or Digital Elevation Models of a high resolution need to be created. This processing would take considerable time to achieve. However, the techniques involved in creating 1m vertical resolutions have been tested by SEQWater in monitoring surface dynamics of storage waters.

However, no single application can be used to detect, map and generate flood events. Furthermore, many of the models reviewed dealt only with the downstream effects and the river systems discharge capabilities. Many were able to show where outbanking or overbanking would occur, but this was done in only a downstream direction. The generation of some of these models is based on simple techniques such as the 8-point direction method which simply determines those neighbouring grids that flow in a downstream direction.

It is assumed, therefore, that some computational analysis would be needed, once it was established that the system being studied had reached the maximum discharge capacity. Water in the catchment would appear to remain stationary or flow in a negative direction. Application of more traditional estimation techniques, based on critical flow analysis, may form a temporary solution until a more robust method is tested.

The computational requirements of detecting flood inundation and modelling the extent of the area have been well documented. The need to subset catchment areas by only those likely to flood is a plausible answer. However, the main area of concern has multiple creeks which criss-cross the catchment's area as well as the Brisbane River which meanders through the area and finally the Lockyer Creek that bisects the lower portion of the shire.

Caution is also indicated by Carpenter et al. (1999) about the use of historical flood events, as these can often be derived from neighbouring sources and this applies also to climatic data for the area. Also of significance are the unstable cross sections which vary over short distances (Carpenter et al. 1999). There are possible examples of this in the Brisbane River outside of Lowood, where the terrain varies as the river takes a sharp bend, adjacent to some hills.

Data input sources for the tasks of simulating/estimating backwater effects are varied. Spatial and temporal resolutions would not be displayed unless they were available in a GIS. The data used for this system would also have multiple ownerships, as the data sources would include, at a minimum, the Bureau of Meteorology, SEQWater, CSIRO[11], Department of Environment and Resources Management and Queensland Department of Emergency Services. Furthermore, future analysis of the interactions between flood mitigation, water supply, catchment management and water quality poses a problem that needs to be solved.

Due to the scale of the area concerned there are very few options open for mapping flood events over many thousands of square kilometers. Being able to select features such as roads and bridges in views that can be toggled on and off is a major advantage of using a GIS for flood inundation and estimation of backwater effects.

Servicing the needs of multiple users such as members of the public, local councils and State Emergency Service volunteers by utilizing the ability to use multiple themes is a key factor in the use of a GIS for flood inundation mapping. Further portability is gained by the application available to display maps via the World Wide Web.

CONCLUSION

Despite the apparent lack of progress in establishing a system to model backwater effects, it is assumed with further advances in hydrological modelling and terrain modelling, that future programs would include back-building as a key feature. In the meantime, developing a system to monitor flood inundation, collect records of historical flood events and provide access to catchment management would provide the platform to adapt to the new emergent hydrological modelling programs.

The availability of remotely sensed data, prediction of flood heights, and estimation of discharge rates from river systems will combine to aid in flood monitoring. Spatial temporal concerns will be reduced, as more satellites are brought into service, sensors gain increased resolution, and detection algorithms improve for currently sensed images. Integrating the system with either a disaster management system or a water resources monitoring system would greatly enhance the benefits of the data acquisition and storage. The ability to acquire AVHRR images four times a day and other images, on a similar temporal resolution, would enable accurate post flood event analysis including the water resources issues of pollution and erosion in the catchment area.

It can be deduced that, as much of the data needed to commence such an endeavour is available for research purposes, the creation of a system to model flood inundation and backwater effects is possible. The ability of a GIS to combine a flood modelling package with a disaster management system means that this is a project worth undertaking with considerable secondary benefits. Expanding the capabilities to include water resources monitoring, salinity monitoring in the catchments and erosion studies is possible within a G.I.S.

With the South East and South West regions of Queensland experiencing severe and long term drought conditions, combined with a rapidly expanding population growth, estimations of backbuilding in these study areas will have even more impetus, with the increasing population at risk and the development in areas that may be affected by road closures and short term flash flooding.

---

[11] CSIRO. Commonwealth Scientific Research Organisation.

References

Ackerman, C. T., Evans, T. A., and Brunner, G. W. (1999). HEC-GeoRAS: Linking GIS to Hydraulic Analysis Using ARC/INFO and HEC-RAS. Accessed 23.04.09 http://proceedings.esri.com/library/userconf/proc00/professional/papers/pap737/p737.htm

Bates, P. D., and de Roo, A. P. J. (2000). A Simple Raster Based Model for Flood Inundation Simulation. *Journal of Hydrology, 236*, 54-77.

Bates, P.D., Wilson, M.D., Horritt, M.S., Mason, D., Holden, N. and Currie, A., (2006). Reach scale floodplain inundation dynamics observed using airborne Synthetic Aperture Radar imagery: data analysis and modelling. *Journal of Hydrology*, 328, 306-318.

Brakenridge, G. R., Tracy, B. T., and Knox, J. C. (1998). Orbital SAR Remote Sensing of River Flood Wave. *International Journal of Remote Sensing, 19*(7), 1439-1445.

Carpenter, T. M., Sperfslage, J. A., Georgakakos, K. P, Sweeney, T., and Fread, D. L. (1999). National Threshold Runoff Estimation Utilizing GIS in Support of Operational Flash Flood Warning Systems. *Journal of Hydrology, 224,* 21-44.

Carson, E.C. (2006). Hydrologic modeling of flood conveyance and impacts of historic overbank sedimentation on West Fork Black's Fork, Uinta Mountains, northeastern Utah, USA. *Geomorphology, 75,* 368-383.

Featherstone, R. E., and Nalluri, C. (1995). *Civil Engineering Hydraulics: Essential Theory with Worked Examples* (3rd ed.). Oxford: Blackwell Science.

Garbrecht, J., and Martz, L. W. (1999). Digital Elevation Model Issues in Water Resources Modeling. Accessed 21.07.06. http://gis.esri.com/library/userconf/proc99/proceed/papers/pap866/ p866.htm

Garnaut, R. (2008, October). *The Garnaut Climate Change Review.* Final Report. Melbourne: Cambridge University Press.

Giddings, L., and Choudhury, B. J. (1989). Observations of Hydrological Features with Nimbus 7 37GHz Data, Applied to South America. *International Journal of Remote Sensing, 10*(10), 1673-1686.

Intergovernmental Panel on Climate Change. (2007). Intergovernmental Panel on Climate Change Fourth Assessment Report. Climate Change 2007: Synthesis Report. Summary for Policy Makers. Available at: http://www.ipcc.ch/pdf/assessment-report/ar4/syr/ar4_syr_spm.pdf

Jeffrey, S. J., Carter, J. O., Moodie, K. B., and Beswick, A. R. (2001). Using Spatial Interpolation to Construct a Comprehensive Archive of Australian Climate Data. *Environmental Modeling and Software, (16),* 309-330.

Matgen, P., Schumann, G. Henry, J-B, Hoffmann L. and Pfister, L. (2007). Integration of SAR-derived river inundation areas, high-precision topographic data and a river flow model toward near real-time flood management, *International Journal of Applied Earth Observation and Geoinformation*, *9*(3), 247-263.

Melack, J. M., Hess, L. L., and Sippel, S. J. (1994). Remote Sensing of Lakes and Floodplains in the Amazon Basin. *Remote Sensing Reviews, 10*, 127-142.

Pickup, G., and Marks, A. (2001). Identification of Floodplains and Estimation of Floodplain Flow Velocities for Sediment Transport Modelling - NLWRA Sediment Project 14/01 (CLW12). CSIRO Land and Water Technical Reports. Accessed 22.07.06. http://www.clw.csiro.au/publications/technical2001/

Richard, A., Kraus, P. E., and Dodson and Associates, Inc. (1999). Flood Plain Determination Using ArcView and HEC-RAS. http://gis.esri.com/library/userconf/proc99/proceed/papers/pap808/p808.htm

Roy, D., Dodson, P.E., Xiaojian, Li and Dodson and Associates, Inc. (1999). The Accuracy and Efficiency of GIS-Based Floodplain Determinations. Accessed 21.07.06. http://gis.esri.com/library/userconf/proc99/proceed/papers/pap809/p809.htm

Sanders, R. A. (1999). Flood modelling using airborne synthetic aperture radar data. Accessed 21.07.06. http://gis.esri.com/library/userconf/proc99/proceed/papers/pap569/p569.htm

Sippel, S. J., Hamilton, S. K., Melack, J. M., and Choudhury, B. J. (1994). Determination of Inundation Area in the Amazon River Floodplain Using the SMMR 37 GHz Polarization Difference. *Remote Sensing of Environment, 48*, 70-76.

Sippel, S. J., Hamilton, S. K., Melack, J. M., and Novo, E. M. M. (1998). Passive Microwave Observations of Inundation Area and the Area/Stage Relation in the Amazon River Floodplain. *International Journal of Remote Sensing, 19*(16), 3055-3074.

Sui, D. Z., and Maggio, R.C. (1999). Integrating GIS with Hydrological Modeling: Practices, Problems and Prospects. *Computers, Environment and Urban Systems, 23*, 33-51.

Townsend, P. A., and Walsh, S. J. (1998). Modeling Floodplain Inundation Using an Integrated GIS with Radar and Optical Remote Sensing. *Geomorphology, 21*, 295-312.

Wang, Xinhao, Huang, Shaoli and Du, Changming. (1999). Development of a Flood Warning Information System. Accessed 21.07.06 http://gis.esri.com/library/userconf/proc99/proceed/papers/pap814/p814.htm

In: Meltdown: Climate Change, Natural Disasters…
Editor: Kathryn Gow
ISBN 978-1-60876-153-1

Chapter 13

BUSHFIRE DISASTER PREPAREDNESS: A RURAL SURVEY

Francine Pritchard, Kathryn Gow, Iraphne Childs and David Chant
Queensland University of Technology, Australia

ABSTRACT

The implications of climate change scenarios for increased wildfire activity is directly related to the need for people living in those countries already most affected by wildfires (Australia, Southern France, and California in particular) and which have experienced significant increases in temperature in the past century (Australia by 1˚C and the state of Queensland by 1.2˚C) to prepare for the onslaught of forest, grass, bush and firestorms. This chapter focuses on a study undertaken in several bushfire-prone shires in rural South East Queensland which was conducted to assess people's preparedness to fight future bushfires (wildfires). This area had suffered severe bushfires in the past decade. The data from 713 respondents is set out and focuses on the factors of risk, exposure and preparedness. Pertinent findings included lack of available water resources, heavy vegetation danger, lack of access or take up of the internet and other preparedness prevention issues, such as inability to afford the equipment or services, or have the time or human resources, to adequately prepare.

Keywords: Bushfires, wildfires, disaster preparedness, natural disasters, temperature increases

INTRODUCTION

> Preparedness is a special issue. As warnings and forecasts of particular events improve, opportunities for saving lives and property will depend increasingly on preparedness.
>
> (NOAA, 1994) [1]

The IPCC (2007) has warned that there will be more severe heat waves with Climate change, and the UN Report (2000) indicated that in Australia, the frequency of days when bushfires would be at their peak would increase from 4% to 25% by 2020. Together with the increased temperature, Garnaut (2008) reported the likelihood that the number of droughts will increase and be more frequent, indeed, as frequent as one in every 1-2 years. Pittock (2005) reported that the period 1995 to 2004 was the warmest decade on record. Thus it is likely that hot dry days, with little or no water available to fight fires with in rural areas, will become the norm in future summers. At the time this chapter was being written, Queensland was heading into a summer for which the worst bushfire season in 15 years was being forecast; indeed shortly afterwards, firestorms ravaged Victoria leaving hundreds dead and thousands homeless with heavy losses to farming animals, wildlife, forests and infrastructure.

Normally disaster preparedness is a vexed issue with rising numbers of people being affected in some way by natural hazards, whether that be via tsunamis, hurricanes, earthquakes or prolonged drought, heat waves, cyclones, floods, wind storms and bushfires, with which Australia is more familiar. Australia has invested considerable effort in educating and promoting bushfire preparedness. It is against this backdrop of preparedness that the study was undertaken.

Prelude – Dry Crackling

As South East Queensland prepared for another long dry hot summer in 2005/2006, warnings were being transmitted about fire preparedness. Already in winter and spring (not the usual seasons of concern) bushfires had been reported. Along with most states in Australia, South East Queensland had been suffering from the longest drought on record and residents had need to feel worried. Fire authorities gave instructions about methods to prepare houses and properties in order to prevent or mitigate damage to livestock and destruction of property. (See the Rural Fire Service website[2] and the Queensland Fire and Rescue Service[3] website.)

These warnings and instructions to prepare generally included the following advice: (1) Clear scrub and undergrowth around your property; (2) Keep grassed areas well trimmed and watered; (3) Clear your gutters of leaves and rubbish; (4) Store flammable items well away from the house; and (5) Make sure you have hoses and reserve water supplies available. As Gow, Pritchard and Chant (2008) indicated, these strategies are not new; they have been consistently relayed to the public via television, radio, local newspapers and letter drops, as well as through real estate business mail outs.

---

[1] http://www.nws.noaa.gov/oh/hrl/surveys/alberto/contents.htm.
[2] www.ruralfire.qld.gov.au.
[3] www.fire.qld.gov.au.

Indeed, similar warnings were made before, during, and after the severe summer fires of 2004/2005, and thus it is interesting to note whether a year after these severe fire episodes, residents living in areas which had suffered devastation from those bushfires (grass fires, forest fires, firestorms etc.), were sufficiently motivated to ensure that during the following summer months, they would be as best prepared as they could be.

A portion of South East Queensland was chosen for this study (based on its history of bushfires) to ascertain residents' preparedness following previous bushfire disasters. This study is important, as it tapped a time frame in which there had been no rain since the 2004-2005 fires and at a time before the floods engulfed a large area of Queensland in early 2008 and then storms and floods late in 2008. Thus the attitudinal confound of 'time passes and we have had rain and everything will be all right' had not encroached on the population from which the data were being collected.

Disaster Management

Disaster preparation encompasses many different levels of society and involves infrastructure development, large scale emergency evacuation plans, dissemination of information to the public on the steps householders and businesses can take, reviewing council zoning for hazard prone areas, health department and emergency services and others working in concert to be prepared in the event of a catastrophic event. At the national level, the Department of Transport and Regional Services (DOTARS) explains issues regarding jurisdiction: research, education, and information that is supplied by Emergency Management Australia (EMA) to the public and to each state or territory emergency services division. At the state and local government levels, awareness campaigns including media briefs, mitigation planning and support to voluntary emergency services organizations are maintained.

At a localized or individual level, disaster preparation involves having an evacuation plan for the family (Victorian Department of Sustainability and Environment, Emergency Management Australia (VIC DSE and EMA), provision of supplies, hazard or fuel reduction for bushfires, moving stock in flood or drought prone areas, and keeping gutters and drains free from debris. Online resources such as the Emergency Management Australia (EMA) website or accessing the individual state government websites, such as the South Australian Government website can also be helpful.

Because the location of this research is located in Queensland, the nature, utility and readability of the Queensland Government website is pertinent (see Gow, Pritchard and Chant, 2008). The Ipswich City Council, Disaster Management webpage (Ipswich is the city closest to the rural areas under study) describes the role of disaster management in the community as: the identification of hazards and why they occur, evaluation of risks, likely occurrences and consequences and preparation of plans to reduce that risk. The local shires in the study area – Esk, Crows Nest, Gatton, Kilcoy and Laidley all had some Disaster Management Plan or Strategy in which they were involved, although to differing extents and emphases.

BUSHFIRES - A NATURAL HAZARD

Bushfires have many causes including lightning strikes, escaped controlled burns, cigarettes being thrown out of car windows, arson, and other vehicular ignitions (Borger, 1999). Often they can start off as slow burning grassfires that only need a change in one of the main conditions to escalate to bushfire status. Fires of all types follow the Triangle of Fire as described by all fire fighters; Heat, Oxygen and Fuel. Major variables used as predictors of bushfires are wind speed, humidity, soil moisture, and fuel load. An increase/decrease in one of these often leads to an escalation of the bushfire threat; for instance, an increase in wind speed can often change a slow moving grass fire into a fast moving bushfire front that moves from treetop to treetop sparking spot fires that start in front of the original bushfire line.

Regardless of the origin of the bushfire, once it has reached the stage of creating its own energy, a fire will pose a serious/extreme threat to residents and their houses in its pathway. This is why it is essential to ascertain the criteria for establishing regions of particular susceptibility to bushfires and assessing knowledge and actions of residents to prepare for fires. In the case of the area chosen for this research project, its susceptibility was proven time and again with high fuel load, drought conditions, frequency of high winds and heat wave weather conditions. What had not been gauged before was the residents' preparedness to prevent and fight encroaching bushfires and whether or not they had the tools and the skills to do so.

It is important to obtain an indication of the level of awareness of the householders to sources of information, such as those available on the World Wide Web (Rohrmann, 2002) and how they access bushfire threat updates; what steps landholders take to protect their property and to reduce the possibility of wildfire and severe fire storm damage; and what their fire management strategies are, such as fuel reduction (Brandes, 2003).

Susceptibility to Bushfire Hazard

Vulnerability to bushfires is made up of a number of factors: Geographic Location, mobility of household residents, access to infrastructure, actions of residents, awareness of bushfires and the dangers they pose, and seasonal variations. Vulnerability for this project will be specified as 'awareness' combined with already identified geographical indicators of susceptibility.

How can we tell which areas are susceptible to bushfires? Generally, if we speak about bushfire susceptibility in identifying areas prone to bushfires based on geographic indices, then a Bushfire Susceptibility Index (BSI) is useful. The BSI is based on previous history of fires, the topography of the area, land cover types, vegetation types and fuel loads throughout the seasons. The basis for a susceptibility indicator as a need in preparation is demonstrated in Borger (1999). The BSI for the area under study was pre-determined – it was a disaster just waiting to happen. Figure 13.1 displays the factors involved in determining bushfire susceptibility.

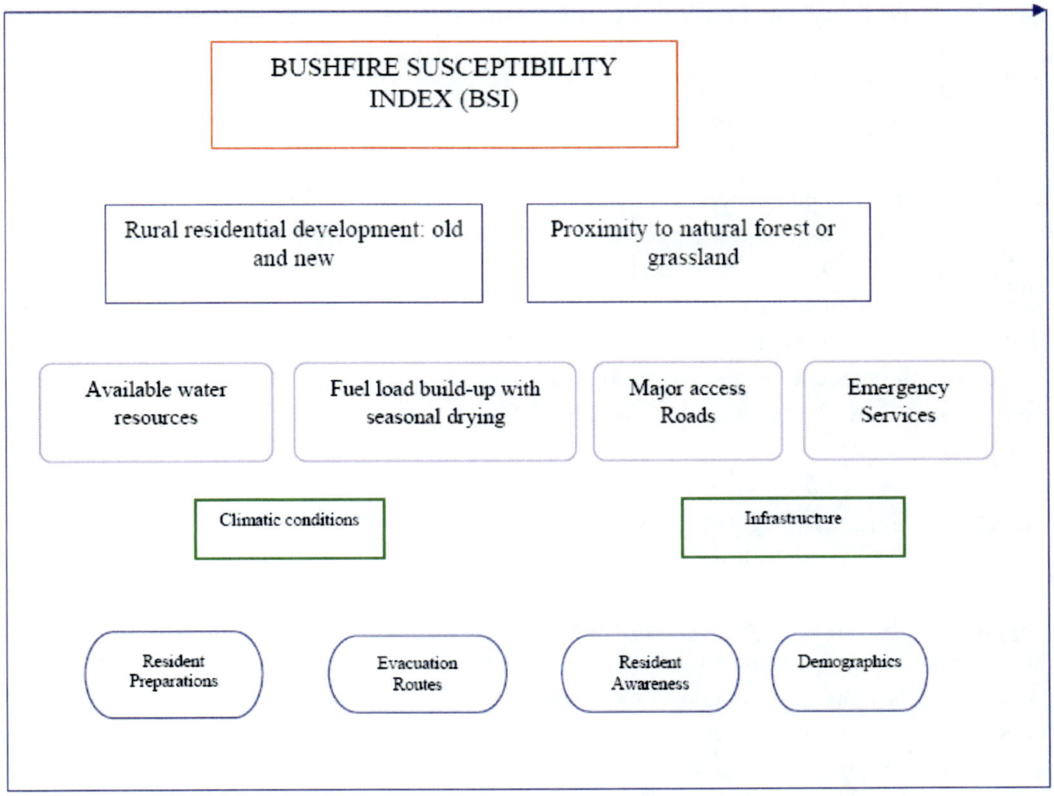

Figure 13.1. Determining Bushfire Susceptibility.

Thus from the factors in Figure 13.1, the areas chosen for the study were established rural areas, in natural forest and grassland regions, and had had fuel load build-up before the last fires, and the majority of the areas around those burnt out areas had not yet been back burned. So serious susceptibility to bushfires still remained. While there was major road access to the areas, within the forests, both private and public access was severely restricted and the grasslands were tinder dry. Although the Emergency Services had been stretched to breaking point, they had gone on fighting fires in the previous fateful summer and were on standby for repeat bushfire occurrences. At the time of the study, the climactic conditions remained the same: very dry, very hot and with no signs of significant rain. Infrastructure had not changed, but work was underway with councils to update disaster management plans and strategies, and emergency services reviewed evacuation routes for future fires. Demographically, a few hundred more people moved to the potential disaster areas where cheaper housing was more available. We had no way of knowing to what extent both the awareness and preparedness of residents had changed.

In examining the BSI of the studied area, we already knew that some of the properties chosen for the study had suffered serious bushfires in the previous few years, culminating in major losses to houses, outbuildings, human lives, properties, farm and domestic animals, crops, wildlife and forests of the summer of 2004/2005.

In studies of this kind, distance measurements should also be calibrated once the study area is chosen, to investigate if a household's distance to the potential source of a bushfire influences the level of preparedness. This was done and the specific outcomes of micro differences in distance and preparedness have been published in an article in *The Australasian Journal of Disaster and Trauma Studies* (see Gow, Pritchard and Chant, 2008).

When attempting to prevent future hazard scenarios leading to loss as evidenced in the bushfires seasons in the last decade in Australia, it is useful to explore residents' participation in disaster mitigation/preparation and their knowledge of the resources available to them. It is also necessary to include the perception of risk (Paton, 2003) and the motivation demonstrated in acting on that perception (see Bushnell and Cottrell, 2007 on preparation for bushfires in North Queensland undertaken at approximately the same time that this study was carried out; see also Childs, Pritchard, Gow and Hastings, 2006 for the preliminary findings of this study). For instance, even if a person's values indicate a strong reason for preservation of lifestyle, the perception that a risk does not exist lessens the person's motivation to initiate preparations (Gow, 2006; Paton and Burgelt, 2005).

Bushfire Disaster Preparedness Resources

There are several resources that the public can access that would assist them in preparedness for disasters and some of these are particularly indicated for bushfires. Each of the websites, such as the Red Cross, BOM, EMA and various state government web pages, include sections on how to cope emotionally when looking at individual and community preparation to aid and foster recovery; however we are particularly interested here in preparedness (see Pritchard and Gow (2008) for discussions about psychological and mental health issues). Rohrmann (2007) has recently evaluated the perceived value of websites created and maintained by fire authorities in Australia. One of the research questions he pursued was the degree to which the information on the websites increases awareness, preparedness and coping in real bushfire situations. Because the internet may be seen to be a valuable information source (see Pritchard and Gow, 2008) for the public, it is essential to assess the extent to which rural residents have ready access to the Internet and with appropriate technology.

Resources for South East Queensland. The Queensland Department of Emergency Services provides leaflets during campaigns, such as "don't be fooled by a flood" which proved critical in the early 2008 floods. Rural fire service website provides information on how to prepare a house in the urban fringe areas for Bushfire safety. This information is reproduced on the Queensland Fire and Rescue Service: Bushfire Safety webpage and similar information is found either diagrammatically or in lists on the Emergency Management Australia's webpage. The effectiveness of this web information is that it caters for people's different strategies for learning by describing what to do and then providing a picture of a house and highlighting what an occupant can do and where to do it.

There is further advice on long term planning, such as landscaping and vegetation planting, precautions to take as the fire season approaches and then providing a checklist to use. Information is also provided for the immediate approach of a fire and what to do, what to have ready when the power goes out, or if the water supply is affected and if advice is given to evacuate. The difference is highlighted between previous occupants of the land; for example, in urban fringes it is typically farmers whose range of equipment was large to that of the rural residential owner, who may only have hand operated tools. Thus in this study it was of interest to ascertain if rural residents had the tools - even basic tools - to fight fires with. There is also a reference to manuals being available to occupants on the construction of houses in rural urban fringe areas. So it would appear that the resources are there, but are they accessible and are they utilized and valued by the population?

Aims of the Study

The main aim of this research study was to ascertain the level of preparedness of residents in rural areas west of Brisbane, and near Ipswich and Toowoomba.[4] The second aim was to determine their level of susceptibility based on their nearness to hazardous materials (in this case land that was high on the BSI). The third aim was to analyse what factors prevents these residents from being adequately prepared for future bushfire events.

Methodology

This was a cross sectional study using survey methodology to tap rural residents' perceptions and risk behaviours about bushfire danger preparedness. The survey was designed to capture certain susceptibility factors and Table 13.1 outlines general data categories under investigation and the reason for including them in this survey.

The Questionnaire Survey

Five questions in the survey focused on home ownership, size and usage of property, vegetation, age, education, occupation and family and pets/animals and length of residency in area.

A single score, often called a factor score, measuring preparedness for future bushfires was created from the standardised responses to 16 questions by taking their first principal component, normalised to have an overall mean of 100 and a standard deviation of 15. Nine other items focused on home ownership, size and usage of property, vegetation, age, education, occupation and family and pets/animals and length of residency in the area. Risk factors were measured by six items. Exposure to bushfires was captured by four items. One item had asked if they had encountered a bushfire in or near their home, while actual personal experience was taken from another item which asked how close the fire was to them. A copy of the questionnaire can be obtained from the lead author.

[4] To see a map of the area of this study, refer to: http://www.socialchange.qut.edu.au/conferences/socialchange/2006proceedings.jsp.

Table 13.1. General Profile of survey sample and expected output

Data Item	Explanation Of Output And Rational For Community Profile Usage
Age Groupings	Gives overall indication of the population; Standard Demographic Survey Question
Education Levels	Standard Survey Question. Leads to a general understanding of the types and level of education campaigns and the relevance of the information sources
Tenure Type	To gain an estimate of home ownership rates in the area
Information Sources	Such as Internet Usage to estimate the proportion of the population that is able to access information sources; radio usage for ND alerts, evacuation points and other news items in emergency situations.
Length of Residence	Although an ideal data source would be a mobility survey, there is much discussion about the scarcity of available information at the small area level. Approximate length of residence would be a useful source of information during emergency situations, to gauge experience in disasters and knowledge of how bushfires behave in the area.
Vehicle Ownership	This should provide a snap shot on the overall ability of residents to evacuate in an emergency and includes access to vehicles.

The 32-item survey was mailed out to 4850 rural residents in the shires of Crows Nest, Esk, Gatton, Kilcoy and Laidley over the November 2005 - January 2006 time period. Reply paid return envelopes were supplied with the questionnaires to respondents. Seven hundred and thirteen questionnaires were returned (13.4%) about 6% less than anticipated for a rural mail out, but understandable given the Christmas/New Year period. There was evidence that several hundred surveys were, in fact, not delivered. In addition, several participants contacted the researchers directly by phone to talk about their concerns and, as outlined in this article, many offered suggestions about reducing the bushfire danger.

Results and Discussion

Firstly, the results address demographic and susceptibility influences; then risk factors and exposure factors are explored in relation to preparedness. Tables 13.2 to 13.4 outline the descriptive statistics for community demography, level of preparedness, and deterrents to capacity and readiness.

The majority of the respondents (72.6%) were in the age range from 35 to 64, with just 1% being in the 18 to 24 category, 8.3% were aged from 25-34, with 12.5% falling into the 65-74 category and 5.6% were aged 75 or more.

Most people had lived in their locations for more than 5 years (68%), while 24.9% had resided there for less than 5 years and 7.1% for less than 12 months. For 25.7 % of them, it was the first time they had lived in a rural residential area or in an area close to forest, grassland or scrub, but the majority of them (74.3%) were no strangers to rural living.

Table 13.2. Demographic and Susceptibility Details

Data Item	Descriptive Statistics
Age Groupings	18-24 (1%); 25-34 (8.3%); 35-64 (72.6%); 65-74 (12.5%); 75+ (5.6%)
Education Levels	Up to Year10 (10.6) Year 10 (23.2) Year 12 (10.8) Trade (11.6) Traineeship (1.1) Certificate (14.2) Diploma (9.2) Degree (3/4 yr) (9.2%) Postgraduate (Diploma, Masters, PhD) (10.1%)
Home Owner	Own home (92.8%); renting (6.5%); with friends/relatives (0.72%)
Time in Area	Under 1 yr (7.1%); < 5 yrs (24.9%); > 5 yrs (68%)
Time in Residence	Under 1 yr (9.9%); < 5 yrs (31.7%); > 5 yrs (58.4%)
First time rural	Yes 25.7%; No 74.3%
Usage	Residential (68.4%); Grazing (14.5%); Crops (2.2%); Orchards (0.4%); Mixed (14.6%); NR (2%)
Vegetation Proximity	<3 m (3.3%); <10 m (3.6%); <55 m (32.06%); < 1.5 klm (50.8); >1.5klm (10.2%)
Size of Property	Average: 130.5 acres; Range: .1 to 19,760 acres; Median indicates 50% < 10 acres
Own Vehicle	Yes: 95.65%; No: 4.35%
Access to Vehicle	Yes: 67.56%; No: 32.44%
Home phone	Yes: 96.5%; No: 3.50%
Mobile	Yes: 85.5%; No: 14.5%
Internet Access	Yes: 60.6%; No: 39.4%

The size of their properties ranged from as small as 0.1 of an acre to as large as 19,760 acres). The average size of the properties was 130.5 acres. The main usages of their properties were as residences (68.4%), grazing (14.5%), crops (2.2%), orchards (0.4%) or mixed farming (14.6%) (2% gave no response to this question).

When determining how serious their vulnerability would be when fires hit the area, it was reported that 3.62% of them had heavy vegetation within 10 metres of their houses, and 3.3% of houses were located as close as 3 meters to this vegetation; 32.06% indicated that it was about 55 metres away and for 50.8%, it was as far away as 1.5 kilometres, and the remainder (10.2%) responded that it was under 1.5 kilometres.

The responses to the questions about preparedness indicated that rather than the communities in these areas being totally unprepared, they were in fact equipped and prepared to do battle with the next bushfire season.

Table 13.3. Level of Preparedness

Data Item	Descriptive Statistics
Evacuation Plan	Yes (56.9%); No (43.1%)
Smoke Alarms	Yes (75.7%); No (24.3%)
Know Emergency Phone No.	Local FB (81.6%); SES (30.6%); Ambulance (83.3%); Energex (50.5%); Police (0%)
Hazard Info.	Radio (56.2%); TV (50.6%); Internet (8.70%); Phone (40.7%); Neighbours (54.9%)
Gutter Cleaning	3 Monthly (49.3%); 6 Monthly (17.0%); Yearly (23.0%); Never (10.7%)
Fire fighting tools	Shovels (85.8%); Axes (62.7%); Working Pump (42.9%); Hessian Bags (37.7%); Pump Packs (32.1%); Fire Blankets (19.1%); Fire Hoses (14.0%)
Buckets/hoses	Yes (94.3%); No (5.7%)
Storing flammables	Garage (25.5%); Shed (75.6%); Under House (6.0%); Verandah (7.9%); In house (0.8%)
Have Water Sources	Yes (97.1%); No (2.9%) Tank (83.7%); Town (28.1%); Bore (37.5%); Dam (42.4%)
Know additional water sources	Yes (70.7%); No (29.3%)

Table 13.4. Deterrents to Capacity and Readiness

Data Item Deterrents	Descriptive Statistics Percentage
Need Assistance	40.3
Lack skills	43.8
Lack equipment	36.0
Cost of equipment too high	39.1
Competing priorities	42.3

Most people (94.3%) indicated that they had buckets and hoses to fight fires with and knew their location: garage (25.5%); shed (75.6%); veranda (7.9%); in the house (0.8%); and under the house (6%). Additional equipment they nominated having for fire fighting included: shovels (85.8%); hessian bags (37.7%); axes (62.7%); pump packs (32.1%); fire blankets (19.1%); fire hoses (14%); and working water pumps (42.9%).

As reported in their chapter on bushfire preparedness, Pritchard and Gow (2008) found that while some of these residents never cleaned their gutters (10.7%), most did either at 3-monthly (49.3%), 6 monthly (17%) or 12 monthly (23%) intervals. Again while residents admitted to storing flammable liquids in their garages (25.5%), sheds (75.6%), verandas (7.9%) and under the house (6.0%), almost 70.7% of them knew where to find additional water supplies, other than their own, to fight fires with in the form of tanks, dams, rivers or bore water.

Interestingly these rural respondents were aware of some things (see Table 13.4) that might prevent them (somewhat - to a great deal) from being better prepared for the bushfire seasons such as: they did not have the necessary skills (43.8%); they needed assistance (40.3%); they did not have sufficient equipment (36.0%); or the cost of equipment was too high (39.1%); and in terms of the time it would take to prepare, they had other competing priories (42.3%).

From the perspective of being prepared for normal fire hazards, such as house fires, 75.7% had had smoke detectors fitted to their houses. This higher percentage may have been due to the new State law that had just been passed that year about every home having smoke detectors installed.

Perhaps one of the reasons why 75.7% of the property owners had an evacuation plan was that these residents had lived through scenes of fire devastation, season after season, and this may have persuaded them to take some preparations, such as making evacuation plans and being aware of equipment to fight fires with and where to seek help and who to contact in case of bushfire emergencies.

In terms of access to information and communication channels, this survey demonstrated that in this sample, the residents obtained information about the threat and reality of fires from radio (56.2%), television (50.6%), internet (8.7%) phone (40.7%) or neighbours (54.9%). Generally, nearly all those surveyed knew the phone numbers of the fire brigade (81.6%) and the ambulance (83.3%). Surprisingly, Energex (power supplier) was recognised by 50.5%, but no-one ticked the police or emergency services (SES) boxes (this may have been due to the fact that they would not consider phoning the police or emergency services rescue teams personally).

To what extent the internet can be utilized in rural communities is questionable. While 60.6% stated that they had access to the internet in their homes, only 8.7% of those surveyed in these rural communities indicated that they would search the web for such information to keep up to date with fire threats. This is consistent with anecdotal information from rural residents.

Next we explored the level of preparedness and a range of factors that were inherent in the questionnaire design. The actual process of the factor analysis that produced the factors is described in Gow, Pritchard and Chant (2008). Following the principal components analysis that was designed to tap underlying factors, certain items only were found to have adequate loadings on the three factors: risk factors, exposure, and preparedness. The 17 preparedness items reduced to the essentials of knowledge about emergency phone numbers, evacuation plans, having phone access, fire fighting equipments and tools, and access to water resources.

For ease of understanding, first the analyses relating to the demographics as outlined in Table 13.1 are explored, then susceptibility, risk factors, and exposure factors. For the purpose of the direction of the research outlined in this article, post hoc analyses utilising Anova techniques were undertaken on certain aspects of the demographic data relating to (i) time spent in the area/property; (ii) size of property; (iii) the closeness of vegetation, as well as (iv) compounding escape concerns and how this affected their level of preparedness.

For example, we were interested in whether time spent in the area/property affected the level of preparedness? The answer was yes; preparedness (where “a greater index value" means "more preparedness”) increases monotonically with time spent in the area (see Table 13.5) ($F = 3.37$, $df = 3, 696$, $p = .02$).

Table 13.5. Time spent in area and level of Preparedness

Time spent in area	*n*	*m* (*s*)
less than 12 months	50	96.0 (11.6)
1-5 years	174	98.4 (14.9)
5-10 years	103	99.7 (16.1)
> 10 years	573	101.6 (14.6)

Table 13.6. Percentage Endorsement and Level of Preparedness for Compounding Escape Concerns

Compounding Escape concerns	Percentage	Preparedness: *m* and *sd*
Elderly	8.42	102.25 (14.35)
Infirm	1.68	101.65 (11.45)
Special Needs	1.82	99.48 (10.60)
Young Children	16.13	98.34 (13.05)
Pets	70.41	100.56 (14.67)
Horses	26.79	102.01 (15.50)
Farm Animals	33.94	101.23 (15.50)
Caged Animals	23.42	103.82 (13.20)
Any of the above	85.55	100.25 (14.82)

However, neither the size of property nor the closeness of vegetation affected the level of preparedness in any systematic way.

We were also interested in whether other risk factors that did not load onto the final principal components iteration of items that were labelled risk factors, affected the level of preparedness. Clearly they had not been mathematically considered to be key risk factors in the same way as the items which really measured actual prevention issues, such as lack of skills, assistance, insufficient equipment (possibly because it was expensive) and other priorities competing for their time and energy (see Table 13.4). These other items related to responsibilities for people and animals on the property. On closer inspection after the Principal Components Analysis (PCA), it was considered that they should more aptly be labelled as Compounding Escape concerns. Ostensibly, one would think that these would affect the level of preparedness in terms of a better escape plan, and perhaps more overall preparedness. Indeed this was the case (see Table 13.6).

Interestingly, the factor that was called Exposure to bushfires remained constant from one iteration to the next and related to respondents' perceived likelihood of being caught in a fire in the short and intermediate term, their viewing of bushfires on TV, their own eyewitness accounts and their own close encounters with fires on or near their properties (Table 13.7 refers).

Susceptibility via Vegetation

One of the major risk factors that showed up in the descriptive statistics is that 7% of respondents had heavy vegetation within 10 meters of their house and a further 32% within

55 meters. What was more of a concern was that there was little awareness of the danger of this in bushfire prone areas. While this item did not load onto the risk factor, as can be seen in Table 13.2, quite a number of residents had heavy vegetation relatively close to their houses. One would have thought that the closer the heavy vegetation to one's house, the greater would be one's level of preparedness. However the Anova for Preparedness showed that this was not significant; this factor did not affect their preparedness, which is interesting (F = 1.60, df= 4,517, p =.17).

Another susceptibility factor is that of first time rural experience. So would it matter if it were the respondents' first time in a rural area? It did matter quite significantly (F = 8.85, df =1, 62, p = .003). In Table 13.8, we see that the "first timers" had lower preparedness than "long timers".

Similarly did owning a home versus renting a home make a difference? Yes it did (F = 11.07, df= 1,687, p<.001), home owners were better prepared. [However the small cell sizes in the renting category may account for some of this statistical variation.]

Table 13.7. Exposure to Bushfires

	less than .5 L'hood	.5 L'hood	more than .5 L'hood		p-value
Likelihood of being caught in BF in next 12 months?	50.83%	23.83%	25.34%		$X^2 = 91.56$, $df.= 2, p<.001$
Likelihood of being caught in BF in next 5 years?	42.20%	20.34%	37.46%		$X^2 = 51.92$, $df = 2, p<.001$
	None	1-5	5-10	10+	
How many BFs seen on TV in past 12 months?	4.50%	42.73%	29.54%	23.24%	$X^2 = 202.32$, $df. = 3, p<.001$
How many BFs seen with own eyes past 12 months?	36.90%	52.24%	7.38%	3.47%	$X^2 = 458.19$, $df. = 3, p<.001$
	Yes	No			
Have you encountered BF near your home?	68.02	31.98			$X^2 = 92.64$, $df. =1, p<.001$

Table 13.8. Frequency distribution and mean of first time living in area and preparedness

Arrival	Number	Level of Preparedness		
First Time	N	%	Mean	Std Dev
Yes	163	25.7%	97.4	14.3
No	472	74.3%	101.4	14.7

Table 13.9. Home Ownership and Level of Preparedness

Ownership	Number	Level of Preparedness		
Q1 Own/rent	N	%	Mean	Std Dev
Owner	644	93.47	100.7	14.8
Renting	45	6.53	93.2	12.5

Table 13.10. Level of Preparedness and Usage of Property

Usage	*n*	*m (sd)*
Residence	478	98.0 (14.5)
Grazing	101	102.3 (16.2)
Mixed	102	103.3 (14.1)
Crops	15	106.2 (12.7)
Orchard	3	116.1 (6.2)

Table 13.11. Level of Preparedness and Age

Age Group	*n*	*m (sd)*
18-24	7	87.8 (17.5)
25-34	58	95.6 (13.2)
35-44	150	99.5 (14.5)
45-54	170	99.9 (13.8)
55-64	187	103.7 (14.4)
65-74	87	99.4 (15.9)
75+	39	95.7 (17.6)

In terms of the size of property, did it change the outcome (No: $F = 0.03$, $df = 1{,}687$, $p{=}.87$); but its usage did (Yes: $F = 4.06$, $df = 4, 694$, $p = .003$). Interestingly, orchard growers were the most prepared and those who had no farming investments the least prepared. However, as you can see, the very small number of orchard growers in the survey may call that finding into question.

Asking questions of the simple demographics of age and education in terms of preparedness, we found that overall education did not make a difference ($F = 1.46$, df = 8, 697, $p = .17$). Age did make a significant difference ($F = 4.29$, df = 6,691, $p < .001$); however, only two individual between-level comparisons are formally statistically significant at the .05 level after post-hoc adjustment for multiple comparisons: (i) the 55-64 and 75+ age groups; and (ii) the 55-64 and 25-34 age groups (see Table 13.11). While the numbers in the table for the youngest age group indicate that it might be significantly different from the others, this is not the case.

From these figures, we can deduce that the youngest age group (18-24) were the least prepared, and this could be explained by the fact that they were most likely living in someone else's house and or other older people had the responsibility for hazard reduction.

They were significantly different from all other age groups. It seems that those in the age group 55-64 were most prepared and this could be that they were home more often and had the time to do some of the hazard reduction activities. Similarly for the 25-34 and 75+ age groups, someone else may be responsible for these activities. Again, they were significantly different from all other age groups.

With respect to occupation, those who described themselves as farmers, graziers or mixed, were no different from all others taken together; nor were they different from the subgroup that identified themselves as pensioners.

The other risk factors that did not survive the second iteration of the PCA were items relating to owning or having ready access to a motor vehicle, having a telephone in the house, having mobile phones, and having internet access (see Tables 13.2 and 13.3). We had considered them as risk factors because without them escape may be impossible in a bushfire. Three of the items relate to communication, but are key information resources. If you see smoke, you can always jump in a car and drive to where the smoke is coming from to check the fire's location, intensity and velocity. You can also check the fire breaks, clearings, dams and break waters between the fire and your own property.

As indicated earlier, several of the respondents had opinions about what could be done in bushfire emergencies and these are now outlined.

Recommendations from Participants

Some recommendations obviously came from participants who knew a great deal about fire hazards, and these are intermingled below with non-hazard trained respondents.

- Rural fire trucks are only small and carry only a small water load; we need large trucks due to lack of available water.
- We need to educate rural people including fire volunteers to hose the base of fire not tree tops. Fires burn slowly down hill and race up hill.
- During the bushfire season, those people with full dams should put signs outside on the road for the fire brigade, so they know where water is available, as this saves time.
- All valuables should be kept in a fire proof safe including insurance papers.
- It would be helpful if there were 'Fire Awareness Meetings" before the fire season each year.
- Fire breaks need to be put in place for all rural areas. This gives emergency vehicles better access to properties; also more controlled burn offs are needed.
- Homes need to be fitted with a wind driven ventilator, and fitted with approved wire mesh to prevent sparks and embers being sucked into roof ceilings.
- Trees and grass need to be cleared 25 metres from the house.
- Residents need to remove rugs from horses during fires.
- More disaster management resources are needed with the growing population in South East Queensland and South West Queensland.
- Do people check if the fire they see is under control? Checking could save a life.
- With respect to residents knowing their local fire brigade number, the Rural Fire Service urges home owners to use '000' when reporting a fire. The fire is then in the system and support is on standby.

- We need to make the public more "bushfire aware" and use photos of heavy vegetation, demonstrating actual distances that are too close. We should give them: (a) reasons for cleaning leaves from gutters, (b) sample evacuation plans, and (c) recommended storage for inflammables.

One Rural Fire Services (RFS) officer was concerned about the value of education for preparedness and asked:

- Will education address the following?
  - Criteria for decision making – flee or fight;
  - Methodology for fighting ember attack on dwelling;
  - Training for RFS Officers/particularly back burning safely;
  - Funding for RFS to be made proactive in inspection and advice.

Following The Inquiry on Bushfire Mitigation and Management (Ellis, Kanowski and Whelan, 2004, p. 240), action has been taken on implementing the indicative national bushfire principles which are listed below.

1. Bushfires are understood, accepted and respected
2. Shared responsibility
3. Decisions within a risk management framework
4. Integration of learning and knowledge
5. Manage fire according to the landscape objectives
6. Consistency of purpose and unity of command
7. Protection of lives as the highest consideration
8. Monitoring performance

Communications

The present survey, undertaken in rural areas where bushfires had caused severe damage within the previous 12 months, demonstrated that the indicative national bushfire principles are framed within the context of emergency management and government instrumentalities. However, if one looks at them through the eyes of residents, they are still important principles. It may be, however, that there are several facts that must be taken into consideration to prevent further loss of life and property. Many rural people do not have access to the internet and where reception is limited, even mobile phones may not be an option for emergency communication when landlines are cut off, or when people are isolated by bushfires away from homes and businesses.

Water Resources

In terms of resources with which to fight fires, we now have proof that most rural areas are without the water resources they need in the form of piped water, hydrants, dams and bores, as many of these dried up in the long drought and may do so again once the flash flood effects dissipate, and in some areas the flood waters recede. Without ongoing rain, the susceptibility to bushfire hazards in rural areas will be high again, with or without the promised recycled water access. Thus it is always important in these studies on bushfire preparedness to obtain an estimate of the water sources available (to ensure enough for the dampening of embers typically remaining for many hours after a fire or for re-hydration of residents after the fire front has passed). It remains for other researchers to explore the effects of severe fire on pollution of water supplies after a bushfire event.

To Go or Stay

Interestingly, only one respondent who was a local fire fighter raised the issue of "flee or fight the fire" in connection with education for bushfire awareness. Proudley (2008) has given interesting first hand insights into this issue from the Wangary fire in South Australia. The issue of "stay or go" is central to bushfire policy in most states (Watson, 2006) and is the position of the Australasian Fire Authorities Council (AFAC, 2005). Emergency management organizations have shifted their focus in addressing community safety towards encouraging greater *community preparedness* and *self-reliance* (Enders, 2001; Hodges, 1999). The ultimate objective is to increase resilience. In terms of bushfire hazard, an important aspect of the desired resilience is the capacity of individual property owners to be able to decide whether to flee a bushfire at an appropriately safe juncture or to stay and fight the fire, thereby defending their property. Clearly, this aspect of preparedness needs to be addressed in further research and events such as firestorms have to be factored into decision making processes and policy reviews.

Conclusion

Key findings from this research enhance our understanding of factors relating to bushfire vulnerability and preparedness in rural settings in South-East Queensland. These include length of residence in an area, first time rural experience, home ownership, usage of the property, age and education. Interestingly, neither actual size of the property nor the closeness of vegetation to their house made a difference, nor did a person's occupation exert an influence on their preparedness.

With the increasing number and intensity of bushfires accompanying climate change, this survey has highlighted the fact that rural residents need assistance to help them prepare for bushfires, whether that be in terms of time, skills training, supply of necessary equipment or just an extra pair of hands because they are very busy with other competing priorities. There is no doubt that with the continuing drought, and threats of more to come with increasing temperatures, rural communities in treed and grassland areas will need to ensure that more people are given special training in mitigating and fighting bushfires; in this way, they can assist both the paid and volunteer Emergency Services groups to save people and their homes, farm animals, business and infrastructure, and our wildlife and bush landscapes.

With temperatures predicted to increase over the next few decades (with or without climate change mitigation action), it is to be hoped that in all countries, and not just in those which have been listed as the most vulnerable to bushfires, wildfire preparedness plans will be implemented in every rural area where there is danger to human life, ecology and infrastructure. Countries which have not yet commenced such planning could benefit from accessing proven strategies those from countries such as the USA, Australia and France. Not to take preventive action in full cognisance of climate change indications for increased bushfire activity across many countries in the world, can only feed the destructive powers of Vulcan, the Roman Fire God.

REFERENCES

AFAC (Australasian Fire Authorities Council). (2005). *Position paper on bushfires and community safety*. AFAC Ltd., Melbourne. http://www.afac.com.au

Borger, B. (1999). Wildfire Threat Analysis: Literature Summary. Appendix E. (Accessed 10.08.07)
http://nrfa.fire.org.nz/projects/wta_docs/Analysis_Phase_Appendices/AppendixF.PDF

Brandes, I. (2003). "Burning Questions: Australia's Bushfire Policy, The Drawing Board http://www.econ.usyd.edu.au/drawingboard/digest/0302/brandes/html (Accessed 30/03/ 2004).

Bushnell, S. and Cottrell, A. (2007). Increasing community resilience to bushfire - implications from a north Queensland community case study. *The Australian Journal of Emergency, 22* (2), 3-9.

Childs, I., Pritchard, F., Gow, K., and Hastings, P. (2006, October 27). Bushfire preparedness of rural-residents in selected Southeast Queensland shires. Social Change in the 21st Century Conference. Accessed 01.02.09 http://www.socialchange.qut.edu.au/ conferences/socialchange/2006proceedings.jsp

Ellis, S., Kanowski, P., and Whelan, R. (2004). *National Inquiry on Bushfire Mitigation and Management*, Commonwealth of Australia, Canberra.

Enders, J. (2001). Measuring community awareness and preparedness for emergencies. *Australian Journal of Emergency Management*, 16(3), 52-58.

Garnaut, R. (2008). *The Climate Change Review*. Cambridge, UK: Cambridge University Press.

Gow, K. (2006, September). "How psychological theories can help make sense of residents lack of preparedness for Bushfires". Paper presented at the APS/NZPS Conference, Symposium: Addressing Disasters: From threats to risk communication to preparedness, Auckland: SKYCITY Auckland Convention Centre.

Gow, K., Pritchard, F., and Chant, D. (2008). How Close Do You Have To Be To Learn The Lesson? Fire Burns! *The Australasian Journal of Disaster and Trauma Studies*, 2, (http://www.massey.ac.nz/~trauma/issues/2008-2)

Hodges, A. (1999). *Towards community safety*. Paper presented at the AFAC Sixth annual Conference. AFAC Ltd., Melbourne. http://www.afac.com.au

IPCC (Intergovernmental Panel on Climate Change). (2007). Intergovernmental Panel on Climate Change Fourth Assessment Report. Climate Change 2007: Synthesis Report. Summary for Policy Makers. Available at: http://www.ipcc.ch/pdf/assessment-report/ar4/syr/ar4_syr_spm.pdf

NOAA. (1994). Tropical Storm Alberto Floods of July 1994 Disaster Survey Report: U.S. Department of Commerce, National Oceanic and Atmospheric Administration, National Weather Service, Silver Spring, Maryland. http://www.nws.noaa.gov/oh/hrl/surveys/alberto/contents.htm

Paton, D. (2003). Disaster preparedness: a social-cognitive perspective, *Disaster Prevention and Management, 12* (3), 210-216.

Paton, D. and Bürgelt, P.T. (2005). "Living with bushfire risk: residents accounts of their bushfire preparedness behaviour", paper presented at the AFAC/Bushfire CRC Conference, Auckland, October. http://www.bushfirecrc.com/research/downloads/Paton-Bushfire-Warnings_wcover_1.pdf

Paton, D., Kelly, G., Burgelt, P.T. and Doherty, M. (2006). Preparing for bushfires: understanding intentions. *Disaster Prevention and Management, 15*(4), 566-575.

Pittock, A.B. (2005). *Climate Change: Turning up the heat.* Collingwood, Victoria: CSIRO Publishing.

Pritchard, F. and Gow, K. (2008). Bushfires: Preparation, Resilience and Recovery. In K. Gow and D. Paton (Eds.). *The Phoenix of Natural Disasters: Community Resilience* (pp. 209-233). New York: Nova Science Publications.

Proudley, M. (2008). Fire, Families and Decisions. *The Australian Journal of Emergency Management*, *23*(1), 37-43.

Rohrmann, B. (2002). "The utility of the WWW for fire preparedness of residents" Study Report. University of Melbourne Australia, December 2002. http://www.psych.unimelb.edu.au/staff/br/brafw.pdf.

Rohrmann, B. (2007). Exploring information for residents on websites of fire authorities-practical experiences. *The Australian Journal of Emergency Management*, *22*(2), 10-15. http://www.ema.gov.au/agd/EMA/rwpattach.nsf/VAP/(1FEDA2C440E4190E0993A00B7C030CB7)~10-15rohrmann.pdf/$file/10-15rohrmann.pdf

Watson, K. (2006). *Building community self-sufficiency for fire safety. The Australian Journal of Emergency Management*, *21*(1), 43-46.

Contributed by Kathryn Gow, Queensland.

Photo. A typical Rural Bushfire Warning Signs in shires under study.

In: Meltdown: Climate Change, Natural Disasters... ISBN 978-1-60876-153-1
Editor: Kathryn Gow

Chapter 14

ADAPTING TO SEA-LEVEL RISE: LESSONS FOR THE FUTURE FROM REDCLIFFE, AUSTRALIA

Peter A. Hastings and Iraphne R.W. Childs
Queensland University of Technology, Australia

ABSTRACT

This longitudinal study tracks and analyses a local government's planning and adaptation to potential sea-level rises, over the twenty-year period 1987-2008, and identifies future concerns regarding inundation. The case study location, Redcliffe, is a small bay side city in Southeast Queensland that is potentially prone to coastal inundation. Over the period of study, there have been significant scientific, policy and public awareness developments relating to climate change at the international, national and state levels which local governments have had to take into consideration in decision-making. The Redcliffe study reveals continuity in the range of challenges faced by this local council including: dealing with scientific uncertainty; policy application at the local government level, public awareness, political response, planning and land use adaptation. While progress has been evident over the past two decades in addressing these issues, some obstacles still persist and some major future challenges remain unresolved. These challenges include funding for adaptation, legal liability and governance issues.

Keywords: Climate Change Adaptation; Sea-level Rise; Local Government Planning; Redcliffe, Queensland

INTRODUCTION

In Australia, the three levels of government, federal, state and local, have specific and different responsibilities in managing land use and the environment. Local government, administered by a council, is responsible for a broad range of functions, including capital works and essential services. These councils also have mandatory responsibilities for land use planning schemes that duly consider the environment, settlement patterns and economic activities within their communities. Thus, in this context, local government is the vehicle via which practical policy and planning adaptation to climate change occurs at the community level. This local government responsibility poses a range of challenges for decisions concerning land use and land use planning. The fundamental question is whether local councils are effectively responding to the evolving science and policy frameworks relating to climate change to benefit their constituents.

Over two decades, the researchers have tracked the progress of one Australian local council's planning and adaptation in response to potential climate change-induced sea-level rise. The following report describes this progress and analyses on-going and future concerns over this issue in the case study of Redcliffe, a small coastal city near Brisbane, Queensland. This community is susceptible to inundation from sea-level rise and storm surge and hence it provides insight into the practical problems faced by local government in the management of the issue in future scenarios of sea-level rises, induced or exacerbated by climate change factors.

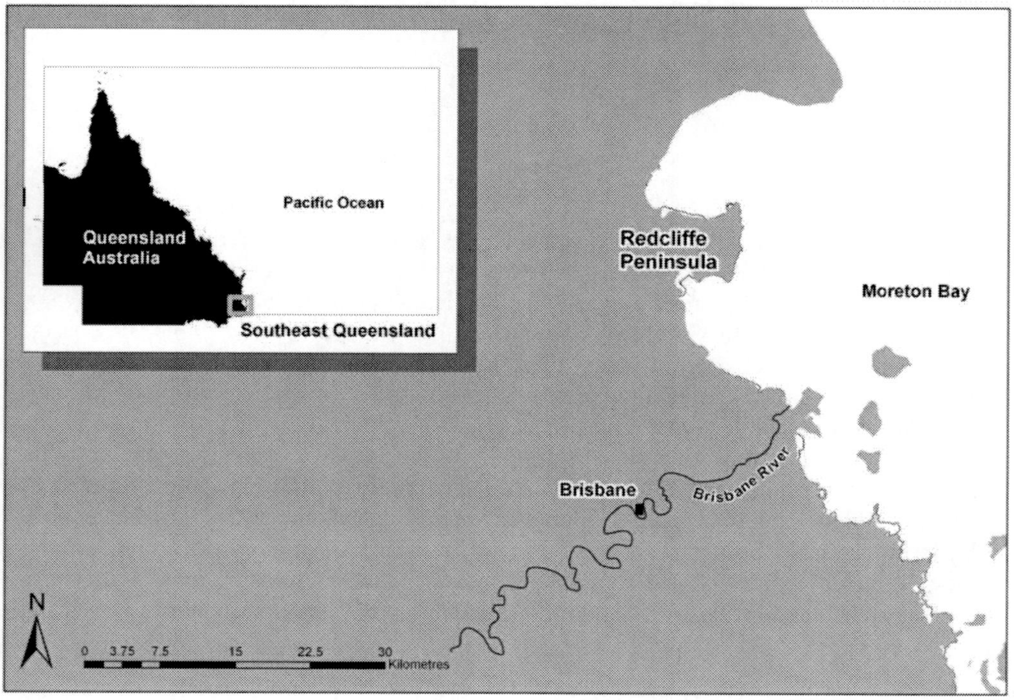

Figure 14.1. Redcliffe Peninsula location map.

The Case Study: Redcliffe City

Redcliffe City is located on a peninsula on the western shores of Moreton Bay, a subtropical locality situated approximately 30 km north of Brisbane (see Figure14.1). In 2006, the city's population was 52,000 (Australian Bureau of Statistics, 2006). Redcliffe is a readily identifiable community that has evolved from a quiet seaside retirement area and dormitory suburb serving Brisbane into a more diverse community featuring considerable high-value waterfront real-estate developments. Virtually all of the elevated land in the city has now been developed and expansion has occurred on the remaining low-lying areas in the form of reclamation and residential canal developments. Redcliffe has seen several changes of local government administration over the period 1987-2008; although once a local government area in its own right, as of 2008, the city is now part of the larger amalgamated Moreton Bay Regional Council (MBRC).

In the late 1980s, climate change was emerging in Australia as a topic of concern in the scientific community, but was not yet an issue of public debate. In 1987, a group of researchers embarked upon a pioneering study of potential impacts from climate change at Redcliffe (Childs, Auliciems, Hundloe and McDonald, 1988); the key problem identified was additional flood risk, from the combination of sea-level rise and storm surge, to the existing developments in the low-lying areas around the Redcliffe peninsula. Both the potential inundation risks on land use and infrastructure, and the associated likely economic, social and managerial consequences were examined and found to be significant. A further aim was to identify potential decision-making within the local council with regard to adjustment to this hazard. At that time, awareness of "greenhouse warming" (as the issue was then commonly termed) and its potential impacts at Redcliffe was found to be limited to a small number of concerned professional officers in the local council and to only a select few in the wider community.

The researchers returned to Redcliffe a decade later in 1996 to find much greater awareness of "greenhouse" issues, both within the local Council and in the general community (Childs, Hastings and Auliciems, 1996). Progress in planning and adapting to potential sea-level rise by the local council was documented; comparisons were made with results from the earlier survey (Childs et al., 1988); and on-going themes were analysed, with a particular emphasis on the sea-level rise issue.

More recently in 2008, the research team again revisited Redcliffe, to investigate how the council is now negotiating appropriate planning and adaptation for the community. Two professional officers (engineering/ hazard management) and two elected representatives (councillors) were interviewed by the researchers. The aims were to review current activities of land use adaptation and to elicit the concerns regarding the management of potential sea-level rise into the future. The Council now operates within the context of a range of evolving international initiatives, national and state government climate change policies and a greater mass media exposure of the issue.

Sea-Level Rise: Key Aspects of Concern

In 2008, the researchers determined that the major concerns in Redcliffe regarding potential sea-level rise were comparable to those raised in the previous two studies. These have been categorised under the headings of:

- scientific uncertainty
- policy frameworks and application
- public opinion on sea-level rise and political response
- planning and land use adaptation

The following discussion presents these issues within a longitudinal framework as well as the situation in 2008 and projected concerns for the future; Table 14.1 maps some concerns.

Table 14.1. Selected Predictions or Scenarios for Sea-level Rise Released Over the Study Period

Year	Sea-level Rise Prediction/Scenarios	Source Organisation	Comments
1987	0.4m – 1.4m by 2050	CSIRO	The Childs et al. (1988) study was based on this scenario together with storm-surge scenarios
1989	0.2m - 1.4m by 2030	Queensland State Government	
1989	0.1m to 0.5m by 2050	CSIRO	
1992	0.05m to 0.35m by 2030 and 0.1m to 0.8m by 2070	CSIRO	Childs, Hastings and Auliciems (1996) found that the Redcliffe Council had settled upon this scenario as a bench-mark for policy-making.
2001	0.09m and 0.88m from 1990 to 2100	IPCC Third Assessment	
2007	0.18m to 0.59m by 2100	IPCC Fourth Assessment and CSIRO	
2007	0.3m to 0.17m by 2030 and 0.3m to 0.5m by 2070 (Queensland coast)	LGAQ reporting CSIRO	

Sources: Pearman (1988); Childs et al. (1988); Childs et al. (1996); Marsh (2002); IPCC (2001, 2007) CSIRO (2007); LGAQ (2007); Pyper (2007).

Scientific Uncertainty about Climate Change Effects

Uncertainty has surrounded the magnitude and timing of sea-level rise associated with climate change, and therefore the nature and scale of inundation risk has been an ongoing issue. This is in addition to the more fundamental debates concerning the reality, mechanisms and magnitude of climate change itself. Many scenarios of sea-level rise have been released since the 1980s by several organisations. Table 14.1 lists a selection of those identified by the Redcliffe studies, and the most recent predictions of the International Panel on Climate Change (IPCC) and Commonwealth Scientific and Industrial Research Organisation

(CSIRO). The variability of the magnitudes, ranges and lead times of the scenarios, reflecting the inherent uncertainty over the study period, is evident (see Table 14.1).

By the mid-1990s, Childs et al. (1996) confirmed that the Redcliffe Council had settled upon the 1992 scenarios from CSIRO as a reference bench-mark for policy-making. This scenario, however, still reflected notable uncertainty with a 0.7m range between the low and high scenarios of the 2070 predictions.

Most recently, CSIRO (2007) and IPCC (2007) suggested a projected global average sea-level rise of 0.18m to 0.59m by 2100 across several emissions scenarios. Uncertainty, however, remains despite the apparently narrowing ranges between the high and low predictions. CSIRO and IPCC report that additional contributions from Polar ice sheets may add 10cm to 20cm to the upper projection, or much more if there are further, but as yet unquantifiable, changes to the ice sheets. In any case, CSIRO notes that mean sea-level rise on the east coast of Australia may be greater than the global mean sea-level rise (CSIRO, 2007; Pyper, 2007) (see Table 14.1).

At Redcliffe, it is the combined effect of sea-level rise and tropical cyclone-induced storm surge that is critical in defining the inundation risks (Childs et al., 1988). Again, the nature of any changes to the magnitude and frequency of tropical cyclones, and therefore storm surge, with the climate change impact, is unclear. A popular view over the past two decades was that there would be more frequent incursions of intense tropical cyclones to higher latitudes, placing Redcliffe (at approximately 28°S) at higher risk. CSIRO (2007), however, cites Australian studies that respectively suggest both a maintenance of, and a decrease in, tropical cyclone occurrence off the eastern Australian coast. Geoscience Australia (then Australian Geological Survey Organisation, 2001) calculated tropical cyclone and storm tide risks for Redcliffe based on historical data, specifying the 100 Average Recurrence Interval (ARI) at 2.5m Australian Height Datum (AHD). The risk calculations, however, did not incorporate climate change scenarios.

The Future

The lack of scientific certainty over impacts of climate change at the local scale, including the magnitude of potential sea-level rise was continually raised by Redcliffe study participants in 1987 and 1996, as the key difficulty in addressing the issue. While published projection ranges for sea-level rise have narrowed (except for the addition of Arctic / Antarctic sea-ice melting), the future remains unclear. Current predictions are qualified by statements of uncertainty due to modelling limitations; even basic information such as tide gauge data is still subject to inaccuracies.

The 2008 Redcliffe study confirmed the continuing problem of uncertainty for local government decision-makers. Professional officers interviewed recognised that they were still working with "best-guess" scenarios and inadequate data, but were prepared to take steps to manage the risks with the information available; that is, a precautionary approach was adopted. The view of another study participant, however, was one of scepticism about the whole climate change issue. "Proof" was called for in advance of Council funding any substantial mitigating or adaptive action. Existing measures to mitigate against inundation were considered to be adequate.

When uncertainty manifests in people's different subjective appraisals of available evidence, it can delay their decision-making, either due to a lack of consensus, or because of the temptation to defer decisions until greater certainty is achieved. Acting in the face of uncertain risks is especially difficult for local government, which is charged with balancing duty of care for the community, weighing risks, and disbursement of limited local government funds.

In this context, the emerging politicisation of climate change "media hype" surrounding the issue and some more assertive advocacy of sceptical viewpoints is exposing and augmenting public (and scientific) perceptions of uncertainty and disagreement regarding climate change (e.g. *The Great Global Warming Swindle,* 2007). This could reinforce conservative viewpoints and diminish council's propensity to apply a precautionary approach and make proactive decisions.

EVOLVING POLICY FRAMEWORKS AND APPLICATION

In Australia, policy development occurs within each of the three tiers of government, each having particular responsibilities and jurisdictions. In 1987, key Redcliffe stakeholders called for coordination between all levels of government in addressing potential "greenhouse"- related inundation (Childs et al., 1988).

International and national cooperation and guidance on the issue became more concerted with the advent of the IPCC and its publications in the early 1990s, and the *National Greenhouse Response Strategy for Australia* in 1992 (Commonwealth of Australia, 1992). The latter was followed up in Queensland by the *Queensland Government Response Strategy* (Queensland Government, 1995), newsletters from Queensland Government departments, and nationally by the *Greenhouse 21C* document (Commonwealth of Australia, 1995). Despite these developments, Childs, Hastings and Auliciems (1996) reported a continuing perception by Redcliffe decision-makers and professional officers that there was inadequate policy or legislative guidance, from federal and state governments, to direct local planning decisions in response to potential sea-level rise, and that information on the hazard was difficult to access. The emphasis in the documents cited above was on determining and managing carbon dioxide emissions.

In the late 1990s, the development of substantive federal and state policies gained momentum. In 1998, the *National Greenhouse Strategy* specifically advocated the adoption of planning strategies that take into account potential sea-level rise (Commonwealth of Australia, 1998). State governments, including Queensland, responded with their own policy framework and implementation plans (e.g. Queensland Government, 2001; 2004). These documents established the basis for the most recent iteration of state policy, *ClimateSmart 2050* (Queensland Government 2007a, 2007b) (under review at the time of writing). Related documents synthesize the science and potential impacts of climate change (e.g. Queensland Government, 2008). Again, however, the emphasis is on carbon emissions rather than potential sea-level rise. Nevertheless, regional plans (i.e. the *Southeast Queensland Regional Plan*) and the *State Coastal Management Policy* do encourage consideration of sea-level rise in planning schemes (Queensland Government 2003, 2005).

The recently released *Climate Change Adaptation Actions for Local Government* (Commonwealth of Australia, 2007) and the LGAQ's *Adapting to Climate Change, A Queensland Local Government Guide* (LGAQ, 2007) more specifically aim to provide guidance to local governments dealing with climate change, including potential sea-level rise. The greater practicality and specificity of these documents for local government appears to address the policy gap identified by the earlier studies.

The 2008 Redcliffe study revealed that the Council's professional officers were generally aware of the contemporary climate change policies and documents at federal, state and regional/local levels. They specifically identified the LGAQ's (2007) document as being particularly useful in their professional roles. The officers considered that the specific guidelines presented in this document could readily be applied to strategic corporate planning and land use planning schemes. One officer had attended a workshop based on the LGAQ guidelines and described it as not simply a "talkfest", but as providing practical direction on how to amend planning to comply with newly recognised climate change imperatives.

LGAQ (2007) provides a "snapshot" of climate change scenarios across Queensland and clearly links the issue to specific sections of a typical local government corporate plan; for example, asset and growth management, and community development. Climate adaptation is broadly contextualised, with facilitating tools and capabilities outlined (for example, regulate, advocate, engage, design) and applies a four-step risk assessment process to each of the delineated elements of climate change: rainfall, temperatures, sea-level rise, storms and cyclones. In the case of sea-level rise, impacts and links across core local government functions are defined, a scheme for risk prioritisation suggested, and specific guidance in selecting response and adaptation measures provided in the contexts of information/ awareness, planning, design, risk diversification and geography. This follows through to framing the development of an action plan and identifying relevant internal and external roles and linkages.

The Future

The translation of state and regional policies, such as those previously outlined, into local planning schemes is required by the Integrated Planning Act (IPA) (Queensland Government, 1997). Redcliffe's most recent Town (land use) Plan under the Act was released in 2005. The professional officers interviewed in 2008 described local government strategic planning as an ongoing process; they suggested that the MBRC would begin appropriate amendments in line with relevant policies and guidelines, but that operationalising these would realistically take around five to seven years. Implementation of such amendments, however, can be significantly influenced by the degree to which public perception of the climate change issue is driving political response.

The enthusiasm with which the LGAQ (2007) guidelines were received by the council professional officers suggests that a detailed, locally applicable and integrated risk management approach could provide the way forward in dealing with potential sea-level rise. In essence, uncertainty is conceptualised in terms of risk and analysed in a systematic process relevant to local management. Such guidelines, with adequate revision and extension, are needed to bridge the gap between over-arching policy and local application. The existence of

LGAQ (2007) now affords opportunity for increased regional consistency in the assessment and treatment of climate change.

PUBLIC OPINION OF SEA-LEVEL RISE AND POLITICAL RESPONSE

Public awareness of climate change is a key catalyst triggering political action and in achieving community acceptability of adaptation measures. In 1987, only a moderate level of awareness of the "greenhouse effect" was reported among Redcliffe community representatives. Sixty-one percent said that they had not heard of the term and most survey participants thought that sea-level rise would not_become a major public policy issue within the next twenty years (Childs et al., 1988).

A period of heightened public interest in "greenhouse issues" and intense media attention on the Redcliffe case (after the publication of the Childs et al. (1988) study) followed the *Greenhouse '87* conference in Melbourne. One of the key aims of this conference was to raise public awareness of climate change issues (Pearman, 1988). Beginning with a spate of letters to the editor and articles in the local weekly, *The Redcliffe Herald*, media interest in the Redcliffe case study extended to the metropolitan Brisbane daily, *The Courier Mail,* and eventually to a national television special *Beyond 2000 (Beyond 2000 – Climate in Crisis, 1989*) which featured local Redcliffe residents and school children, politicians and other prominent citizens giving their views on "The greenhouse problem". This program was eventually shown around the world by BBC-TV. This media exposure coincided with a national greenhouse awareness campaign in Australia, *Greenhouse '89,* representing the peak of public pressure on politicians in the late 1980s to "do something about the greenhouse problem".

It was a different story in the 1990s however. Although there had been mounting public concern over broad environmental issues, there had been no accepted published proof clearly identifying deleterious environmental effects which could be attributed to climate change. In particular, there was still no apparently visible evidence of sea-level rise reported in Australia. The 1990s was a period of prolonged drought, with relatively few tropical cyclones affecting the southern Queensland coast. While the scientific community argued about possible effects and time horizons, the general public became sceptical, interest waned and climate change dropped off the political agenda (Enviro-Futures, 1994; Childs and Hastings, 1991).

This lack of public interest was one of the most important problems reportedly faced by Redcliffe professional officers in 1996 when trying to implement further risk management strategies (Childs et al., 1996). Climate change was not then viewed as an urgent threat, either by the media or by the public, and therefore there was little pressure on politicians to implement change. A search of locally produced newspapers - *The Redcliffe Herald* and the *Peninsula Post* - over the period January to May 1996, yielded no articles at all that addressed climate change (Childs et al., 1996); this situation contrasted greatly with the high level of media interest in the late 1980s (Childs and Hastings, 1991). In 1996, Redcliffe Council officers had expressed the view that raising the profile of potential risks of climate change in the public arena was essential to achieve appropriate and strategic policy and land management changes. Thus, public awareness and opinion continued to be a formidable factor influencing political decisions regarding short and long-term planning for climate change.

By 2008, media and political attention to climate change effects (both real and tenuous) such as extreme weather events, carbon-trading and rising fuel prices had increased considerably. One elected Redcliffe representative, however, reported no particular expressed concerns from constituents over sea-level rise, with lobbying from only a minority of rate-payers on general environmental issues such as pollution, canal maintenance and tree-planting. There were concerns with drainage issues, particularly in zones adjacent to the immediate shoreline because of saltwater intrusion during high tides and storm surges, but not in the context of potential sea-level rise. Other representatives asserted that the current "media hype" surrounding climate change was simply not translating into community pressure on Council to act. In addition, professional officers at Redcliffe concurred that, although the issue of sea-level rise is now commonly discussed at professional meetings, it had not been raised with them very often by the public.

The Future

Direct public pressure is a key driver of policy development and action at the local government level; in its absence, as reported by participants in the 2008 Redcliffe study, local government may be less inclined to further adaptive change. It is unclear as to why the apparent inconsistency between media saturation of the climate issue and the lack of perceived public pressure to act has emerged. It may be that: (a) the public is simply assuming that Council is already managing the risks in the community's best interests; or (b) that there is still a high degree of community scepticism regarding the risks of sea-level rise; or (c) that lead times expressed in decades (e.g. 2020, 2050) are beyond individuals' decision timeframes concerning their foreseeable futures. In any case, closer engagement between the community and its representatives is desirable to ensure inclusive decision-making and community ownership over any land use adaptations.

The 2008 Redcliffe study suggested that local government sees the provision of measured, positively presented, locally applicable and factually-based information and education as the best way to enhance community engagement, gain a political mandate and hence facilitate appropriate local government decision-making.

Planning and Land Use Adaptation

In 1987, some Redcliffe community stakeholders were aware of the potential impacts of sea-level rise (Childs et al., 1988). Nevertheless, there was general unwillingness to implement local climate change adaptation via land use planning or zoning changes, given the scientific uncertainties of the risk scenarios at the time. It was acknowledged, however, that the existing ten year local government planning and decision-making time horizon was inadequate to cope with the "greenhouse problem". The need to consider longer-term planning horizons arose from a recognition that office-bearers were responsible for structures that could be affected in 20-50 years.

As a local council dealing with climate change issues, Redcliffe's response in the early 1990s was to initiate some significant planning and policy changes, particularly in the area of building regulations. The impetus for at least some of these developments is reasonably assumed to have been connected with the involvement of the Redcliffe Council and community in the 1988 Redcliffe study and the subsequent publicity that arose. In the 1990s, Council engineers regularly produced reports on scientific advances in greenhouse information incorporating findings from the latest available literature (Marsh, 2002). Such reports set the context of scientific advice to elected Councillors in their decision-making deliberations.

In 1996, Redcliffe was using the CSIRO (1992) sea-level rise estimates. Deliberations concerning adaptation took into consideration the median value in the range of estimates from published scenarios. Risk planning for possible sea-level rise was achieved basically through development controls; for example, raising the base level of minimum kerb level (i.e., height) required for development of new buildings around the Redcliffe Peninsula. The nomination of minimum kerb levels is a key mechanism available to local councils to manage the risk of inundation. Decisions regarding levels are frequently based on historical analyses of tide heights. In the case of Redcliffe, further adjustments have been made in response to available scientific information concerning potential sea-level rise, as outlined below. At the time, this represented a progressive planning initiative. Table 14.2 summarizes the changes in kerb levels at Redcliffe.

In 1964, the minimum kerb level at Redcliffe was set at 1.86m AHD. This level remained until 1974 when, because of the effects of tropical cyclone Pam (i.e., combined tide and storm surge), Council decided to raise the minimum kerb level to 2.01m AHD. This was a reactive measure to a physical occurrence of high water. In 1987, a proactive change to the minimum kerb level occurred, this time on the basis of greenhouse-related publicity and scientific information. On the advice of the Council's professional engineers, the minimum kerb level was raised to 2.16m AHD to give the administration "some breathing space" and in the hope that the scientific community would firm up its scenarios and predictions to allow a more measured and accurate response in the near future (Marsh, 2002).

A significant test for the Council's commitment to its new raised development levels (i.e., minimum kerb levels) occurred in 1989, when an extension proposal for the already established Newport waterways canal estate was lodged. The development consultants were required to consider sea-level rise in the mandatory Environmental Impact Statement for their proposal. On the basis of updated scientific information, Council engineers initially recommended a new minimum kerb level for the Newport extensions of 2.81m AHD (Marsh, 2002).

The developers, however, argued for a lower kerb level on the basis of alternative sea-level rise scenarios from other sources (Childs et al., 1996). The Redcliffe administration eventually compromised by setting the minimum kerb level for new developments at 2.51m AHD, while leaving the level for redevelopment at 2.16m AHD. Interestingly, this was not changed in the 2005 Redcliffe planning scheme.

This is a clear example of how uncertainty in scientific information and confusion surrounding differing publicised scenarios can pose practical difficulties of policy setting at the local government level. In this case, a relatively conservative scenario was used by the developer to exert pressure on the Redcliffe Council to lower its preferred minimum kerb level. The Council was effectively placed in the difficult position of having to evaluate the alternative sea-level rise scenarios and was seen by developers and the community as being unnecessarily obstructive towards development.

Despite the uncertainty, precautionary responses have been taken in Redcliffe planning initiatives. In the 2008 Redcliffe study, the professional officers reported that the last low-lying Greenfield development site on the peninsula was earmarked for an extension of the Newport estate. Residential properties in this new development are at 3.5m AHD, above the Council's requirement, with minimum kerb levels at 2.51m AHD. Furthermore, a new major bridge development linking Redcliffe to Brisbane across an estuary also illustrates response to sea-level rise concerns. The 2008 study participants believed that the bridge height and placement was changed late in the design phase, following further reflection on potential climate change and sea-level rise impacts in light of extreme weather events occurring in other countries.

Table 14.2. Summary of Recent Changes to Minimum Kerb Level at Redcliffe

Year	Minimum Kerb Level (m AHD)	Comments
1964	1.86m	Level based on the highest tide recorded in Moreton Bay
1974	2.01m	Response to Tropical Cyclone Pam impacts
1987	2.16m	Reflecting "greenhouse" sea-level rise concerns
1990	2.51m	Following further "greenhouse" concerns given new scenarios, and significant negotiations between the Council and developers. For new developments.
2008	2.51m	No change in policy to the present

Sources: Childs et al. (1988); Childs et al., (1996); Marsh (2002); Redcliffe City Council, pers. com. (2008).

In the context of increased public awareness of wider environmental concerns over the past two decades, the Redcliffe City Council was successful in promoting a number of incidental initiatives benefiting adaptation to climate change and sea-level rise, although this was not the primary motivation in the decision-making. These initiatives included expanding public recreational land use (e.g. parklands, bikeways and a saltwater swimming lagoon) in low-lying foreshore areas, effectively creating buffer zones between the sea and residential and commercial properties. Recreational zoning is an excellent choice of land use for these relatively high-risk areas.

Local government has commissioned several studies that assess natural hazard risk on the shorelines and which now, at least, incorporate consideration of "climate change"; for example, a current beach erosion study and a new storm tide hazard study - both contributing to a management plan. The professional officers suggested that the inclusion of climate change considerations is now commonplace in environmental, engineering and related studies relevant to the region.

The Future

Determining the nature and timing of any adaptation to sea-level rise through planning schemes and building codes, even with guiding policies and documents, is problematic for councils on several fronts. Throughout the Redcliffe study, diversity of opinion was noted on how adaptation should proceed, ranging from "do nothing at the moment" to "proactive actions now".

The Council's professional officers interviewed in 2008 did not envisage any change to the current minimum development levels given existing circumstances. They did not perceive any particular pressure to alter them (in either direction) from developers, the wider community, or from internal sources. They did, however, concede that new pressures were likely if regulations and codes changed and would lead to higher, and potentially more costly, standards. Pressure to reduce minimum kerb levels could also result from a perceived lowering of sea-level rise predictions. The persistence of contrasting predictions / scenarios remains highly significant, as it potentially facilitates challenges to existing or proposed local standards, as was seen in Redcliffe with the Newport development application in the early 1990s.

With Greenfield development sites now virtually exhausted at Redcliffe, Council's attention has turned to re-development issues. Specifically, concerns centre on the risk exposure and piecemeal pattern of small-block re-developments on private property on low-lying sites along the coastline. Historically much of the peninsula was settled on the basis of lower minimum level requirements, and it is therefore now difficult for the Council to raise the minimum kerb levels on existing properties. Presently, the minimum kerb level applicable to redeveloped sites is 2.16m AHD, which is notably lower than that for new developments, but still higher than some existing properties where sea water inundation is a high risk. At such sites, Council seeks to enforce its codes on any re-development application to at least 2.16m AHD minimum level. Still, there remains a geography of contrasting risk and standards across the community. This leaves the Council with inherent inconsistencies in standards.

The protection and maintenance of existing infrastructure against the risks of potential sea-level rise was an issue raised by some participants throughout the Redcliffe studies. Professional officers recognised that progressive sea-level rise, contributing to saltwater intrusion, could increase maintenance costs of community lifeline infrastructure such as sewerage, electricity and telecommunications. In the future, the incremental exposure of such infrastructure may also be critical for individual hazard events (e.g. storms, tropical cyclones) with gradually increasing risks of failure in emergency situations.

WHERE TO FROM HERE?

Participants in the 2008 Redcliffe study raised several new issues in terms of addressing climate change adaptation into the future, including the availability of funding, the need to resolve possible legal liability issues, and governance within the local councils.

Funding Availability

Burton (2007), in reviewing contemporary responses to climate change (regarding greenhouse gas emissions) by South East Queensland local governments, reported concerns that local councils had neither the resources nor technical know-how to properly address the problem. This was also reinforced by the 2008 Redcliffe study in relation to adapting to potential sea-level rise, particularly with regard to funding. Councillors suggested that, if the financial burden of such adaptations were left with local councils they simply could not fund significant mitigation works or adaptation. In a worst case scenario, adaptation could necessitate resumption of private property and other land acquisition by local councils for the purpose of hazard protection. Although state and federal funding is currently available on a project-by-project co-funding basis, a more reliable and substantive framework is being sought.

One suggestion to augment local councils' resources was the introduction of a levy specifically to fund climate change adaptation. This, however, may be politically difficult to initiate due to the many factors discussed above. In particular, obstacles identified include: on-going scientific uncertainty, public perception of the risks versus financial costs of adaptation, and the present lack of direct public pressure on local politicians for action.

Resolution of Possible Legal Liability

The contrasting risks between individual properties which now exist in Redcliffe because of historical differential approved developments (see above) have resulted in concerns that legal liability issues may arise in the future. That is, if Council did not amend planning schemes and building codes adequately, even given the uncertainty of current climate change scenarios, they may be held accountable for any future impacts. If they do adapt, however, is this an admission that previous standards were inappropriate? What is to be done with development that no longer complies with current standards? How might this be compensated? Changes in Council's policies and regulations, or choices not to change, may also affect the availability of insurance to the community, imposing greater costs or hardship. England (2007), nevertheless, concludes that local governments do have the means for protection from liability in circumstances related to climate change, but suggests that this is a rapidly evolving legal area in Australia. [A further account of the development of such legal frameworks can be found in England (2008)].

Governance within Local Councils

The recent amalgamation of Redcliffe Council with two neighbouring councils to form the larger Moreton Bay Regional Council (MBRC) is seen as bringing positive benefits in general for policy implementation, through greater access to resources and opportunities to achieve consistency across a broader geographical region. Councillors did note, however, from their political viewpoint, that a diversity of opinions and priorities within the larger membership and constituencies of MBRC needs to be managed; for example, within the newly-created MBRC, there is an inland constituency that would probably not view mitigation against sea-level rise as a priority issue for rate-payers. As outlined above, the researchers also note that diversity of opinion exists within the Redcliffe component of the MBRC, that is, significant scepticism still exists in terms of committing local government funds for unproven hazards. Therefore, there is a potential for tension over possible responses to climate change and sea-level rise within the MBRC, although this was not reported to be noticeable at the time of the 2008 Redcliffe study.

CONCLUSION

As an Australian local council dealing with the climate change issue, and particularly with the potential for sea-level rise around its shores, the Redcliffe case study provides insights into the processes of risk management and the factors that facilitate or obstruct land use adaptation. Over two decades of this longitudinal study at Redcliffe, persistent issues have included: dealing with the uncertainty of risks of sea-level rise; applying climate change policy from an evolving national and state context to the local level; as well as balancing public opinion, economic and political imperatives with longer-term community stewardship.

Future resolution of these on-going issues is now also linked to additional concerns, clarified in the 2008 Redcliffe study, such as: the availability of funding for adaptation to potential sea-level rise to local councils; concerns regarding legal liability; and the processes of governance in the newly-amalgamated regional councils in Queensland.

The significance of potential sea-level rise is an issue which local government will increasingly need to address in the future and is underscored by a recent ruling in the Civil and Administrative Tribunal in the state of Victoria regarding an application for residential development in a coastal location (Victorian Civil and Administrative Tribunal, August, 2008). The ruling was as follows:

> In the present case, we have applied the precautionary principle. We consider that increases in the severity of storm events coupled with rising sea levels create a reasonably foreseeable risk of inundation of the subject land and the proposed dwellings, which is unacceptable. (p. 13)

In light of this ruling, lessons learned from the Redcliffe case study may be useful for other Australian coastal councils dealing with the need to address liabilities and local government responsibilities in the context of planning for sea-level rise.

Some of the lessons learned from the Redcliffe experience, that may be applicable to other local government areas dealing with similar risks relating to climate change, include the following:

- The desirability of a proactive, integrated risk-management approach to adaptation in which scientific uncertainty about climate change and its impacts can be conceptualised in terms of risk;
- The necessity to develop specific guidelines on applying risk management approaches to local council planning and decision-making, and to bridge the gap between over-arching climate-change policies and their local application;

- The need to ensure that the actions and opinions of the public are being accurately interpreted and translated into pressures on council decision-making as appropriate;
- The recognition that councils may face complex negotiations with developers who can use scientific uncertainty to argue for more conservative risk standards;
- The requirement for councils to consider the consistency of risk standards across both Greenfield and re-development of existing sites established under earlier building standards; and
- The recognition of the emerging issues of attracting adequate funding, resources and intra-council cooperation to ensure that they are meeting their responsibilities to the community.

Thus the extended work of the Redcliffe City Council, in coming to terms with sea level rise, may now assist other councils in their planning for the future impacts of climate change.

ACKNOWLEDGMENTS

The authors wish to acknowledge the valuable assistance of professional officers and councillors of the Redcliffe City Council and Moreton Bay Regional Council who participated in this research.

REFERENCES

Australian Bureau of Statistics. (2006). Redcliffe (C) Local Government Area, 2006 Census QuickStats. Accessed 08.08.2008. http://www.abs.gov.au/websitedbs/D3310114.nsf/Home/Census+Data

Australian Geological Survey Organisation (AGSO). (2001). *Community risk in Redcliffe city – a multi hazard risk assessment*. AGSO, Canberra, Australia.

Beyond 2000 - climate in crisis. (1989). CEL Arts and Thames Video Collection. Beyond International Group, Australian and BBC TV, UK in association with WTBS and TVE/UNEP.

Burton, D. (2007). *Evaluating climate change mitigation strategies in South East Queensland*. Research Paper 11. Urban Research Program, Griffith University, Brisbane, Australia. Available from: http://www.griffith.edu.au/environment-planning/urban-research-program/publications

Childs, I.R.W., Auliciems, A., Hundloe, T.J. and McDonald, G.T. (1988). Socio-economic impacts of climate change: potential for decision making in Redcliffe, Queensland. In Pearman G.J. (Ed.), *Greenhouse, planning for climate change (pp.* 648-664). CSIRO Division of Atmospheric Research, Brill, Leiden, Melbourne, Australia.

Childs, I.R.W. and Hastings, P.A. (1991). Public awareness of greenhouse: developments in Queensland since 1987. *Environment of Institute of Australia Newsletter,* April 1991, 9-10.

Childs, I.R.W., Hastings, P.A. and Auliciems, A. (1996). Redcliffe ten years on: A longitudinal study of policy responses to potential greenhouse induced storm surge in a local government area. In R.L. Heathcote, C. Cuttler, and J. Koetz (Eds.), *Proceedings of NDR96 conference on natural disaster reduction*, Surfers Paradise, Australia October 1996. Institution of Engineers Australia, 211-222.

Commonwealth of Australia. (1992). *National greenhouse response strategy*. Australian Government Publishing Service (AGPS), Canberra, Australia.

Commonwealth of Australia. (1995). *Greenhouse 21C, a plan of action for a sustainable future*. Department of Environment, Sport and Territories, Canberra, Australia.

Commonwealth of Australia. (1998). National greenhouse strategy - strategic framework for advancing Australia's greenhouse response. Australian Greenhouse Office, Canberra, Australia.

Commonwealth of Australia. (2007). *Climate change adaptation actions for local government*. Australian Greenhouse Office, Department of the Environment and Water Resources, Canberra, Australia. Available at: http://www.climatechange.gov.au/impacts/publications/local-government.html

Commonwealth Scientific and Industrial Research Organisation (CSIRO). (1992). *Climate Change Scenarios for the Australian Region*. Climate Impact Group, CSIRO Division of Atmospheric Research, Melbourne, Australia. Available at: http://www.cmar.csiro.au/e-print/open/CIG_1992a.pdf

Commonwealth Scientific and Industrial Research Organisation (CSIRO). (2007). *Climate change in Australia*. Technical Report, CSIRO Marine and Atmospheric Research, Melbourne, Australia. Available at: http://www.cmar.csiro.au/e-print/open/2007/pearcekb_a.pdf

England, P. (2007). *Climate change: What are local governments liable for?* Issues Paper No. 6, Urban Research Program, Griffith University, Brisbane, Australia. Available at: http://www.griffith.edu.au/environment-planning/urban-research-program/publications

England, P. (2008). Climate change law for planners, developers, local government and greenies: A quick stock take and some ideas for the future. Research Paper 16, Urban Research Program, Griffith University, Brisbane, Australia. Available from: http://www.griffith.edu.au/environment-planning/urban-research-program/publications

Enviro-Futures. (1994). *Greenhouse action and local government, the new directions*. Final report prepared for the National Environmental Law Association Limited, Australia.

IPCC. (2001). *Climate change 2001: Synthesis report*. A Contribution of Working Groups I, II, and III to the Third Assessment Report of the Intergovernmental Panel on Climate Change [Watson, R.T. and the Core Writing Team (Eds.)]. Cambridge University Press, Cambridge, United Kingdom.

IPCC. (2007). *Climate change 2007: Synthesis report*. Contribution of Working Groups I, II and III to the Fourth Assessment Report of the Intergovernmental Panel on Climate Change [Core Writing Team, Pachauri, R.K and Reisinger, A. (Eds.)]. IPCC, Geneva, Switzerland. Available at: http://www.ipcc.ch/ipccreports/ar4-syr.htm

Local Government Association of Queensland (LGAQ). (2007). *Adapting to climate change, a Queensland local government guide*. LGAQ. Available from: http://www.lgaq.asn.au/lgaq/publications/ClimateChange.html

Mallon, T. (1996). Deputy City Engineer, Redcliffe City Council. Personal communication.

Pearman, G.J. (Ed.). (1988). *Greenhouse, planning for climate change*. CSIRO Division of Atmospheric Research, Brill, Leiden, Melbourne, Australia.

Pyper, W. (2007). Preparing for sea-level Rise. *Ecos, 137*, 4-17.

Queensland Government. (1995). *Queensland Government response strategy*. Department of Environment and Heritage, Brisbane, Australia.
Queensland Government. (2001). Queensland greenhouse policy framework. Queensland Government, Brisbane, Australia.
Queensland Government. (2002). *State coastal management plan – Queensland's coastal policy*. Environmental Protection Agency, Brisbane, Australia.
Queensland Government. (2004). *Queensland greenhouse strategy*. Queensland Environmental Protection Agency, Brisbane, Australia.
Queensland Government. (2005). *South East Queensland regional plan 2005-2026*. Queensland Department of Local Government, Planning, Sport and Recreation, Brisbane, Australia. Available at: http://www.dip.qld.gov.au/regional-planning/regional-plan-s.html
Queensland Government. (2007a). ClimateSmart 2050, Queensland climate change strategy 2007: A low carbon future. Queensland Government, Brisbane, Australia. Available at: http://www.climatechange.qld.gov.au/response/adaptation_plan.html
Queensland Government. (2007b). *ClimateSmart adaptation 2007-2012, an action plan for managing the impacts of climate change*. Queensland Government, Brisbane, Australia. Available at: http://www.climatechange.qld.gov.au/response/pdfs/climatesmart_plan.pdf
Queensland Government. (2008). *Climate change in Queensland: What the science is telling us*. Queensland Government. Available at: http://www.climatechange.qld.gov.au/downloads/ index.html
Redcliffe City Council. (2002). *Sea levels and their relationship to development levels in Redcliffe*. Memorandum from Coordinator Contracts, 11 April, 2002.
The great global warming swindle. (2007). UK Channel 4, March 2007. Information available at: http://www.channel4.com/science/microsites/G/great_global_warming_swindle/
Victorian Civil and Administrative Tribunal. (2008). Red Dot Decision Summary. Sc and Ors (No.2) VCAT 1545 Gippsland Coastal Board v South Gippsland Shire Council. Accessed 19 August 2008. http://www.vcat.vic.gov.au/CA256902000FE154/Lookup/decisions/$file/gippsland_coastal_board_v_south_gippsland_sc_and_others.pdf

Contributed by Tim McCrorey, Toowoomba.

Photo. Rising Sea Levels – a Surety in the Future?

In: Meltdown: Climate Change, Natural Disasters… ISBN 978-1-60876-153-1
Editor: Kathryn Gow

Chapter 15

CLIMATE CHANGE IMPACTS AND PLANNING IN AFRICA

Workineh Kelbessa
Addis Ababa University, Ethiopia

ABSTRACT

The aim of this chapter is to analyse climate change impacts, and climate risk management in Africa. Although African emissions have played a minor role in the build-up of greenhouse gases in today's atmosphere, Africa has been suffering from rising temperatures and evaporation, widespread water stress, increased frequency and severity of droughts and floods, crop loss, rising sea level, decline in biodiversity, higher level of disease, and conflicts over access to land and water. Unless the greenhouse gases emissions are reduced significantly, Africa will continue to suffer from the adverse effects of climate change, as it has low capacity to cope with such phenomena. In spite of this, national governments and the local people in Africa have used various strategies to tackle the adverse consequences of climate change. African countries have tried to address the problems of climate variability and change by introducing national and regional mitigation strategies. Peasant farmers and pastoralists have been coping with environmental hazards, and adapting their livelihoods to changing conditions for generations. However, indigenous mechanisms are not sufficient by themselves to deal with the effects of climate change; they should be complemented by modern strategies. This chapter, thus, suggests that policy makers, modern scientists and peasant farmers have to work together to address the adverse effects of climate change.

Keywords: Africa, climate change, climate variability, climate risk management, indigenous coping strategies, international response, regional cooperation

INTRODUCTION

Africa has had multiple problems that have contributed to its underdevelopment. Both internal and external factors (slavery, colonialism, neocolonialism etc.) are responsible for the current situation of Africa. Africa has been incapacitated by poverty, uncontrolled disease including the tragedy of HIV/AIDS, environmental degradation, inequitable land distribution, overdependence on rain-fed agriculture, the debt crisis, the declining competitiveness of its raw materials in the international market, an unjust international trade system, exploitation of its resources by external powers, limited institutional capacities, poor management, corruption, ethnic conflict, lack of political will on the part of the leaders, and massive human rights violations. Africa has become structurally irrelevant to the present global economic order; it is no longer very important to the major actors in the world economy. In short, because of its weak economy, inherent historical biases, feeble capacities for negotiations, its deplorably low and declining share of global trade and the aggravation of the brain drain, Africa has very little influence and voice in today's global policy-making forums.

The list of challenges facing Africa is so long and oft repeated that in contemporary scholarship there is a tendency to try to somehow escape any real responsibility for effectively addressing such problems. I might note here without dwelling on the facts, or attempting to prove the obvious that Africa is not a major player in the globalised and supposedly 'competitive' 'free market' economy, and that, quite simply, the global economic system is not fair, based on any remotely viable form of justice.

Moreover, climate change has further aggravated the problems of the African people. Recurrent droughts, floods and storms have had devastating impacts on Africa. Thus, it is imperative to understand the nature and impacts of global climate change on Africa, and to propose viable solutions that can address climate change impact, vulnerability, adaptation and other related problems in Africa.

There is a dearth of serious academic studies of climate change impacts on African countries. This is partly due to the two key scientific constraints on African climate science: lack of reliable data and scientific capacity (Washington et al., 2006). Additionally, there are very few climate scientists in Africa and moreover, Africa has low capacity to plan and invest in adaptation to climate variability; it lacks sufficient and functioning meteorological stations. Currently, the density of its stations is estimated to be eight times lower than the minimum recommended by the World Meteorological Organization (WMO) (Giles, cited in Hellmuth, 2007a, p.10). "One upshot of the relative scarcity both of African scientists and of African observations is that we have little to no systematic understanding of the basic state of the circulation over critical parts of the continent, including the central African convective region (which at certain times of the year is the largest on the planet) adjoining Angola and much of the Sahara" (Washington et al., 2006, p.1358). A combination of factors including a lack of locally calibrated models, a shortage of observational data in computer-readable form and an exclusion of African scientists from accessing the results of global climate and climate impact models, explain the paucity of detailed climate change impact case studies for Africa (Hulme et al., 1995; see also Hulme et al. 2005).

In this chapter, I examine climate change impacts in Africa and how Africans have responded to these impacts. First, I explore past climate impacts in Africa by citing some examples from different parts of the continent; then, I discuss projected climate change impacts on Africa. Second, I shall examine climate risk management in Africa; finally, I offer some concluding remarks. The material in this chapter is based on secondary sources about the subject, and interviews I conducted in Borana, South Ethiopia, Ambo, Central Ethiopia, and Illuababorra, West Ethiopia.

CLIMATE CHANGE IN AFRICA

Past Climate Change Impacts in Africa

Africa, the continent that has contributed least to causing climate change has nevertheless been facing countless climate change related problems. This is despite the fact that Africa's contribution to greenhouse gas emissions is the least of any continent. For instance, it is estimated that each year Africa produces an average of just over 1 metric ton of the greenhouse gas carbon dioxide per person (U.S. Department of Energy's International Energy Annual, cited in Fields, 2005, p. A535). "Africa's share of global CO_2 emissions rose from 1.9% in 1973 to 3.1% in 2002, representing an emission of about 747 Mt of CO_2 in 2002" (IEA, cited in Manful, 2005, p. 297). Thus, Africa contributes less than 4% to global emissions of GHGs (greenhouse gases), 75% of which are due to land use changes (deforestation, land degradation).

While industrialised countries, such as South Africa, produce 8.44 metric tons per person, the least developed countries, such as Mali, produce less than a tenth of a metric ton per person. Thus, the largest proportion of greenhouse gas emissions in Africa belongs to South Africa, which is the fifteenth largest industrial emitter of greenhouse gases globally and the seventh largest 'developing' country emitter (Justice et al., 2005, p. 174). "For example, it is estimated that in 1997 South Africa generated 2% of global CO_2 emissions and 3-7% of the global total emissions of NO_x" (Justice et al., 2005, pp. 173-174). In general, the following countries, taken collectively, share about 95% of Africa`s total C02 emissions: South Africa, Nigeria, Kenya, Zimbabwe, Sudan, Ethiopia, Angola, Ghana, Ivory Coast, Equatorial Guinea, Senegal, Botswana, Tanzania, Cameroon, and Congo Brazzaville (Unmüßig and Cramer, 2008).

By comparison, the United States pumps out about 25 percent of all greenhouse emissions. Whereas the G8 countries emit about half the world's total output, the entire African continent emits less than four percent of the world's total output.

Recently, changes in rainfall have been observed in different parts of Africa. Among others, the African Sahel suffers from very variable annual rainfall and a prolonged drought (Taylor, 2001), which continues today. Annual rainfall during the period 1931-60 had been between 20% and 40% greater than during the most recent three decades (Hulme, cited in Hulme et al., 1995, p. 6). "Individual years, such as 1984 and 1990, have seen rainfall totals drop below 50% of those typical of the last decades of the colonial era prior to 1960. The human suffering that has accompanied this prolonged drying out provides a stark indication of the vulnerability of parts of African society to climate change" (Hulme et al., 1995, p. 6).

A fairly recent study also shows that there was a dramatic decline in average rainfall conditions in all West African drylands for the period 1960-1990 (Put et al., 2004, p. 27). "Some of the regions in the northern zone with semi-arid conditions in 1930-60 had clearly become arid (on average) in the 1960-1990 period, with unsuitable conditions for millet or sorghum production in most years. A considerable part of the sub-humid zone in the period 1930-1960 had become semi-arid in 1960-1990 with considerable drought risks, certainly for crops which are less adaptable to drought stress (e.g., maize, cotton)" (Put et al., 2004, p. 27). About 40% to 60% of West Africa's major river systems have dropped since the early 1970s (Niasse et al., and Niasse, cited in Nkomo et al., 2006, p. 15).

Furthermore, some studies have shown that El Niño-Southern Oscillation (ENSO) has had a negative impact on climate in Africa. The ENSO phenomenon and the temperature of the Indian Ocean affect Eastern and southern Africa whereas the conditions in the Atlantic Ocean affect West Africa. The interactions between the Pacific tropical ocean and the global atmosphere drive El Niño phenomenon, which can bring large year-to-year changes for some parts of Africa (Basher and Briceño, 2005, p. 274). Drought is a recurrent feature in most parts of Southern Africa, with five recent major episodes, in 1980-1983, 1987-1988, 1991-1992, 1994-1995, 1997-1998 and 2002 (Basher and Briceño, 2005, p. 272). Severe drought has reduced the amount of rain over parts of southern Zimbabwe and southeastern Botswana during the rainy season of 1991/92 largely owing to the (ENSO) phenomenon, the periodic warming of the tropical Pacific Ocean and related shifts in the atmospheric circulation which brings climatic disruption to many low-latitude areas (Glantz et al., cited in Hulme et al., 1995:8). The severe drought that affected nearly 100 million people in south-eastern Africa in 1991-1993 coincided with an ENSO event (Cane, Eshel and Buckland, 1994, cited in Lovett et al., 2005, p. 168). Thus in 1997/98, the El Niño-induced high rainfall killed 1000 people, displaced 150,000 people, and damaged trunk and rural road infrastructure estimated at USD 400 million in Uganda (Eriksen et al., 2008, p. 16). The devastating floods from tropical cyclones in 2000 reduced Mozambique's annual economic growth rate from 8% to 2.1% (World Bank, cited in Washington et al., 2006, p. 1355), and droughts in the same year substantially reduced hydroelectric power output in Kenya, leading to a $72 million (U.S. dollars) emergency loan from the World Bank (World Bank, cited in Washington et al., 2006, p. 1355). During the floods of 2000, two million people were displaced, and 350,000 jobs were lost in Mozambique (Nkomo et al., 2006, p. 3). South Africa lost cereal crops worth billions of dollars during the 1990s because of droughts (Clay et al. 2003, cited in Washington et al., 2006, p. 1355).

Another study suggests that vegetation productivity and desertification in sub-Saharan Africa may be influenced by global climate variability attributable to the North Atlantic Oscillation (NAO) and ENSO (Oba et al., 2001). According to Oba and his co-authors, "[f]ully 75% of the interannual variation in the area covered by the Sahara Desert was accounted for by the combined effects of the NAO and ENSO" (p. 343). However, most variance is attributable to the NAO. Much of the interannual variability in vegetation productivity in the Sahelian zone and Southern Africa is the outcome of the combined indices of NAO and ENSO. This conclusion should be supported by further research.

According to the database of the EM-DAT, flood, drought/famine, windstorms, extreme temperature and wildfire killed 15,713 people, and affected over 136 million people in Africa between 1993-2002 (Basher and Briceño, 2005, p. 273). Floods accounted for the most deaths, whereas droughts affected over a hundred million people over the decade.

Moreover, glaciers on Mount Kilimanjaro on the border of Tanzania and Kenya are melting. According to Ahmed Djoghlaf, UNEP's Executive Secretary, "[a]lthough the snow and ice cover of the mountain has been shrinking for more than a hundred and fifty years, it has accelerated in recent decades due to higher temperatures and less precipitation. It is widely anticipated that the remaining glacier[1] will disappear entirely within 20 years" (UNEP, 2007, p. 1). The melting of the glaciers on Mount Kilimanjaro will have negative implications for some local rivers that rely on them in Kenya (Odada, cited in Migiro and Nutt, 2007, p. 12).

My field research also confirmed the negative impacts of environmental degradation in Ethiopia; soil erosion, deforestation, overcultivation, and overgrazing have contributed to this environmental degradation. Inappropriate economic policies, inappropriate political decisions and actions of the state, poor infrastructure, and insecure land tenure have exacerbated the environmental crisis in Ethiopia (Kelbessa, 2008).

According to interviewees, environmental degradation has had negative consequences including climatic change, ecological imbalance, the spread of human and animal diseases, and increasing shortage of critical resources such as pasture land and arable land. It also involves air and water pollution, drought, reduction of genetic diversity, poverty, famine, migration of the people and animals, lack of construction materials, reduction in land productivity, reduction of fallow years (period of being untilled or unseeded), and so on. Environmental degradation forced the poor people to live in exposed and marginal areas (i.e. floodplains, exposed hillsides, arid or semi-arid lands). Such people can thus suffer from even minor climate changes.

In the past, there were long rainy seasons in both Illuababorra and Borana. Particularly in Illuababorra, in the past till about 1976, it was only for one week in January that the region lacked rain. Currently, there is no rain from January to March; likewise, there has not been sufficient annual rainfall in Borana. The interviewees lamented that temperature has been increasing over time, and the volume of surface and underground waters has decreased in both Illuababorra and Borana; some springs, wells and permanently flowing rivers have dried up. Thus, with changes in land use and climate, some areas have become drier, and water sources are becoming intermittent or disappearing. As in Australia, drought has appeared there every two years, whereas in the past, drought occurred only every ten years. Some peasant farmers and pastoralists have been migrating to neighbouring districts in search of grass and water since 1995; for instance, about 14 Peasant Farmer Associations have migrated from Liben district to Adola and Wadara, Hagere Mariyam and Odo Shakisso districts in Borana. Unfortunately, after the moves, many animals have died because they could not adapt themselves to the new climate and environment (Kelbessa, 2008), and thus this disaster shows that climate change can create environmental refugees by forcing relocation because of droughts and consequent lack of food supplies and income raising farming pursuits.

Environmental degradation and competition for natural resources have given rise to massive internal and external migration of people and has been putting new pressures on host countries in different parts of Africa. As Cardy (2003) notes, the causes of migration include overexploitation of natural vegetation, pollution of water and soils, depletion of water resources, changes in land use, loss of 'reserve lands' and spread of diseases and decreased resistance to diseases.

The UN convention has defined a 'refugee' as a person who crossed national borders and escaped persecution. Later, a UNEP publication used the title 'environmental refugee' (El-Hinnawi, cited in Cardy, 2003 p. 142). Environmental refugees are people who were forced to

[1] http://www.cbd.int/doc/bioday/2007/ibd-2007-booklet-01-en.pdf.

leave their homelands, temporarily or permanently, because of a marked environmental disruption (natural and/or triggered by people) that has had a negative impact on their life (El-Hinnawi, cited in Cardy, 2003, pp. 142-143; see also Myers, 2005). In this definition, 'environmental disruptions' means any physical, chemical and/or biological changes in the ecosystem (or the resource base) that render it, temporarily or permanently, unsuitable to support human life. The concept 'environmental refugees' does not include people displaced for political reasons or by civil strife, and migrants seeking better jobs purely on economic grounds.

El-Hinnawi (2003) proposed the existence of three broad categories of environmental refugees: those who have been temporarily displaced because of an environmental stress and who can return to their original habitat once the environmental disruption is over. The second category comprises those who have to be permanently displaced because of permanent changes and re-settled in a new area. The third category of environmental refugees consists of individuals or groups of people who migrate from the original habitat, temporarily or permanently, in search of a better quality of life because of the deterioration of the quality of their original habitat. People in the third group may cross national borders and may be officially defined as 'refugees' if they establish 'persecution' at home (El-Hinnawi, cited in Cardy, 2003, p. 143).

It is difficult to determine the number of environmental refugees. According to Myers (2005), of the 25 million environmental refugees recorded in 1995, roughly five million were in the African Sahel. Of eleven million refugees of all types, the Horn of Africa hosted four million; about seven million people migrated in different parts of Sub-Saharan Africa where 80 million people were considered to be semi-starving due largely to environmental reasons. The UNHCR estimated that there are at least 21 million internally displaced people in the world (cited in Cardy, 2003, p. 145).

Some of the negative impacts of migration include: over-exploitation of natural vegetation which may lead to increased soil erosion and gully formation, increased water run-off and siltation of rivers and reservoirs downstream; pollution of water and soils; depletion of water resources; changes in land use from sustainable to unsustainable status; reduction of loss of 'reserve lands' for local populations in drought periods: spreading of, and decreased resistance to, various kinds of diseases between hosts and refugees (Cardy, 2003, p. 149). Large concentrations of displaced people on confined areas often lead to localized deforestation and poor sanitary conditions. In short, the increased flows of refugees across international borders have had impacts on environmental security; environmental refugees create additional pressure and demand on water resources and grazing land.

Environmental refugees are found all over the world. Some of the likely causes of environmentally induced migration include: active volcanoes and earthquakes on the Pacific Rim; cyclones and flooding in Bangladesh; prolonged drought, locusts, land degradation in the Sahel; accidents at nuclear installations and industrial plants, tornadoes and hurricanes in the United States; increasing numbers of landless through sub-division of smallholdings at death; salinisation of irrigated lands; and many other problems (Cardy, 2003, p. 157).

It is worth noting that diseases such as malaria, diarrhoea, tuberculosis, trypanosomiasis, skin infections and other contagious diseases are rampant in most parts of the study sites in Ethiopia. Previously, malaria was common only in Dollo, Malka Guba and Kiltu in Borana, and in lowland areas in Illuababorra. New weeds, leaf blight and what the Illuababorra Oromo called *coleraa* (a disease which makes crops dry out) attack plants.

Likewise, new climate conditions have affected insects and the diseases they harbour in other parts of Africa (Patz, cited in Fields, 2005 p. A537). According to Patz, vector borne diseases - those in which a pathogen is carried from one host to another by pests such as mosquitoes - have increased their range, because of global warming. This enabled malaria to move into higher African latitudes as highlands have warmed enough for mosquitoes to breed.

"Malaria is responsible for about 91% of all mortalities in Africa and Africa accounts for about 85% of all malaria-related deaths in the world" (Nkomo et al., 2006, p. 3). On the other hand, Marlies Craig, a South African scientist, is of the opinion that factors besides increasing temperatures, such as increased resistance to drugs lead to the spread of this vector (cited in Fields, 2005 p. A537). Moreover, Rift Valley Fever, cholera and meningitis have affected different parts of Africa.

Potential Climate Change Impacts in Africa

As many scholars stated, climate change will lead to increasing levels of drought in Africa if temperatures continue to rise. Further, it can lead to floods, starvation, landslides, drought and rising sea levels. According to the United Nations, climate change will affect Africa more than anywhere else in the world due to extreme poverty levels, high rates of population growth, over-reliance on rain-fed agriculture[2] and over-dependence on natural resource-based livelihoods. The projected global temperature increase by the year 2100 will lead to less precipitation in Central Asia, the Mediterranean region, Africa, parts of Australia and New Zealand, associated with a greater probability of droughts (UNEP, 2007). In order to understand the potential impacts of climate change on Africa, it is important to examine how various sectors are susceptible to climate change.

Water Resources and Agriculture

The regional report by Working Group II of the Intergovernmental Panel on Climate Change (IPCC) predicted that because of continued increase in greenhouse gases, 75-250 million people in Africa will be at risk of increased water stress with a 1°C rise in global temperature by the 2020s, 350-600 million people will be affected with a two degree climb by the 2050s, and up to 1.8 billion if global temperature rises by three degrees, which could happen by around 2080 (Boko, Niang, Nyong et al., 2007). Africa's persistent water-stress[3] will be aggravated by climate change. According to the United Nations Environmental Programme (UNEP), as of 1999 "14 countries in Africa are subject to water stress or water scarcity" and "[a] further 11 countries will join them in the next 25 years" (quoted in New Economic Foundation, 2005 p. 13). The decline of rainfall would lead to the deterioration of water for sanitation purposes and industrial effluents would become more concentrated and provide more breeding grounds for water-borne diseases.

[2] Rain-fed agriculture = cultivation relying solely on rainfall category.

[3] Water stress occurs when the demand for water exceeds the available amount during a certain period or when poor quality restricts its use.

African economies are overwhelmingly agriculture-based, and highly susceptible even to minute variations in temperature and rainfall. As Fields states, crop failure in Africa is a catastrophe, but in other continents it causes difficulties of a lesser nature. "About 40% of the gross national product of African countries flows from agriculture, and about 70% of African workers are employed in agriculture, most of them on small plots of land" (Fields, 2005, p. A536). African peasant farmers always depend on rain. If there is no rain one year, there will be no food. Thus, changes in precipitation adversely affects African peasant farmers particularly those who have no opportunity to use irrigation. Droughts and water stress will also affect both pastoralists and their cattle in rangelands, and many rural communities who rely on animals for traction and fertilization, as well as meat and milk.

Further global warming will lead to the further drying up of arid and semi-arid areas. In many African countries and regions, it is expected that climate variability and change will reduce agricultural production, including access to food, in many African countries and regions. By 2020, yields from rain-fed agriculture are expected to reduce by 50% in some countries (IPCC, 2007, p. 10). Agricultural loss up to 7% of GDP is expected in parts of the Sahara and Central and Western African countries are expected to lose up to 4% of their GDP.

Nevertheless, climate change can also have positive impacts on some countries. Higher predicted rainfall and increased temperatures may extend the growing season in places like the Ethiopian Highlands. It may also be possible to grow more temperate crops such as apples, pears and barley in alpine zones like those in and around Mount Kenya and Mount Kilimanjaro because of falls in frost.

Health and Disease Patterns

Africa suffers from climate or water-related diseases, such as yellow fever, cholera, river blindness, bilharzia, malaria and tuberculosis (Hulme et al., 1995 p. 34). In particular, climate change will create more favourable conditions for the spread of malaria in Africa in the years to come.

According to the IPCC Working Group II Fourth Assessment Report (IPCC, 2007), vector- and water-borne diseases may increase in the future. It is estimated that an additional 80 million people will be at risk of malaria epidemics in the 21st century. The highlands of Ethiopia, Kenya, Rwanda and Burundi—currently malaria free—become highly suitable for malaria mosquitoes by the 2080s. Malaria is also likely to rise in the highland areas of Angola and Somalia, areas currently with low levels of malarial cases; AIDS and malaria are the biggest killers in Africa. Moreover, heat waves, floods, storms, fires and droughts will give rise to increased deaths, diseases and injury. Although climate change may lead to fewer deaths from cold exposure in temperate areas, overall the negative health effects of rising temperatures world-wide, especially in 'developing' countries, will outweigh these benefits (IPCC, 2007 , p. 10).

Sea-level Rise and Fisheries

Rising sea levels may have negative impacts on a number of African countries. It can lead to inundation and displacement of wetlands and low-lying coastal zones. The Nile delta, the lagoons and cities along the Gulf of Guinea in West Africa and the large urban centres in East and Southern Africa are vulnerable to the impact of flooding. The IPCC (2001) estimated that as many as 70 million people could be affected by flooding in 2080, compared to 1 million in 1990. According to the IPCC, "[s]ea-level rise and human development are together contributing to losses of coastal wetlands and mangroves and increasing damage from coastal flooding in many areas" (2007, p. 4). It is estimated that the reduction of fisheries resources will have a negative impact on local food supplies. The IPCC (2007) report further revealed that sea-level rises in coastal countries may cost up to 14% of GDP. Adaptation may cost less — between 5% to 10% of GDP.

Rising water temperatures, continued overfishing, alterations in river flows and intrusion of salt waters into lagoons can have negative impacts on fisheries in Africa. For instance, Chidi Ibe (cited in Hulme et al., 1995, p. 36) examined the vulnerability to sea level rise of three coastal cities - Lagos, Banjul and Dar Es Salaam. All three cities face coastal erosion owing to a high proportion of the economic activity of the respective countries; anticipatory planning and action is required to avoid severe social and economic disruption. Ibe (in Hulme et al., 1995) suggests that the following actions should be taken: (1) phased 'disengagement' from the vulnerable areas where practicable; (2) prevention of new developments in the coastal zones; and (3) the installation of low cost, low technology erosion and flood defence measures.

High rates of influx of people from the hinterland, population growth and urbanisation have put heavy pressure on the coastal zone of Ghana, which flanks a coastline of approximately 550 km (Armah et al., 2005, p. 204). The coastal zone in Ghana faced overfishing, poor sanitation, wetland degradation and coastal erosion (Armah and Amlalo, cited in Armah et al., 2005, p. 204). Armah et al. (2005) envisaged that if the current rate of discharge of greenhouse gases into the atmosphere continues, several portions of the coast of Ghana would undergo physical changes in the following areas through: "establishment of permanent connections with the sea for most of the 90 lagoons within the coastal areas; increased penetration of salt water inland into estuaries and open lagoons; increased rates of coastal erosion, resulting in beach recession; decrease in the depth of the water table in coastal areas (i.e. decrease in the thickness of the freshwater lens overlying saline water); and intrusion of salt water into coastal freshwater aquifers" (p. 205).

Conflict

Climate change can further lead to conflict between neighbouring people. Such conflict occurs when resources become scarce and there is no food or water security. In Ethiopia, there has been an intermittent conflict between the Borana Oromo and Ethiopian Somali in Southern Ethiopia about water and grazing land. Furthermore, drought led to conflict, and the death of pastoralists and animals in the remote village of Sambarwawa in the heart of the northern Kenyan district of Isiolo, and in Mai Mahiu area, 90 km north-west of Kenya's capital, Nairobi; some pastoralists murdered others who owned boreholes (Migiro and Nutt,

2007). Lack of water, grazing, and land has become the cause of inter-ethnic conflict in Kenya, for example between the Maasai and others on the Ewaso Kedong river, between Borana and Gabra pastoralists in Kenya's Marsabit district, near the Ethiopian border. Drought has also caused conflict between Ugandan and Kenyan pastoralists (Migiro and Nutt, 2007). Similarly, there have been conflicts over river basin resources in African drylands; for instance, there were conflicts between Senegal and Mauritania in the 1980s over the utilisation of resources in the Senegal basin, between the Niger and Nigeria over the two dams on the Lamido and river Maggiya, and between Nigeria and Cameroon over the Ladoo Dam on the Benue river (Darkoh, 2003, p. 62). There have also been serious disagreements among various countries over the use of the Nile river in Africa.

This conflict issue is common throughout Africa. According to Scholes, "[s]ubtropical dry, arid areas are going to be a huge source of conflict over the next half-century because we still have very, very high population growth rates in those areas, very low economic growth rates, and deteriorating environment" (quoted in Fields, 2005, p. A537). The carbon-intensive lifestyle of the people in the global North triggers conflicts among the poor people in the global South.

In general, climate change will affect water resources, agriculture, fisheries, health, peaceful relationships, and so on. Regional level studies also support the trends discussed above. A case in point is the projected impact of climate change on southern Africa. It is estimated that over the next 60 years, global climate change will have the following impacts on southern Africa: temperature rise of 1.5ºC; modest drying over large parts of the region; decline in grasslands, which are replaced by thorn scrub savannah; expansion of dry forest biomes; expansion of desert areas; 15 to 20 per cent of large nature reserves and national parks to experience a change in biome, which has consequences for biological diversity; variations in runoff, and greater annual variation in the same; crop yields increasing generally, though with some adverse impacts in semi arid regions; changes in the distribution of disease-bearing insects; a decline in the distribution of ungulate species of richness; along with other impacts, for example, sea-level rises - not explicitly considered by the study (Hulme, 1996, cited in Rowlands, 1998, p. 14).

CLIMATE RISK MANAGEMENT IN AFRICA

The local people and national governments in Africa have used various strategies to tackle the adverse consequences of climate change. For example, as I have shown elsewhere (Kelbessa, 2008), Ethiopian peasant farmers, through continuous experiments on their environment, have managed to learn how to control weeds and insects, diversify their crops and animals, classify vegetation types, and cope with climatic and environmental changes. They have observed changes in the regularity, length, intensity, and timing of precipitation, and have developed various strategies to cope with climatic variability. They conserve water resources and avoid unnecessary problems and crisis during dry seasons. They use drought-resistant crops to address problems related to climate variability and drought in particular.

Peasant farmers are well aware of the fact that trees and rain are interconnected. Trees help to prevent soil erosion, influence climate, provide shade for the soil, humans and animals (Kelbessa, 2001), and break the force of the wind. They argue that cutting all available forests will not only deprive the future generation of rare plants and animals, but will also affect climatic conditions leading to drought, excessive wind storms and soil erosion. All of my interviewees confirmed that the rain originates on the forested lands. They specifically mentioned the names of these places. There is no dust in these places. Instead these places are covered by cloud. Springs can only be found around forested areas. My interviewees maintain that trees help retain moisture by sheltering the land from the sun and the wind; they also state that planting trees has influenced the weather of their locality. They plant trees for religious, social and economic reasons. Thus, in terms of ecology ethics, we can say that the preservation of forests is extremely important to peasant farmers for an endless number of utilitarian reasons.

Moreover, pastoralists and peasant farmers have adopted the strategy of spreading risks across locations, time and the diversity of materials they use. They grow the same landrace[4] in different places. In other words, they enhance biodiversity in diverse geographical areas. In so doing, they spread their risks across locations; if something fails in one location, it will do well in another. They also spread risks across seasons. When a landrace does not work in one season, they go to the local market and exchange it with the variety that will grow well on their own location. The former one may grow well in other places in a more appropriate planting season. This knowledge exchange has created a system which has allowed the landrace to be grown on a wide range of locations (for details, see Kelbessa, 2008).

The Oromo people, the largest ethnic group in Ethiopia, use different means to predict the future and prepare for and manage future climate events. The Oromo practise astrology and they use the positions of the stars, moon, and sun to predict weather changes. In fact, the Oromo calendar is founded on lunar rather than solar cycles. "Time-reckoning" experts observe the position of the stars during the night and inform the local people that summer is setting in or winter is coming. A Borana Oromo time-reckoning expert uses his memory to distinguish different seasons. If he is unable to do so, he can then determine different seasons by exploring the relative position of the stars and the moon. This leads Asmarom Legesse to say, quite correctly, that the Oromo "schedule life-crisis ceremonies to a degree that would be inconceivable in the most time-conscious Western cultures" (1973, p. 179). Furthermore, for the Oromo, the physical appearance and conditions of both domestic and wild animals are important indications of future events. The unusual destruction of crops by baboons, pigs, or rats is also an indication of famine in the future. People are thus advised to save crops for the future (Kelbessa, 2008).

The Oromo also use flowering and fruiting trees and the appearance of certain birds to anticipate the behaviour of the natural environment and the weather conditions. When many fruits drop from the newly sprouted trees, or some trees unusually lose their leaves, or the boundaries of rivers dry up in spite of the presence of sufficient rainfall, or some trees blossom before the usual time, when solanaceous tubers are empty, or when grasses fail to grow well in spite of abundant rainfall, then famines, disease and other problems are expected to come in the future (Kelbessa, 2008).

Like the Oromo, some indigenous people in Tanzania, Mozambique, and Zambia have developed their own prediction schemes based on the observation of the behaviour of the surrounding world. The people in the three countries commonly use the following methods as

[4] Landraces are crop plant populations that have not been bred as varieties, but have been adapted through years of natural and artificial selection to the conditions under which they are cultivated.

indicators of local climate change: the appearance of plants, flowering density of certain trees, immature dropping of fruits by certain tree species, dripping of water from the leaves of some trees before the onset of the rains, higher than normal flowering density of certain trees, higher than normal ambient temperatures, wind direction, appearance of insects, appearance of certain animals, mists and rumbling sounds in mountains and hills, birth of babies (birth of many baby girls as indication of good rains), and appearance of the moon (for details see The Pilot Project Research Team (PPRT), 2002). Local communities in rural Ghana use similar coping strategies to cope with climate variability (Gyampoh et al., 2008).

Indigenous peoples interpret climate change in various ways. Their interpretation depends on personal observations, experiences, and local cultural frameworks. "While scientific explanations of climate changes have mainly concentrated on anthropogenic, greenhouse gas emissions, local interpretations of observed climate changes are often much more varied and encompassing" (Salick and Byg, 2007, p. 17). Some people consider adverse weather conditions as punishments for human wrongdoings. They identified various spiritual and social causes of climate change; however, spiritual explanations of climate change are incompatible with scientific explanation. It should also be noted that although some indigenous coping strategies are still useful, others may not in themselves be sufficient and effective to deal with contemporary climate variability and change.

As I have discussed above, people's perceptions also are based on scientific elements. Ethiopian peasant farmers are aware that lack of trees can lead to drought. Modern science also confirms that trees can absorb excess carbon dioxide and we now know that tropical rainforests can serve as the earth's natural storage facilities for CO_2. Like peasant farmers, scientists use a combination of observed soil water conditions and river levels to estimate flood risk; they also use observations of soil moisture and plant conditions to estimate drought risk (Basher and Briceño, 2005, p. 274).

In general, peasant farmers and other indigenous people are not only the receivers of new technology; they are also the initiators of development and sound climate change mitigation strategies. They have attempted to understand the secret of nature and avoid famine, war and other problems by employing different mitigation strategies. They have tried to discover the unknown and the future by considering the reactions of plants, animals and the natural environment to both human-induced change and natural change. The major goal of this effort is to maintain a positive relationship with the natural environment.

On the other hand, however, until recently, African countries have not considered climate as a priority area in development. As Washington et al. have observed, "[t]o a certain extent, development planning within Africa has taken a limited account of climate in an explicit sense (climate has generally been implicit as one of many biophysical considerations), despite the significant vulnerability of the human population to the high levels of climate variability over the continent" (2006, p. 1356). The International Research Institute for Climate and Society (IRI) also found gaps in four main areas: integrating climate into policy; integrating climate into practice; climate services and climate data (IRI, cited in Hellmuth, 2007a, pp. 8-9).

Nonetheless, there have been some attempts to address the problems of climate variability and change in Africa. The Environmental Initiative of the New Partnership for Africa's Development (NEPAD) receives support from the African Union Commission (AUC). AUC, in collaboration with the UN Economic Commission for Africa and African Development Bank started to support a major new initiative, Global Climate Observing

System (GCOS)-Africa Climate for Development (Babagana, 2007, p. v). It aims to integrate climate information and services into development and thereby support Africa's progress towards the Millennium Development Goals.

Regional cooperation is another way of dealing with climate change in Africa; it can be a cost effective way of greenhouse gas mitigation. Coordinated action among neighbouring countries could reduce greenhouse gas emissions in ways that are environmentally, economically and socially beneficial. Regional mitigation strategies have economic (developmental gains), and political advantages (greater bargaining strength in negotiations with external actors) (Rowlands, 1998).

African countries formed regional organisations to deal with climate change related problems. The Permanent Interstate Committee for Drought Control in the Sahel (CILSS) in West Africa was formed in 1974 after an extended drought in the Sudan-Sahelian region from 1970-1974. Following the 1984-1985 droughts in the Horn of Africa, governments in the region created the Intergovernmental Authority on Drought and Development (IGADD) in 1986. The need to address the persistent issues of drought, desertification and food insecurity led to the creation of its successor, the Intergovernmental Authority on Development (IGAD). Likewise, governments in Southern Africa created the Southern African Development Coordination Conference (SADCC), and its successor the Southern African Development Community (SADC) to deal with drought and other related issues (Basher and Briceño, 2005, p. 278).

Some countries have also created the following regional technical institutions to address the consequences of climate hazards: the Africa Centre for Meteorological Applications for Development (ACMAD) in Niamey, Niger; the Regional Centre for Training and Application in Agrometeorology and Operational Hydrology (AGRHYMET) also in Niger and linked to CILSS (Permanent Interstate Committee for Drought Control in the Sahel); the Drought Monitoring Centre at Harare, Zimbabwe (DMC-Harare), linked to SADC; and the Drought Monitoring Centre at Nairobi, Kenya (DMC-Nairobi), linked to IGAD. These institutions receive support from UN agencies such as the World Meteorological Organisation (WMO), Food and Agriculture Organisation (FAO), United Nations Environment Programme (UNEP), and United Nations Development Programme (UNDP) (Basher and Briceño, 2005 p. 278).

In West Africa, the 16-country Economic Community of West African States (ECOWAS) has programmes including a regional meteorological and water resource management programme, and subregional programmes for desertification control, "but there is no designated subregional activity on natural disaster risk reduction or consolidated regional strategy on risk management" (Basher and Briceño, 2005, p. 279).

It is worth noting that since 1997, various regional institutions have organised the Regional Climate Outlook Forum (RCOF) to disseminate early warning information. The purpose of this forum has been to review the latest climate information and to formulate a consensus seasonal forecast for the concerned region (Basher and Briceño, 2005, pp. 279-280). The important lesson here is that a regionally coordinated approach can help African countries to respond to global climate change challenges.

Moreover, various African countries have been trying to address problems related to climate change (for details see, Hellmuth, et al., 2007b). A case in point is Ethiopia's food security policy. Drought, floods and fires have affected Ethiopia for many years. In particular, droughts had a devastating impact on Ethiopia during the periods 1957-58, 1964-65, 1972-73 and 1983-84 (Erkineh, 2007, p. 35).

The relationship between climate change and famine in Ethiopia was not adequately studied. According to one study, the high-latitude cold period of the Little Ice Age (Lamb, cited in Mohammed, 2005, p. 25) experienced the high frequency of famine events from the sixteenth to nineteenth centuries in Ethiopia (Pankhurst and Mwachando et al., cited in Mohammed, 2005 p. 25). The El Niňo Southern Oscillation (ENSO) event partly caused high inter-annual variability of rainfall during the twentieth century and famine in Ethiopia during the beginning of the 1970s and 1980s (WMO, cited in Mohammed, 2005, p. 26). However, there is no definitive evidence that indicates that ENSO event alone is responsible for rainfall variability in Ethiopia. As Mohammed (2005) notes, the accelerated human impact has put pressure on the environment, under conditions of high decadal and inter-annual natural variabilities.

Successive Ethiopian governments took various measures to address food security problems: (A) The military government established Relief and Rehabilitation Commission in 1976; (B) Policy on Disaster Prevention, Preparedness and Mitigation was launched in 1995; (C) and later, the Relief and Rehabilitation Commission was renamed the Disaster Prevention and Preparedness Commission Agency.

In 1976, after the famine of 1973-74, Ethiopia started an early warning system (Erkineh, 2007, p. 36) relating to changes in the climate. The National Meteorological Agency has been supporting the early warning system in Ethiopia. "It collects and analyzes climate data, produces forecasts, and disseminates information to various user groups in the form of regular bulletins" (Erkineh, 2007, p. 36). It includes daily weather reports, 10-day weather summaries and 10-day forecasts, monthly weather summaries and 1-month forecasts, and three seasonal forecasts a year covering the two-rainy seasons.

The measures taken by other African countries vary from country to country. Some African countries have formulated strategies in poverty reduction, flood management, malaria control, drought insurance, and disaster risk reduction, and sectoral strategies in climate-sensitive sectors. However, as has been stated earlier, there are many problems that hinder the attempt to address the problems of climate change in Africa. The presence of other stresses, such as current climate hazards, poverty, unequal access to resources, food insecurity, trends in economic globalization, conflict, and incidence of disease such as HIV/AIDS can exacerbate vulnerability to climate change. These stresses can reduce resilience and adaptive capacity "because of resource deployment to competing needs" (IPCC, 2007, p. 19).

In spite of wide-ranging problems, Africa has a great potential for additional carbon sequestration in biotic sinks, as well as for reduction of emissions caused by clearing and degradation of forests and woodlands (Justice et al., 2005, p. 175). More woodlands can be managed by adopting more efficient wood fuel use technologies (e.g. efficient stoves and charcoal kilns or by zero-tillage agriculture (Justice et al., 2005, p. 175). Africa could provide both a significant increase in the terrestrial carbon pool and greatly benefit soil fertility and productivity by increasing soil organic matter in agricultural systems through minimum tillage or agroforestry methods (Woomer et al., cited in Justice et al., 2005 p. 175). Indigenous environmental knowledge can also contribute to these activities.

Recently, some countries and institutions have made positive interventions to assist African countries regarding development and climate change. Among others, the international research community has been developing a set of regional global change research priorities through the System for Analysis Research and Training (START), by paying attention to issues directly relevant to developing countries and regional natural resource management

(Justice et al., 2005, p. 177). "START http://www.start.org is helping to develop regional research networks within Africa" (for details see Justice et al., 2005 , p. 177). Regional research networks in Africa receive support from several US agencies through such activities as the US Country Studies Programme, the NASA Land Cover Land-Use Change Programme, the NOAA ENSO Applications Programme, (NOAA, cited in Justice et al., 2005, p. 177), and the USAID Central African Regional Programme for the Environment (CARPE) (Justice et al., 2005, p. 177). In particular, CARPE is contributing to the compilation of regional environmental data sets in Central Africa (Justice et al., 2005, p. 178). Although the US Global Change Research Programme and the component research agencies contribute significantly to the international global change research programmes, "they are often unable to support scientific capacity-building in Africa directly" (Justice et al., 2005, p. 178). For instance, in spite of its significant contribution to natural resource management, the USAID traditionally has not supported global change research or scientific capacity-building in Africa (Justice et al., 2005, p. 178).

The U.K. government and the G8 countries made another positive intervention. In the spring of 2004, the U.K. government launched the Commission for Africa, and published "Our Common Interest" in March 2005. The G8 meeting at Gleneagles, Scotland, in July 2005, had addressed two priority items on the agenda: Africa and climate change. The then UK Prime Minister, Tony Blair, who took up the presidency of the G8 at Gleneagles, underlines the danger of climate change for Africa. He says, "[t]he size of its land-mass means that, in the middle of the continent, overall rises in temperature will be up to double the global rise, with increased risk of extreme droughts, floods and outbreaks of disease. It is estimated that African GDP could decline by up to 10% because of climate change" (2005). The G8 countries approved a pledge to forgive about US$40 billion of debt owed by 18 nations, including 14 African countries. They also committed to double annual African aid from US$25 billion to US$50 billion. However, this is not sufficient to address the adverse effects of climate change in Africa.

Furthermore, the Climate Change Adaptation in Africa (CCAA) research and capacity development program was launched in 2006 and is jointly funded by Canada's International Development Research Centre (IDRC) and the United Kingdom's Department for International Development (DFID).

CONCLUSION

Although climate change affects all countries in the world, the world poorest countries face the most severe impact. It is the rich who are using most of the energy that leads to the emissions that cause climate change, while it is the poor who will bear most of the costs. Those who lack the resources to adapt, to find alternative sources of food and healthcare will suffer a lot from climate change. The foregoing discussion shows that African emissions have played a minor role in the build-up of greenhouse gases in today's atmosphere. In spite of this, Africa has been suffering from rising temperatures and evaporation, widespread water stress, increased frequency and severity of droughts and floods, crop loss and rising sea level. Changes in climate have also caused a decrease in forest and cultivation area, a reduction in ecosystem integrity and resilience, a decline in biodiversity and higher levels of disease.

The foregoing discussion also showed how peasant farmers and pastoralists have been coping with environmental hazards, and adapting their livelihoods to changing conditions for generations. However, indigenous mechanisms are not sufficient by themselves to deal with the effects of climate change; they should be complemented by modern strategies. Thus the field research upon which this chapter was based suggests that policy makers, modern scientists and peasant farmers have to work together to address the adverse effects of climate change. Scientists and policy makers who want to address the impacts of global climate change can derive important lessons from peasant farmers' and pastoralists' strategies for addressing weather-related risks, although indigenous strategies have been affected by various internal and external factors. So far, much of the climate discourse in Africa has not involved farmers, as it has been at the national and international levels. It is important to incorporate indigenous knowledge and perceptions into the climate change forum in the world. Logically, it would always be useful to involve local communities in the design and implementation of coping strategies. Local communities should be encouraged to grow drought-resistant and early-maturing crops, introduce agroforestry, plant and protect trees, develop water storage and harvesting, and improve empirically tested coping strategies.[5] Externally generated and imposed strategies cannot help solve the problem. As Salick and Byg persuasively suggested: "[i]ntegration and feedback loops between climate change science and indigenous peoples must be established and employed. Both parties can gain knowledge from the other and support each other in action" (2007, p. 25). Thus, African governments should consolidate indigenous coping strategies, and strengthen community institutions along with modern coping strategies and institutions, in order to effectively respond to climate change.

Western scholars and the public also have ethical responsibilities and obligations to influence their governments to take responsible actions regarding global problems and the marginalized people in the world. However, "one problem is that the public at large in the 'developed' world who have at least formally, a good deal of democratic control over their governments, are not sufficiently informed and far-sighted enough to understand what is going on in other parts of the world or their real interests in defending the global environment and marginalized people from the ravages of corporate driven development" (Kelbessa, 2005, p. 34). Western scholars and other organisations need to facilitate information flows and dialogues between indigenous and marginalised groups in the 'developing' world and majority populations in the 'developed' world.

Global justice requires rich countries to protect not only the interests and rights of their citizens and transnational corporations but also the interests and rights of the people all over the world regardless of race, religion, and background. Humanity is required to build a world where all citizens are fairly treated. Our enlightened self-interest itself should compel all of us to reexamine our current situation, and act collectively to reduce human misery and depredation in our world.

Increasing public education, training in both the formal (school) and informal (community) sectors and public awareness raising in climate change issues especially those that affect the communities can also help Africans to mitigate the impact of future climate

[5] Coping strategies refer to the immediate actions in the face of an event or changes and ability to maintain welfare in contrast to adaptation which refers to long-term adjustments to the framework within which coping takes place.

change. People should be encouraged to understand the importance of global climate and improve their indigenous coping strategies. Media have a vital role in raising awareness and transmitting climate information.

The nature of climate change and its impacts in Africa, and African sensitivity to climate change require further research.

ACKNOWLEDGMENTS

An earlier version of this chapter material was presented at a session of the International Conference on Riding a Moral Storm. The Global Challenge of Climate Change: Science-Economics-Ethics-Politics, in Greifswald, Germany, 30 May to 1 June 2007. I want to thank those who participated in this session for their helpful comments and suggestions. I would particularly like to thank the Alexander von Humboldt Foundation for supporting my research.

REFERENCES

Armah, A. K. (2005). Sea-level Rise and Coastal Biodiversity in West Africa: a Case Study from Ghana, in Low, Pak Sum. (Ed.). *Climate Change and Africa* (pp. 204-217). Cambridge: Cambridge University Press.

Babagana, A. (2007). "Foreword", in "Lessons Learned and Next Steps", in M. E. Hellmuth, A. Moorhead, M.C. Thomson, and J. Williams. (Eds.). (2007). *Climate Risk Management in Africa: Learning From Experience*. New York: International Research Institute for Climate and Society (IRI), p. v. [Available online at http://iriportal.ldeo.columbia.edu/portal/server.pt/gateway/PTARGS_0_2_1171_0_0_18/Climate%20and%20Society%20No1_en.pdf.]

Basher, R. and Briceño, S. (2005). Climate and Disaster Risk Reduction in Africa, in Low, Pak Sum (Ed.). *Climate Change and Africa* (pp. 271-283). Cambridge: Cambridge University Press.

Blair, T. (2005). A Year of Huge Challenges, *Economist*; 1/01/2005, 372(8407), 44-46.

Boko, M., J. Niang., A. Nyong., C. Vogel., A. Githeko., M. Medany., B. Osman-Elasha., R. Tabo., and P. Yanda. (2007). *Africa: Climate Change 2007: Impacts, Adaptation, and Vulnerability: Contribution of Working Group II to the Fourth Assessment Report of the Intergovernmental Panel on Climate Change,* M. L. Parry, O. F. Canziani, J. B. Palutikof et al., (Eds.) (pp. 433–467). Cambridge, UK: Cambridge University Press. Available from http://www.gtp89.dial.pipex.com/09.pdf.

Cardy, W. F. G. (2003). Environment and Forced Migration: A Review, in M. Darkoh and A. Rwomire (Eds.). *Human Impact on Environment and Sustainable Development in Africa* (pp. 141-161). Hampshire: Ashgate.

Darkoh, M. B. K. (2003). Environmental Security in the Drylands of Africa: The Role of Science and Technology, in M. Darkoh and A. Rwomire (Eds.). *Human Impact on Environment and Sustainable Development in Africa* (pp. 55-74). Hampshire: Ashgate.

Dessler, A. E. and Parson, E. A. (2006). *The Science and Politics of Global Climate Change: A Guide to Debate*. Cambridge: Cambridge University Press.

Eriksen, S., O'Brien, K. and Rosentrater, L. (2008). Climate Change in Eastern and Southern Africa: Impacts, Vulnerability and Adaptation, Global Environmental Change and Human Security (GECHS) Report: 2. Oslo: Norwegian Ministry of Foreign Affairs (MFA).

Erkineh, T. (2007). Food Security in Ethiopia, in M. E. Hellmuth, A. Moorhead, M.C. Thomson, and J. Williams. (Eds.). *Climate Risk Management in Africa: Learning From Experience* (pp. 31-44). New York: International Research Institute for Climate and Society (IRI), Columbia University, New York, USA. [Available online at http://portal.iri.columbia.edu/portal/server.pt/gateway/PTARGS_0_2_1171_0_0_18/Climate and Society No1_en.pdf.]

Fields, S. (2005). Continental Divide: Why Africa's Climate Change Burden is Greater. *Environmental Health Perspectives, 113*(8), A534-A537.

Gyampoh, B. A., Amisah, S., and Idinoba, M. (2008). Coping with Climate Change: How Local Communities Use Traditional knowledge in Rural Ghana. Paper Presented at Adaptation of Forests and Forest Management to Changing Climate with Emphasis on Forest Health: A Review of Science, Policies, and Practices, Umea, Sweden, August 25-28, 2008.

Hellmuth, M. E., Moorhead, A. and Williams, J. (2007a). Introduction, in M. E. Hellmuth, A. Moorhead, M.C. Thomson, and J. Williams. (Eds.) *Climate Risk Management in Africa: Learning From Experience* (pp. 8-12). New York: International Research Institute for Climate and Society (IRI), [Available from http://iriportal.ldeo.columbia.edu/portal/server.pt/gateway/PTARGS_0_2_1171_0_0_18/Climate%20and%20Society%20No1_en.pdf.]

Hellmuth, M. E., Moorhead, A., Thomson, M. C., and Williams, J. (2007b). "Lessons Learned and Next Steps", in M. E. Hellmuth, A. Moorhead, M.C. Thomson, and J. Williams. (Eds.). *Climate Risk Management in Africa: Learning From Experience* (pp. 89-97). New York: International Research Institute for Climate and Society (IRI), pp. 89-97. [Available from http://iriportal.ldeo.columbia.edu/portal/server.pt/gateway/PTARGS_0_2_1171_0_0_18/Climate%20and%20Society%20No1_en.pdf.]

Hulme, M., Conway, D., Kelly, P., Mick, Subk, S., and Downing, T. E. (1995). "The Impacts of Climate Change on Africa", CSERGE Working Paper GEC 95-12. Stockholm: SEI.

Hulme, M., Doherty, R., Ngara, T., and New, M. (2005). Global Warming and African Climate Change: A Reassessment, in Low, P. S. (Ed.) *Climate Change and Africa*. (pp. 29-40). Cambridge: Cambridge University Press.

Intergovernmental Panel on Climate Change (IPCC). (2001). Working Group II Contribution to the Intergovernmental Panel on Climate Change Third Assessment Report. *Climate Change 2001: Impacts, Adaptation and Vulnerability*. Geneva: IPCC. Available from http://www.ipcc.ch/ipccreports/assessments-reports.htm

IPCC. (2007). "Summary for Policymakers", in *Climate Change 2007: Climate Change Impacts, Adaptation and Vulnerability*. Contribution of Working Group II to the Fourth Assessment Report of the Intergovernmental Panel on Climate Change, M.L. Parry, O.F. Canziani, J.P. Paultikof, P.J. Van der Linden, and C.E. Hanson (Eds.). Cambridge University Press, Cambridge, 7-22 [Available from http://www.ipcc.ch/SPM6avr07.pdf.]

Integrated Regional Information Networks (IRIN). (2005). "Africa: Climate Change Becoming a Matter of Life and Death", [Available from http://www.irinnews.org/report.aspx?reportid=53719.]

Justice, C. O., Wilkie, D., Putz, F. E., and Brunner, J. (2005). Climate Change in sub-Saharan Africa: Assumptions, Realities and Future Investments, in Low, P. S. (Ed.). *Climate Change and Africa* (pp. 172-181). Cambridge: Cambridge University Press.

Kelbessa, W. (2001). *Traditional Oromo Attitudes towards the Environment: An Argument for Environmentally Sound Development. OSSREA Social Science Research Report Series,* No. 19. Addis Ababa: Commercial Printing Enterprise. [Available from http://www.ossrea.net/ssrr/workneh/toc.htm.]

Kelbessa, W. (2005). "The Utility of Ethical Dialogue for Marginalised People in Africa", International Institute for Environment and Development (IIED), Edinburgh, UK. [Available from http://www.policy-powertools.org/related/docs/Workineh_Kelbessa_final3.pdf.]

Kelbessa, W. (2006). "Remapping Global Realities: The Necessity of Including African and Other 'Third World' Voices", Paper Presented at the Seventh International Conference of the International Development Ethics Association, held in Kampala, Uganda, July 19-22.

Kelbessa, W. (2008). *Indigenous and Modern Environmental Ethics: A Study of the Indigenous Oromo Environmental Ethic and Modern Issues of Environment and Development: Ethiopian Philosophical Studies, I.* Washington, D.C.: The Council for Research in Values and Philosophy.

Legesse, A. (1973). *Gada: Three Approaches to the Study of African Society.* New York: Free Press.

Lovett, J., Midgley, G. F., and Barnard, P. (2005). Climate Change and Ecology in Africa, *African Journal of Ecology, 43,* 167-169.

Manful, G. (2005). Capacity-building Initiatives to Implement the Climate Change Convention in Africa, in Low, P. S. (Ed.). *Climate Change and Africa* (pp. 297-309). Cambridge: Cambridge University Press.

Migiro, K. and Nutt, D. (2007). Kenya: Drought and Conflict, in Church of Sweden and Norwegian Church Aid. *Understanding the Issue: Climate Change Threatens the Fight against Poverty*, 1, 9-15.

Mohammed, U. M. (2005). The Relative Importance of the Different Forcings on the Environment in Ethiopia during the Holocene, in Low, P. S. (Ed.). *Climate Change and Africa* (pp. 23-28). Cambridge: Cambridge University Press.

Mulama, J. (2006). ENVIRONMENT-AFRICA: Changing Climate, Changing Lives, [Available from http://ipsnews.net/news.asp?idnews=35448.]

Myers, N. (2005). Environmental Refuges: An Emergent Security Issue, Paper Presented at 13th Economic Forum, Prague, 23-27 May. [Available from http://www.osce.org/documents/eea/2005/05/14488_en.pdf.]

New Economic Foundation, The (NEF). (2005). Africa – Up in Smoke: The Second Report from the Working Group on Climate Change and Development. Available from http://www.neweconomics.org/gen/uploads/4jgqh545jc4sk055soffcq4519062005184642.pdf

Nkomo, J. C., Nyong, A. O. and Kulindwa, K. (2006). The Impacts of Climate Change in Africa, Final Draft Submitted to The Stern Review on the Economics of Climate Change.

[Available from http://www.hm-treasury.gov.uk/d/Chapter_5_The_Impacts_of_Climate _Change_in_Africa-*5.pdf]*

Oba, G., Post, E., and Stenseth, N. C. (2001). Sub-saharan Desertification and Productivity are Linked to Hemispheric Climate Variability, *Global Change Biology,* 7, 241-246.

Perry, M. (2007). Climate Change sees Fish Grow Faster in Warmer Water. [Available from http://news.yahoo.com/s/nm/20070427/sc_nm/globalwarming_fis…]

Pilot Project Research Team, The [PPRT]. (2002). Promotion and Integration of Indigenous Knowledge in Seasonal Climate Forecasts: Interim Report prepared for Drought Monitoring Centre (Harare, Zimbabwe, Sokoine University of Agriculture, Tanzania. Available from http://www.dmc.co.zw/research/PilotProjects/Promotion%20And%20Integration%20Of %20Indigenous%20Knowledge%20In%20Seasona.pdf

Put, M., Verhagen, J., Veldhuizen, E. and Jellema, P. (2004). "Climate Change in Dryland West Africa", in A. J., Dietz, R. Ruben, and A. Verhagen. (Eds.). *The Impacts of Climate Change on Drylands: With a Focus on West Africa* (pp. 27-32). Dordrecht: Kluwer Academic Publishers.

Rowlands, I. H. (1998). Climate Change Cooperation in the Global Greenhouse, in I. H. Rowlands (Ed.). *Climate Change Cooperation in Southern Africa* (pp. 1-26). London: Earthscan Publications Ltd.

Salick, J. and Byg, A. (Eds.). (2007). Indigenous Peoples and Climate Change. Report of Symposium 12-13 April 2007, Environmental Change Institute, Oxford. [Available from http://tyndall.webapp.uea.ac.uk/publications/Indigenouspeoples.pdf.]

Taylor, C. M. (2001). Feedbacks between the Land Surface and the Atmosphere in the Sahel. Aridlands Newsletter, No. 49, May/June.

UNEP (United Nations Environment Programme). (2007). *Biodiversity and Climate Change.* [Available from http://www.biodiv.org/doc/bioday/2007/ibd-2007-booklet-01-en.pdf.]

UNFCCC (United Nations Framework Convention on Climate Change). (1992). Reprinted in *International Legal Materials*, *31*, 849-73.

UNmüßig, B. and Cramer, S. (2008). Climate Change in Africa. *GIGA (German Institute of Global and Area Studies) Focus*, 2, 1-7.

Washington, R., Harrison, M., Conway, D., Black, E., Challinor, A., Grimes, D., Jones, R., Morse, A., Kay, G., and Todd, M. (2006). African Climate Change: Taking the Shorter Route, *American Meteorological Society*, 87 (1), 1355-1366.

Contributed by Tim McCrorey, Toowoomba.

Photo: The Delicate Balance between Water and Life.

In: Meltdown: Climate Change, Natural Disasters…
Editor: Kathryn Gow
ISBN 978-1-60876-153-1

Chapter 16

DROUGHT IN RURAL AREAS: NOT JUST THE ABSENCE OF WATER

Kathryn Gow
Queensland University of Technology, Australia

ABSTRACT

Water is a precious and vital resource, and is now becoming highly valued after lack of precipitation in many countries has led to serious water shortages, over the past few decades. If the precipitation shortfalls are due to climate change, then the careful harvesting of water becomes of paramount importance. The lack of, and shortfall in, precipitation leads to droughts which affect the whole fabric of rural life; in some case, towns and cities, which lack adequate water supplies, take on the barren look of desert towns. Farmers and their produce are particularly affected by drought, and, as a consequence, whole regions can be deprived of food, and without food, people starve. Without water, people die.

Keywords: Drought, lack of precipitation, water shortage, rural towns, climate change, climate variability

INTRODUCTION

Tear Drops Falling in the Dust

In a country stricken by the worst drought in 160 years (some 160 million tears), El Niño and La Niña have become household words as farmers, graziers, irrigators, and politicians all grapple with severe shortages of drinking and irrigation water supplies. Town and city people learn what others in countries, such as Mexico and Arabia and vast stretches of Africa, have known for centuries - water is the most precious gift of life; "don't waste a drop".[1] Most rural people have learned to live with 'making do' by catching their own drinking water, and using dam and bore water for crop and orchard irrigation and watering their stock. While stories passed down from generations before had told tales of the ground cracking and the rivers drying up and all the animals dying, they had never seen it with their own eyes. When the Government called for the wetlands in the Murray Darling Basin to be drained, conservation outrage died in parched throats, as despair etched deeper and deeper into eyes and faces and hunched shoulders, and ropes hanging from trees told of depression and despair too deep to be rescued.

Is Drought a Natural Disaster?

Floods, fires, cyclones and storms are powerful and damaging events, but drought is quiet, slow, and creeps up over time. If this is the case, can drought then be classified as a natural disaster? It is certainly a natural occurrence, but the effects of drought tend to gather and coalesce over time and its effects may take years to dissipate, if ever. Further, in terms of where it sits in the natural disaster classification system, drought spreads over a much larger geographical area. The WMO (World Meteorological Organization) defines drought as a "prolonged absence or marked deficiency of precipitation" or a "period of abnormally dry weather sufficiently prolonged for the lack of precipitation to cause a serious hydrological imbalance" (1992, p. 198). A meteorological drought 'is a result of the occurrence of persistent large-scale disruptions or anomalies in the global circulation pattern of the atmosphere" (Wilhite and Svoboda, 2000, p. 4).

"Drought is a normal, recurring feature of climate; it occurs in virtually all climatic regimes" (Wilhite, 1992a, p. 5) in both high and low precipitation areas. While drought is temporary, aridity is permanent. The severity of drought depends on several factors: duration, intensity, and spatial extent of the episode and also on the demands of human activities, farming, irrigation, vegetation, energy production, and other factors. The "intensity refers to the degree of the precipitations shortfall and/or severity" (Wilhite, 1992a, p. 5) of impacts associated with the shortfall. Wilhite and Svoboda (2000) point out that the impacts of droughts are "spread over a larger geographical area than are damages that result from other natural hazards" (p. 4), such as floods and hurricanes.

As Government policy makers have found recently in Australia (the most arid country in the world) with their Exceptional Circumstances drought payments,[2] it is difficult to pinpoint when a drought event commences and when it is over (Kenny, 2008). Tannehille, back in 1947 (in Wilhite and Svoboda, 2000), was astute enough to label a drought as a "creeping phenomenon" Defining drought, however, is not a simple matter as Wilhite and Glantz (1985) proved when they attempted to analyse more than 150 definitions. They concluded that it is very difficult, and that therefore much of what policy makers use in their decisions is arbitrary, as "they are not linked to specific impacts in key economic terms" (Wilhite and Svoboda, 2000, p. 4). The renaming of drought to 'dryness' in the Kenny report (Kenny,

[1] This phrase became the focal point of the campaigns to save water in Queensland during the recent severe drought.

[2] The Australian Government has supported farmers and small businesses in drought declared areas for several years. These funds were based on exceptional circumstances; in this case, long and severe drought.

2008) was intended to address a broader concept than the currently accepted terminology could, but this definition, too, falls short[3] of the enormity of what prolonged drought does to both human and wildlife habitations.

According to Wilhite (2000), droughts, more than other natural disasters, cause the most damage. Most researchers on drought emphasise its far reaching effects on economic and social factors in rural communities. Drought is not sudden, it is a slow relentless dying and death, creating a string of domino effects: family stress, breakdown structure changes, then loss of, or changes to, employment and problems with social inclusion (Australian Government Bureau of Rural Sciences, 2008).

Some Historical Drought events. The year 1934 brought the worst drought in the USA, with 65% of the country being drought declared; this figure was matched decades later in Australia in 1992 and 2006/7. In both countries, such severe droughts have led to a focus on the critical issues surrounding the resultant water shortage. I will return to this matter in another section of the chapter. Wilhite and Svoboda (2000) note that every year, an average 12% of the area of the USA (Hawaii and Alaska are not included in these calculations) is in drought; indeed this has been the case for a century and the occurrences are so regular that 30 states in the USA now have a Drought Preparedness Disaster Management Program in place.[4] Hayes et al. (2003) recorded that "57 weather-related disasters caused more than $1 billion in losses in the United States between 1980 and 2003"[5]. Ten of those were drought-related, with the highest cost periods ranging from $48.4 billion to $61.6 billion. FEMA (1995) calculated that, in terms of economic loss, drought in the USA costs about $6-8 billion each year.

Wilhite (1991) emphasises that meteorological drought is a natural event and results from the "occurrence of persistent large - scale disruptions or anomalies in the global circulation pattern of the atmosphere" (p. 4). In Australia, most citizens are now familiar with terms such as the SOI (Southern Oscillation Index) and its role in the El Niño[6] and La Niña events.[7] Only recently have the weather scientists begun to understand the importance of the Indian Ocean Dipole (IOD), a factor highlighted in the Garnaut Report (2008) as leading to the drying out of Southern Australia.[8] A recent report, which claimed a matching across eight climate models, pointed to the trend that has begun, and will continue, with the rain patterns from the Indian Ocean moving south to the Antartica and generally tending more and more to miss the southern states of Australia. Earlier reports in 2001 by the Bureau of Meteorology Research Centre, based on three climate change models, had indicated that the IOD changes appeared to be due to climate variability, rather than climate change.

Lexie Donald[9] (2008), from the Queensland Climate Change Centre of Excellence, points out that our climate is influenced by both natural (e.g., orbital cycles and volcanic activity) and human induced changes (such as greenhouse gases from power stations). He emphasises

---

[3] Dry Spell: "Period of abnormally dry weather. Use of the term should be confined to conditions less severe than those of a drought." (UNDHA, DM Glossary, 1992, p. 31).

[4] They have produced the *Drought monitor.*

[5] *The National Climatic Data Center in Asheville, NC. http://www.ncdc.noaa.gov/oa/reports/billionz.html.*

[6] El Niño: An anomalous warming of ocean water resulting from the oscillation of a current in the South Pacific, usually accompanied by heavy rainfall in the coastal region of Peru and Chile, and reduction of rainfall in equatorial Africa and Australia. *EMA Glossary.* (1998). Dickson, ACT.: Emergency Management Australia.

[7] La Niña: The opposite of an El Niño event, during which waters in the west Pacific are warmer than normal, trade winds or Walker circulation is stronger and, consequently, rainfalls heavier in Southeast Asia. *EMA Glossary.* (1998). Dickson, ACT.: Emergency Management Australia.

[8] For people not familiar with the vastness of Australia, this is a massive area to be moving into dryness.

[9] http://www.longpaddock.qld.gov.au.

that we need to adapt practices to the trends, rather than be blinded by what is causing the changes.[10] Hence the advice to farmers to think in a more flexible manner, in terms of location of vegetables, fruit trees and vineyards and crops for stock feed, as well as being prepared to move grazing enterprises to other areas with more rainfall.

It seems that the drought and rainfall situation in Australia has been a little worse than the majority of the population is aware. According to Donald, in the late 1800s, rainfall was reasonably high, and then declined through to the mid 1940s. However from the 1950s to the 1970s and 1980s, there was an increase overall, but since then there has been a downward trend for most of Queensland, New South Wales, Victoria and South Australia, as well as the central and southern coastal strip of Western Australia.

According to climatologist, David Stephens (2007),[11] 2006 was the driest year on record in the south of Western Australia. Prior to that, it was 1914; both 1969 and 2002 were also dry years, with the wettest being 1917, 1963, 1945 and 1974 in that order. The problem is, Stephens says, that while the average rainfall has only decreased 15% to 20% since the 1970s, which is still significant, runoff has reduced by 60% to70%; a phenomenon that has left dams dry, even after heavy rain falls in some parts of Australia.

Climate model simulations (1000-year long) indicate that natural variability can actually lead to decadal (or longer) dry spells over areas such as south-west Western Australia, but the models say this won't happen until well into the next century - 2100 (Bates, 1999).

Across the world, droughts and lack of water have been of critical concern to many countries in the past two to three decades. In 2008, the drought was so bad in Chile that the government commenced handing out water during its worst drought in decades; the rainfall records showed that it had received its lowest precipitation in 50 years. Previously, the Mexican people had groundwater to fall back on, but it became so dry that the wells dried up in many places.[12] Indeed in Mexico, drought continues to be one of the major environmental threats.

In terms of predicting what might happen in the future with respect to drought in China, Wang Shourong (2008) noted that precipitation had been reasonably high in most parts of China during the 1950s, but had decreased after the 1960s, especially in the northern part of the country. He considers that for China, global warming will be higher than the projected global mean, with unpredictability in precipitation patterns. For China, he says, this could mean that by 2030, the yield from seven main crops would decrease by 4.4% and stock breeding by 6%.

According to Maude Barlow (2009), a senior adviser on water to the President of the United National General Assembly, countries such as Argentina, California, Mexico City, Nevada, Jordan, Kenya, Spain, Pakistan, Northern China, Israel, Iraq and South Asia are all in trouble through droughts and water shortages. Barlow disagrees with the labelling of our droughts as cyclical droughts. She points out that the 'hot stains' in the world include 22 countries in the continent of Africa, many of which have been in recurring drought for the past few decades.

[10] Climate update; Rural Weekly, Southern Edition. p. 23. Friday April 25, 2008.

[11] David Stephens cited by Sharon Kennedy, 03.01.07. Climate dry and drier. http://www.abc.net.au/water/stories/s1821092.htm.

[12] Rodrigo Martinez. Chile Government hands out water in major drought. Friday, 22.02.08. Hurricane Hunters http://hurricanehunters.blogspot.com/2008/02/chile-government-hands-out-water-ib.h...

Prediction and Monitoring of Droughts

Regardless of what is going to happen with the farming and grazing lands, there needs to be a drought plan in place in every country and area that has suffered from drought in the past and is scheduled to suffer again in the future, with or without the climate change trilogy coming into play. If those areas in drought were able to work with a Drought Index, that is, a "computed value which is related to some of the cumulative effects of a prolonged and abnormal moisture deficiency" (UNDHA, *DM Glossary*, 1992, p. 31), then it may well lead to more justified awarding of grants to farmers, graziers and irrigators during prolonged droughts, and also prevent the withdrawal of financial support from those most in need, because of drought definition problems and assessment of properties based on non-representative data.

Such an index would be part of a drought plan. In 1991, Wilhite devised a 10-step plan for coping with drought for the USA. Wilhite and Svoboda (2000, p. 1) emphasise that drought plans should contain three basic components: (1) Monitoring and early warning; (2) Risk assessment; and (3) Mitigation and response. With reference to component 1 of this drought plan, readers may be interested to hear that in poverty stricken African countries, the large aid agencies monitor where rainfall precipitation is low or absent each season, and have models to estimate what crops will fall short, in terms of the food supply, and what aid needs to be diverted to which regions. Such amazing monitoring is possible only through technical software and hardware, such as statistical models that map out the data, as well as having access to this data via the internet.

In America, both academic and government scientists are working together to map the extent of the severity, spatial extent and impact of droughts (Svoboda et al., 2002, p. 1181). This new tool is called *The Drought Monitor.*[13] According to the FEMA (1995),[14] drought is the natural disaster that leads to the most economic loss. It is estimated that once such forecasting and monitoring is set up, then this will lead to better preparation, better management practices and reduce the impact of drought on communities (Svoboda et al., 2002, p. 1181). The Drought Monitor has five levels of magnitude (low to extreme) and three contexts (Agriculture, Water and Fire) (Svoboda et al., 2002, Table 3, p. 1186). Thus with respect to Fire, the fire risk can be: above average, high, very high, extreme and exceptionally dangerous. I will list examples of these levels in two contexts - water and agriculture.

In terms of drought (low or no water), the levels are described as: (i) stream flow being below average; (ii) streamflow, reservoir and well levels are low, with some water shortages evident; (iii) water shortages now common and restrictions imposed; (iv) widespread water shortages now common and restrictions imposed; then finally (v) emergency shortages in all water catchments. In terms of the impact then of the drought on agriculture, for example, the levels range from: (a) a slowing in growth across farm activities and a crop and pasture; to (b) some damage to crop and pasture; (c) likely losses in a crop and pasture; up to (d) major losses in a crop and pasture, and then escalating to (e) exceptional and widespread losses in a crop and pasture.

---

[13] The Drought Monitor (http://drought.unl.edu/dm).

[14] FEMA – Federal Emergency Management Agency – USA.

Experimental forecasts have been conducted and published on the Experimental Rainfall and Crop Predictions (ECPC) WWW site for each region, in consultation with local experts and regional agencies. Their team advises that:

> For longer-period southern hemisphere and tropical fluctuations, we have developed a statistical climate prediction system utilizing the slow eastward propagation of covarying SST and SLP anomalies over the globe (that is, the Antarctic circumpolar wave in the Southern Ocean and the global interannual and biennial SST/SLP wave in the Tropical Ocean) to predict precipitation and crop yields over the globe (that is, Australia, New Zealand, South America, South Africa, Southeast Asia, and India) (Roads et al., 2001).

John Pigram (2006, p. 29) paints a picture of too much and too little water in both flood and drought conditions. He reported that after 91% of New South Wales was drought declared in early 2005, this weather was then followed by floods in June 2005. Unfortunately much of the state remained in drought. Earlier, Pigram had warned that a drought mitigation strategy could not depend alone on building more and larger water storages units. We are being called to account now.

Farming and Food Supplies

As indicated earlier, the continuing drought in Australia has meant that the locations where farmers conduct their farming enterprises may have to change. There has been some discussion about grape farmers having to move their vineyards higher up mountains where it is cooler, and horticulture and agriculture having to move to areas where there is either a consistently higher annual rainfall or permanent underground water. For example, while the Northern Territory is very hot and arid, it does have good supplies of aquifer water in certain areas of the territory. For this reason, some of the larger food production companies have moved their production base away from traditional farming regions. For instance, the Australian Peanut company in Kingaroy, Queensland, which had been in severe drought for more than 10 years, moved its growing operations to the Northern Territory and there were many associated small businesses willing to follow, because the drought had already led to the demise of their own enterprises. Mulgowrie Farms is an example of best practice, in terms of implementation of knowledge of the changing climate change patterns, and chooses different parts of the whole country to grow different crops and foodstuffs. Even after major floods in Queensland, as of 29.04.09, 41% of the state was still drought declared.

In addition to being prepared to change locations, according to Donald (2008), farmers have also responded to the changing climate, through: changing variety and planting dates, using in-crop management techniques such as minimum tillage and control traffic systems, fallows, rotations and improved irrigations systems. For grazing, they are utilising variable stocking rates, rotational/cell/spell grazing systems and increasing the use of tools, such as grazing charts and pasture monitoring.

Also it has been shown that climate change is 'not all bad' for farming and in Australia as elsewhere, farm production will increase by 20% in 2008-2009 because of the higher production of wheat and barley; but in comparison to that, beef numbers will fall.[15]

But 'back at the ranch', one of the major concerns that farmers have is that big operators might push them all out of business,[16] as the private investment leagues buy up large tracts of land for forestry tax deductions, carbon sequestration credits, or just take advantage of good farming land being sold up by banks and finance companies for debt arrears. One company – ITC Forestry - paid $16.5 million for 1725 hectares in far north Queensland. Agforce's Cattle president Greg Brown says primary producers can't compete for land with large forestry companies using the money of investors. "It's a misuse of these funds to be in there buying this sort of land with the advantages they have." "That's what's immoral, the fact that these people have the tax advantages that nobody else in mainstream agriculture has."[17] However, the Managed Investment schemes did not agree that they had better tax advantages than primary producers. Moreover, many farming areas in Queensland, some with the best soil, have been targeted for coal mining leases; this has led many farmers and citizens to ask: "But who is going to feed Australia?"

So do you "work a bit harder, work a bit smarter, and work a bit leaner where you can to ride out these tough times?" asks Gavin Robinson, Managing Director of Bethonga Wholefoods.[18] Or do you take an exit grant[19] and hope you can make a new start in life?

The situation for many farmers across the world is tragic; they can no longer feed their own families in the drought, and have begun to slip, if not sell out at a loss, in economic terms. Rural debt is very high and there is no guarantee that they can recover financially or emotionally. The world has forgotten the farmers, but the farmers have not forgotten who it is that nurtures the citizens of the world and provides our most fundamental need: food. Supposedly, 'God' provided free water, but farmers are increasingly having to pay more and more for water to grow our fruit and vegetables.

In addition to this, because livestock contributes about 14% of the total greenhouse emissions, they have to change[20] the whole business of grazing, but there is good news for farmers as the value of the deeply grassed paddocks are being considered as assets in terms of their sequestering carbon[21]. In 2001, Roger Hegarty, in his report on "Greenhouse Gas Emissions from the Australian Livestock Sector" had indicated that the "principle greenhouse gas arising from agricultural livestock management is methane (99%) with a small quantity (1%) of nitrous oxide (N_2O) being released from degradation of nitrogenous compounds in faeces and urine" (p. 4).

[15] Note that these forecasts were made before the impact of the global financial crisis became clear.

[16] Blue's Country Magazine, December 2007, p. 6; Farmers pushed to the sidelines.

[17] Adam Stephen, 06/11/2007 Forestry says tax breaks don't fuel land purchases http://www.abc.net.au/rural/qld/content/2007/s2083370.htm- farmers can't compete.

[18] Troy Bilsborough. "Discovering gold in barren times". Blue's Country Magazine, December 2007, pp. 12/13.

[19] Exit Grants were made available, via the Federal Government, in 2008 to assist farmers to leave their property and start a new life.

[20] ABC Rural. Million dollar grant will fund research on livestock emissions. 07.04.09 http://www.abc.net.au/rural/news/content/200904/s2537002.htm.

[21] ABC Rural. Athol Klieve. Farmers search for new solutions for agriculture's future. 21.04.09. http://www.abc.net.au/rural/news/content/200904/s2548260.htm.

Just when the farmers think that the rain might save them, they are warned about massive locust invasions,[22] which have tended to become more frequent in the past decade.

The Death of Small Towns

One of the most obvious outcomes of the ongoing drought in Australia, and indeed in the USA and several other countries, is that of the fate of small rural towns which are dependent on the prosperity of farmers and graziers. Already beset by vagaries of 'the weather', the livelihood of small country towns ebbs and flows in sync with the income and expenditure of the farmers' bank accounts. Rogers (2003), a rational, level headed thinker, says that "many rural communities face a declining resource base as solids become saline and water resources become scarce" (p. 1). The loss of farming communities is hard to estimate in terms of social and spiritual contributions to humanity, as they are in touch every day with the four fundamental elements of life: fire, water, air and earth.

When livelihoods and whole community frameworks are crumbling, then mental health suffers, along with emotional and physical health. Current research cites some conflicting data on suicide rates, with certain Australian States indicating increases in particular rural areas, provoking alarm with evidence of increased risk during the severe prolonged drought. The drought resulted in the cessation of water rights to some farmers for months at a time and a major reduction in supply to others; supply of water was restricted to rural businesses and towns, with a subsequent cascading loss of employment and income to many rural people whose livelihoods depend on water availability. The most recent critical concern has arisen over the cessation of irrigation rights to the farmers and businesses surrounding the Murray Darling Basin. Depression follows drought for those farmers who do not receive rain or lose their water access or supply.

Not all was lost however. Many farmers and irrigators learned to harvest water in ways they had not done before and many new ways of conserving water and cutting back on water wastage were explored and innovated. Trials of desalinisation and recycled water were introduced in dire circumstances which under normal situations would never have been contemplated, let alone put into practice. New mining fields were opened and new communities built up around them (but it remains to be seen how many will survive the world financial crisis).

Musicians held concerts all around Australia to raise funds for the bush; new mental health programs emerged and were promulgated across rural regions; a Vietnam Veteran, Brian Egan and his wife Nerida, set up 'Aussie Helpers'[23] to take practical assistance personally to the drought stricken and cash strapped farmers. Suddenly the farmer was not just a person or a family who lived a long way away; they were real live people on the brink of bankruptcy and needed 'a fair go'.[24] Songs, such as "You'd better thank a farmer while you can"[25] were sung across community radio stations.

---

[22] ABC Rural. 14.04.09. Farmers urged to prepare for locusts. http://www.abc.net.au/rural/wa/content/2009/04/s2542662.htm.

[23] Aussie Helpers website: Brian and Nerida Egan. http://www.aussiehelpers.org.au/.

[24] 'A Fair Go' is an expression that encapsulates the major value of fairness held in the Australian Community.

[25] "You'd better thank a farmer while you can". From the album "Cool songs for a Warming Climate" by Dermot Dorgan, 2008, Brisbane: djdorgan@bigpond.com.

It is, however, not just water that has been in short supply, but certain crops and meat. Africa and Australia are not alone in their dry years and aridity problems. In the USA Upper Midwest farming area, because of poor water quality, declining biological diversity and decaying rural communities, Jackson (2008) argues that it is now essential to redesign the agricultural landscape.

Berry, whose interviews on farming and conservation (spanning the period 1973 to 2006) have been collated into a book titled: "Conversations with Wendell Berry", has been cited as saying that "society is interested in growth and production" and that "the only economic virtue is competitiveness"[26] and most aptly for the economic downfall in 2008/2009, he noted that "the single goal of the industrial economy, from the beginning of the industrial Revolution, has been the highest possible margin of profit. That is to say that its single motive has been greed."[27] He is an advocate for small scale farming.

Large Scale Agricultural Emergency Disease Outbreaks and Chemical Poisonings

On top of coping with drought, the farmers in Australia were beset in 2007 by an equine flu outbreak. This followed a prior concern about bird flu in the poultry world and SARS for all citizens. Again in 2008, as bird flu broke out and the whole poultry stock of China and other regions were destroyed,[28] Bovine flu and the Hendra virus beset the animal world, leading to the quarantine for many weeks, not just of animals, but of whole communities. After the Swine flu emerged as a threat, doubt was cast on the pig industry which had already had very difficult times.

Following the floods in 2009, water that had spilled over from mines in the north west of Queensland and moved into the river systems was found to be contaminated, worrying organic producers.[29] One of the issues about floods is that they cause a great deal more trouble than is seen by the eye. Brian Williams, an environmental reporter for The Courier Mail, reported on the 27th February 2009, that, after the flood, water had leaked out of 10 of the mines in Queensland (and there are now many such mines). This had contaminated water supplies resulting in unsafe drinking water, dead fish, and acid damage to steel picket fences, which in turn had threatened the whole organically certified beef industry in those areas. Unbelievably, this discharge had polluted water as far as 1300 kilometres away in other states. Thus streams were now unsafe, leading to more problems for cattle and wildlife which drink from them. The fear was that as the rain continued, more polluted runoff would occur and the whole ecology would be contaminated. Primary producers were said to be considering legal action over the contamination.[30]

---

[26] From a 2006 interview; accessed 17.02.09. http://www.americanscientist.org/bookshelf/pub/the-mad-farmer.

[27] From another 2006 interview; accessed 17.02.09. http://www.americanscientist.org/bookshelf/pub/the-mad-farmer.

[28] Jenner (2007) writes about the effect that such slaughter has on emergency responders who often are vets or who have veterinarian training (p. 27).

[29] ABC Rural. Organic producers worried about contamination from flooded mines. Tuesday, 07.04.09. http://www.abc.net.au/rural/news/content/200904/s2537411.htm.

[30] ABC Rural. 29.04.09. Graziers considering legal action over mine contamination during floods. http://www.abc.net.au/rural/nsw/content/200904/s2549552.htm.

In research undertaken by Blacksmith Institute and Green Cross Switzerland on water related diseases, Ericson, Hanrahan and Kong et al. (2008) demonstrate the importance of access to clean water: "[a]lmost 5 million deaths in the developing world annually are due to water related diseases, much of this being preventable with adequate supplies of safe water" (p. 15). They clarify that "[f]resh drinking water makes up only 6% of the total water on Earth and 100% of the world's population relies on it" (Ericson et al., 2008, p. 21). But if the water from icecaps and glaciers is removed from the overall figure, then "only 0.3% of the water on Earth is useable for drinking, and the majority of that is groundwater" (Ericson et al., 2008, p. 21).

Recent findings of high levels of toxic chemicals in the waterways in villages in China where new factories have burgeoned to meet the technology demands of both its own noveau riche and the first world countries rang bells of concern about cancer results in people. This has been matched on an ongoing basis by the findings of high levels of pesticide concentrations in rivers such as the Mae sa Noi Watershed in northern Thailand. This is not a new phenomenon, but in rural areas, it is increasing because of the availability of agrochemicals, pesticides and herbicides. It is no surprise that the increase of suicide in farmers in some countries has been attributed to the use of certain pesticides in their farming methods, although whether the suicides are due to the effects of poisoning, or to the stress of paying high prices for such chemicals, has not yet been determined.

Food Shortages

Food shortages are classified as a secondary hazard.[31] There is no doubt that as the world warms, and droughts increase in length and intensity, more hunger will occur (Smith, 2007). According to the IPCC (Intergovernmental Panel on Climate Change), more hunger, drought, and hurricanes are on the way. The IPCC report noted that with a global increase in temperature of 2˚C, crop yields would crash in Africa, Asia, Southern USA, and Southern Europe. However, at the opposite end of the world, harvests would improve by up to 15% for example, in Britain, Canada, Northern Europe, and parts of the USA and Russia.

Already we have witnessed the effects of droughts on food production in 1996. The USA drought resulted in serious losses in crop productions and livestock and also increased the number of forest fires and wildfires (Wilhite and Svoboda, 2000).

Even if a citizen could inure themselves from media bombardments about global warming and natural disasters that are wreaking widespread havoc, it is difficult for them to ignore the fears that have an impact on them directly in their everyday lives. As news of fuel price increases spell doom for rural businesses and farming enterprises, a resultant outcome of continuing drought and floods is a downturn in food production. Additionally with emerging countries, such as China and Vietnam, being able to pay more for food, the actual price of some food has doubled in some regions, putting even basic necessities such as rice outside the reach of the poor and very poor households in third world countries (Gerstein, 2008). At the 2008 UN Trade and Development Conference in Ghana, the UN Secretary - General Ban Ki-

[31] Secondary Hazards: "Those hazards that occur as a result of another hazard or disaster, i.e. fires or landslides following earthquakes, epidemics following famines, food shortages following drought or floods." (UNDHA, DM Glossary, 1992, p. 66).

moon - indicated that the price rises had led to food riots in Africa and Asia.[32] The fact that Climate Change would increase the risk of hunger in places like Asia was noted by newspaper 'China Daily' quoting from an unknown source.

Gone is the United Nations' dream of halving world poverty by 2015, and such food shortages, coupled with other influential factors, could lead to more mass migration across the world, as yet more countries go into famine. The addition of the major effects of the global financial problems can only make the situation worse. Indeed, Oxfam has warned[33] that the number of people affected by natural disasters would increase by 54% to 375 million each year up to 2015, if the trend since 1980 continues. Nicholas Stern (2006), a senior economist for the British Treasury, gave examples of severe droughts and millions of refugees moving away from the drought affected regions.

In preparation for a seed shortage, the Svalbard Global Seed Vault has created a seed bank with 200 million seeds representing 400,000 individual samples.[34] The CSIRO in Australia are also devising ways to increase crop yields which they believe will help farmers adapt to drought and climate change.[35] According to Dr Richards, wheat yield in particular may be improved, so that it can perform better in dry climates.

EFFECTS OF DROUGHT ON THE ECOLOGY

La Niña did not do her normal dance in 2008 and the drought in Australia continued in most parts of the country unabated, leaving South Australia devastated and some parts of Queensland and New South Wales and Victoria with no rain or a mere 40 ml.[36] Dust storms blew across the nation when the drought had led to loss of ground cover and the rains did not come. Even in April 2009, after the drought had broken in Queensland and some parts of New South Wales, red dust cloaked The Riverina.[37]

It was during this harsh period of drought, that the Murray-Darling Basin (which has certain parts of it world heritage listed) became a household word. School children heard about the endangered species in that river system and learned how to carefully scrape off the encrusted salt from the carapaces of land turtles, affected by the high levels of salinity in places where these unique creatures had always taken shelter.[38] Millions of hectares (5.7m) were marked by the National Land and Water Resources Audit as being at risk of salinity, and in the next 40 years, it is projected that this area will increase by another 11 million hectares.

The Murray Darling Basin has been confronted with an ecological disaster that has affected both the farming and business communities in the Basin area and annihilated so many species of wildlife and flora that this major national crisis has become one of international significance. As of 18.04.09, from the period commencing in June 2008, the

[32] ACCRA, 2008. Rural Weekly, p. 23. Friday April 25, 2008. Southern Edition.

[33] Hundreds of millions will be hurt by climate change, Oxfam warns. The Australian. 21.04.09. http://www.theaustralian.news.com.au/story/0,25197,25364665-11949,00.html.

[34] http://news.theage.com.au/breaking-news-national/world-support-growing-for-seed-v...29/04/09.

[35] Kate Evans. 29.04.09. Scientists working to secure Australia's food future. Http://www.abc.net.au/news/stories/2009/04/22/25/2549953.htm.

[36] This changed markedly in early 2009 with powerful flood waters sweeping across Queensland and parts of New South Wales.

[37] ABC Rural. Dust storm envelopes Riverina. http://www.abc.net.au/rural/nsw/content/2009/04/s2543674.htm.

[38] Georges and Roe paint the plight of these and other turtles during the fierce drought in Chapter 11.

lowest ever inflows were recorded - just 1720GL; just one quarter of what should be expected, based on long term averages.[39] As of 22.04.09, the basin was 17% full and had 3807GL, out of a total capacity of 22621GL.[40]

Water has thus become a hot topic in the literature, in the media and in government circles following droughts. This chapter turns now to a discussion of water shortages.

WATER AS A GIFT OF LIFE OR A COMMODITY?

Water, water, everywhere and not a drop to drink
(Samuel Taylor Coleridge, 1798) [41]

But what actually is water? According to Michael Johnson (in Alexandra and Eyre, 1993),[42] "the abundance of water on earth has given rise to an abundance of organisms which depend on water for their life" (p. 7). The chemical notation H_20 does not sound very natural, with one atom of oxygen and two parts of hydrogen describing mathematically and chemically what is the most important gift of life in the world. We are told that over half the weight of a human is water and that a 70 kg person, for example, contains 42 litres of water in their system. We are also warned that increases in the environmental temperature lead to loss of water, and that heatwaves not only deprive us of water and the natural nutrients it carries, but upset the whole natural balance of the body. Without water, people die; without clean water, people become ill and may also die.

Is it true that the world is short of potable water? Many scientists point to this probability and warn that with further global warming, the situation will become worse. One of the people who are avid about conserving water is Maude Barlow. In her 2007 book, *Blue Covenant*, she posed the question: are we running out of water? But further delving into this book removes the question mark- it is a clear resonant statement: We are running out of water, she insists (p. 2). Barlow 'makes no bones' about the planet drying up.

> Desalinisation plants will ring the world's oceans, many of them run by nuclear power; corporate-controlled nano-technology will clean up the sewage water and sell it to private utilities, which will in turn sell it back to us at a huge profit; the rich will drink only bottled water found in the few remaining uncontaminated parts of the world, or sucked from the clouds by corporate-controlled machines, while the poor will die in increasing numbers from a lack of water. (Barlow, 2007, p. 2)

Quick thinking is commendable, but action without considering all the consequences is fatal. So what if we take the fresh water from the bottom of the oceans? What will that do to the whole equilibrium of the planet earth (and perhaps beyond)? What if we pump back the salt from desalinisation plants into the oceans, what happens to the quality of the sea water and the marine life over time? Gleick (2006) warns us about the side effects of desalination, none of which are pleasant.

---

[39] Siobhain Ryan, Murray Darling Basin inflows slow to a trickle. The Australian, April, 08, 2009. http://www.theaustralian.news.com.au/story/0,25197,25306185-2702,00.html.

[40] http://www.mdba.gov.au/water/waterinstorage.

[41] Poem: The Ancient Mariner, Lyrical Ballads (1798).

Barlow is quite clear that we are in a situation of declining water supplies and, along with others, is not averse to pinpointing the profiteering involved in the new water trade.[43] Barlow also names a number of private firms such as SUEZ, Thames Water, Veolia and a range of other international companies for allegedly failing, sometimes, to deliver clean water and sanitation, and tracks their successes and failures in countries across the continents. The Victorian Women's Trust (2007) are concerned about the way water will be allocated in Australia following the 'unbundling' of water from ownership of land since the 1st July in 2007 and that the share of the water market will be taken up by speculators who have no real stake in how that water is apportioned.

Barlow, a respected activist in food provision, clean water and globalisation, refers to the parts of the world that are 'hot stains' - a particularly visual, graphic and kinaesthetic description of places in many countries, including the USA and Australia, along with China, and the long known dry spots of Mexico, the Middle East, Africa, Asia and South America. That is a vast section of the earth to be short of potable water. Fred Pearce (2007) notes that 100 Chinese cities were facing water shortages during 2005. Water shortage in India, with its one billion people, was chronic. Overpopulation in the countries drawing on the Nile River would lead to fights about both domestic water and irrigation rights. While Vietnam is proactive in building dams along several rivers, the damming will lead to a reduction in river flows, which in turn will impact on the fishing and agricultural resources in the countries of Cambodia, Laos, Thailand and Vietnam (Pearce, 2007, p. 30).

Most Australians are still unaware, but less so since the recent great drought, that Australia is the most arid country in the world. We have always known that Africa had water problems, but the question remains: did it really always have them, and if not, what caused the current problems?[44] Other authors in this book take up the answer to that question.

If, as Global Water Intelligence reports, the desalinisation industry will triple by the year 2015 (in Barlow, 2007, p. 73), then it is no wonder that large financial institutions in Australia were seen buying up as much of the bottled water trade, tanks, and pipes across Australia, as they could purchase. Indeed, according to Barlow, the international water market might reach a value of US$66 billion by 2010 and double that by 2015.

Based on the data from Gleick (2006) with acknowledged sources tables, we can assess countries' and sectors' use of water and then view further pages on access to safe drinking water from 1970 to 2002. The consumption of bottled water was highest in the USA (59.4% in 1997 up to 97.5% in 2004) compared to Africa/Middle East and Oceania which used non-bottled water more - up from 2.6% in 1997 to 4.4% in 2004. Europe's use of bottled water also rose from 47.8% in 1997 to 74.7% in 2004. Barlow's warnings that only the rich will have clean water in the future, and the poor will have little access to clean water can be seen in these figures – economic growth generally wins the day, but not in underdeveloped countries.

In response to the swift move to privatise world water supplies, the Global Water partnership promoted its stance on 'Water for All'; however, this is not what it at first appears. It is not insisting that water for third world countries be made available to them at a cost they can afford, but rather that public funds should be directed to private companies in a

[42] Water in Australia, Chapter 8.
[43] [Editor's note: Kelbessa takes up this point in Chapter 15 on Africa.]
[44] [Editor's note: Kelbessa will also assist us in answering this question in Chapter 15.]

way that insures that they make a profit. There is nothing wrong in making a profit – it is how businesses operate and grow; the reality might be that using normal business principles is a way of ensuring that the poor do get water, when the world's supplies are bought up by profiteers. Nevertheless, water is a natural resource that all peoples across time have taken to be the Creator's free gift to mankind.

A significant goal for the world's poor is poverty alleviation and sustainable development. The question could be put: In what way will having to buy water from international private corporations change the lives of the poor in this world? How will they have easier and better access to water which they can afford? [45]

Michael Johnson (in Alexandra and Eyre, 1993),[46] in a review of the water reforms in England and Wales, concluded that the initiatives did not lead to the proposed outcomes, in terms of improving the economic efficiency of the water industry, nor did advances in service quality or improved environmental standards eventuate. Byrne et al. (2006, p. 51) warned that free trade in water could severely harm agriculture in Australia. They insist that primary allocations have to be made by the government and not by private corporations. There is no need to have water trade to boost agricultural output "because they already move to where the water is" (p. 33).

On the topic of one of the world's heritage listed sites, which has been decimated by the long drought in Australia, John Pigram (2006) addressed at length the trouble ahead for healthier rivers, management of environmental flows and river health, while showcasing the Murray Darling Basin; he pointed to the need for the two States - Queensland and New South Wales - to work together better to share water resources.

Speaking on behalf of the Murray Darling Basin commission, ecologist Lee Benson (2004) in *The Science Behind the Living Murray Initiative (Part 2)* concluded in his report (page xi) that the need for urgent action on the river system was too hasty, because the river system had not collapsed within the specified time frame time. Unfortunately since then, it has; it is indeed now a major ecological disaster. Benson's point was that there had been a lack of consultation with the people in the community, and this overall lack of consultation is now coming back to haunt the governments, as many different stakeholders make their claim for water allocation from the great, but thirsty Murray. "Knowledge-based management should replace the perceived need for urgency" (p. 59) said Benson, but perhaps it is already too late.

It is not just the water from the rivers that is being monitored and metered now, it is groundwater. Jacol Burke and Marcus Moench (2000) warn that groundwater usage has to have a long term management view; these issues are the same across the world (p. 145) and additionally any strategies must be participatory for all users and affected people. In *Groundwater and Society: Resources, Tensions and Opportunities,* Burke and Moench note that some of the factors to be taken into consideration are groundwater depletion and degradation; social conditions and policy failures are other factors to be cognisant of in managing groundwater now and in the future, if we are to keep healthy flows of ground water.

If worry about how to obtain enough potable water was not enough, Peter Gleick (2006) raises our blood pressure over terrorism in our water sources. The report titled *The world's*

[45] These and other difficult questions are taken up further by Kelbessa, and Guille and Evans.
[46] Water in Australia, Chapter 8.

water 2006-2007: The Biennial report on Freshwater Resources adds another conundrum to our water worries: that of water-related terrorism. These mostly focus on attacks of our water systems in the past and the necessity to put in place measures to reduce the vulnerabilities of future attacks. Water security policies need to be set in place in every country, after potential vulnerabilities are assessed, followed by the creation of detection and emergency response systems, he advises. What could this mean? Water contamination, the blowing up of dams, the diversion, or letting go, of waters. These are fears that Gleick is careful not to stir up; however such threats have been taken very seriously across the world (see Coleman, 2005; NATO, 2006).

Water theft[47], like stock theft[48], is quite common in rural areas in different countries of the world, where water trucks pull up and empty people's tanks or suck dry their dams and water holes. In the city suburbs, when the city councils started to put meters into residents' properties, it was not uncommon to hear that neighbours were stealing other people's water to fill their swimming pools and once the water restrictions came into law, stories of residents stealing their neighbours' water to sprinkle their own gardens were not uncommon; alas, in the end, the long drought won and most gardens dried up and died.

Conflict about water is not new; it is just new to some generations who have never encountered the problem before in their own countries, and who may never have travelled to other countries where there was a drought on at the time. It is Gleick who gives us a fascinating extended table about water conflict chronology, commencing with the era of Noah in 3000bc right through to 2006 (in Yemen). Water is critical to life and living, and becomes a sacred resource when drought visits and decides to settle in a region.

One fundamental question needs to be asked and answered by all of us: Do the people of the world have a right to clean water? Along with the international aid organisations that have been attempting to achieve this aim of the provision and affordability of clean water to third world countries, Barlow (2009) adds her significant voice; and she has been known to hint at possibilities, such as the concept of water pipes running across the floors of the oceans to different stations in the world, which she believes is now highly feasible.

Conclusion

There is no doubt that once a country goes into serious and prolonged drought, steps have to be put in place to buffer the effect on farming communities and rural areas. The shortage of ground water at a time of deficit precipitation leads to chronic debt problems over time,[49] from which these communities cannot recover in the short or medium term, if ever. Thus, in terms of disaster management and preparedness, drought plans need to developed and become part of state and regional governments' emergency management policies and strategies.

---

[47] ABC News. 01.07.08. Irrigation water theft increasing. Goulburn Murray Water (GMW) is investigating 240 cases of water theft from the past irrigation season. http://www.abc.net.au/news/stories/2008/07/01/2290988.htm.

[48] Daily Telegraph. 16.05.07. Australia: Rural water theft on rise. http://www.waterconserve.org/shared/reader/welcome.aspx?linkid=75078andkeybold=environment%20AND%20%20crime%20AND%20%20police.

[49] ABC Rural. 22.04.09. Lack of credit strangling drought-stricken farmers. http://www.abc.net.au/rural/news/content/200904/s2549477.htm.

A major social issue is that farmers, including small lot farmers across the world, need to be valued for the role they play in feeding the peoples of the world; this does not currently appear to be the case. Politically and socially, rural areas, as a whole, also need to be acknowledged for preserving the natural environment for the rest of the people in their countries and indeed for the whole world. From a humanitarian (feeding the hungry) point of view and also from self interest (having food to eat), every person needs to heed the words of song writer Dermot Dorgan: You'd better thank a farmer, while you can". [50]

REFERENCES

Alexandra, J. and Eyre, D. (1993). Water and Environment, in M. Johnson and S. Rix. (Eds.). *Water in Australia: Managing Economic, Environmental and Community Reform* (pp. 17–27). NSW: Pluto Press.

Australian Government Bureau of Rural Sciences (2008). Country Matters. *2008 Social Atlas of rural and regional Australia.* Drought in rural Australia- exploring the social impact in three case study regions. www.brs.gov.au/socialatlas

Barlow, M. (2007). *Blue Covenant: The global water crisis and the coming battle for the right to water.* Melbourne: Black Inc.

Barlow, M. (2009). Notes for Opening Keynote. Australian Water Summit, 1 April 2009. Maude Barlow, Senior Advisor on Water to the President of The United Nations. Http:// www.canadians.org/about/Maude_Barlow/UN/KEYNOTE-AWS2009.pdf

Barton, S.B., and Ramirez, J.A. (2004). Effects of El Nino Southern oscillation and Pacific Interdecadal Oscillation on water supply river basin. *J. Water Resour, Ping. And Mgmt., 130* (4), 281-289.

Bates, B.C., Kundzewicz, Z.W., Wu, S., and Palutikof, J.P. (Eds.). (2008). Climate Change and Water. Technical Paper of the Intergovernmental Panel on Climate Change. Geneva: IPCC Secretariat.

Bryne, P., Eagle, O'Brine, J., and McDonald, D. (2006). *High and Dry: How free trade in water will cripple Australian agriculture.* North Melbourne: Freedom Publishing.

Bureau of Meteorology Research Centre. (2001). *Indian Ocean Climate Initiative. Second research report-Towards understanding climate variability in south western Australia.* Research Reports on the Second Research Phase of the Indian Ocean Climate Initiative. East Perth, WA: Indian Ocean Climate Initiative Panel.

Burke, J. J., and Moench, M. (2000). *Groundwater and Society: Resources, Tensions and Opportunities.* New York: UN DESA and ISET. http://www.wrc.org.za/downloads/ watersa/2009/Paper%203.pdf

Coleman, K.G. (2005). U.S. Dams: Is Security Seeping through the Cracks? Dec 1, 2005, http://govtsecurity.com/mag/us_dams_security/

Ericson, B., Hanrahan, D., Kong, V. et al. (2008). *The World's Worst Pollution problems: The Top Ten of the Toxic Twenty.* Blacksmith Institute and Green Cross Switzerland. http://www.worstpolluted.org/reports/file/World's%20Worst%20Pollution%20Problems %202008.pdf

[50] "You'd better thank a farmer while you can". From the album Cool Songs for a Warming Climate by Dermot Dorgan, 2008, Brisbane: djdorgan@bigpond.com.

FEMA. (1995). *National Mitigation Strategy: Partnerships for building safer communities.* Washington, D.C.: Federal Emergency Management Agency.

Garnaut, R. (2008, October). *The Garnaut Climate Change Review.* Final Report. Melbourne: Cambridge University Press.

Gerstein, J. (2008). Food rationing confronts breadbasket of the world. *The New York Sun, 21*st April. http://www.nysun.com/national/food-rationing-confronts-breadbasket-of-the-world/74994/

Gleick, P.H., Cooley, H., Katz, D. Lee, E., and Jason, E.L. (2006). The World's Water 2006-2007. The Biennial Report on Freshwater Resources. Washington, D.C.: Island Press.

Hayes, M.J. Svoboda, M.D., Knutson, C.L. and Wilhite, D.A. (2004). *Estimating The Economic Impacts Of Drought.* 14th Conference on Applied Climatology. Seattle, WA. http://ams.confex.com/ams/84Annual/techprogram/paper_73004.htm

Hegarty, R. (2001). *Greenhouse Gas Emissions from the Australian Livestock Sector: What do we know, what can we do?* Canberra: Australian Greenhouse Office.

Jackson, L.L. (2008). Who "designs" the agricultural landscape? *Landscape Journal, 27*(1), 23-40.

Jenner, M. (2007). The psychological impact of responding to agricultural emergencies. *The Australian Journal of Emergency management, 22* (92), 25-31.

Kenny, P. (2008, September). It's About People: Changing Perspective. A Report to Government by an Expert Social Panel on Dryness. Drought Policy Review Expert Social Panel 2008. Report to the Minister for Agriculture, Fisheries and Forestry. Canberra: Commonwealth of Australia.

Pearce, F. (2007). *When the rivers run dry: What happens when our water runs out?* London: Transworld Publishers.

NATO. (2006). "A Completely New Dimension" or "An Additional Hazard. Springer Netherlands. Large Dams and the Terrorist Threat. Protection of Civilian Infrastructure from Acts of Terrorism. NATO Security through Science Series, 2006, pp 103-124.

Pigram, J.J., and CSIRO. (2007). *Australia's water resources: From use to management.* Collingwood, Vic: CSIRO Publishing.

Roads, J., Chen, S., Han, J., Cui, Y., Miller, A., Auad, G., Gershunov, S., White, W., Somerville, R., Iacobellis, S., Schneider, N. and Yulaeva, E. (2001). ECPC progress report OGP/CDEP (P. Arkin): Jan. 1 - Dec. 31, 2000. University of California, San Diego. http://ecpc.ucsd.edu/psreports/200009.awork.htm

Rogers, M. (2003). Small towns: Big picture- social sustainability indicators and the Arts. Social Dimensions of the Triple Bottom Line in Rural Australia- Bureau of Rural Sciences. http://www.latrobe.edu.au/smalltowns/social%20dimension%20TBL.htm

Stern, N. (2006, October). Stern Review: The Economics of Climate Change, October 2006. HM Treasury and Cabinet Office.

Svoboda, M., LeComte, D., Hayes, M., et al., (2002). The Drought Monitor. *Bulletin of the American Meteorological Society, 83*, 1181-1190.

Tannehille, J.R. (1947). *Drought: Its causes and effects.* New Jersey: Princeton University Press.

United Nations Department of Humanitarian Affairs (DHA). (1982). *Glossary: International Agreed Glossary of Basic Terms Related to Disaster Management* (DHA/93/36).

Geneva, Switzerland, UN DHA, December 1992, 81 pages. Available at: http://www.em-dat.net/glossary.htm; also available at http://www.reliefweb.int/rw/lib.nsf/db900SID/LGEL-5EQNZV/$FILE/dha-glossary-1992.pdf?OpenElement

Victorian Women's Trust. (2007). Water as a common good. http://www.watermark australia.org.au/watermark_pdf/Part1/19bp_waterasacommongood.pdf

Wang Shourong. (2009). Strategies for Climate Change and its impacts on sustainable development in China. National Climate Center. Beijing. http://www.bom.gov.au/climate/cli2000/wShourng.html

WHO. (2003). *Climate Change and Human Health- Risks and Responses. Summary.* Geneva: World Health Organization.

Wilhite, D.A. (1991). Drought Planning: A Process for State Government. *Water Resources Bulletin, 27*(1), 29-38.

Wilhite, D.A. (2000). Drought as a natural hazard: Concepts and definitions. In D. Wilhite (Ed.) *Drought: A global assessment.* (Vol 1, pp. 1-18). New York: Routledge.

Wilhite, D. A. and Glantz, M. H. (1985). Understanding the drought phenomenon: The role of definitions. *Water International,* 10, 111-120.

Wilhite, D.A. and Svoboda, M.D. (2000). Drought Early Warning Systems in the Context of Drought Preparedness and Mitigation. In D.A. Wilhite, M.V.K. Sivakumar, and D.A. Wood (Eds.). *Early Warning Systems for Drought Preparedness and Drought Management* (pp. 1-16). Proceedings of an Expert Group Meeting, Lisbon, Portugal, September 5-7, 2000. World Meteorological Organization, Geneva, Switzerland.

Williams, B. (2009). Toxic flood in Outback: Leaking mines poison water for people and stock. *The Courier Mail 27*.02.09, p. 3. Friday February 27.

World Meteorological Organization (WMO). (1992). *International Meteorological Vocabulary* (2nd ed.). Geneva, Switzerland: World Meteorological Organization.

Contributed by Francine Pritchard, Ipswich.

Photo: Atkinson Dam: Empty of Boats and Turtles.

In: Meltdown: Climate Change, Natural Disasters... ISBN 978-1-60876-153-1
Editor: Kathryn Gow

Chapter 17

THE SOUTH EAST QUEENSLAND WATER GRID – USING THE DROUGHT TO ENGINEER THE MARKET

Howard Guille and Hannah Evans
Queensland University of Technology, Australia

ABSTRACT

The South-East Queensland Water Grid is an extensive engineering works to 'drought-proof' the region. It is also an experiment in creating a competitive urban water market with 'common carriage' arrangements very similar to those applying in electricity. The water grid will facilitate urban users choosing from which company to buy their potable water, and to which one to sell their sewage. This would be a very large, real world application of the abstract models beloved of liberal economists and consultants. The privatisation of water in England and Wales is criticised for creating ten vertically integrated private monopolies without market competition. In contrast, the South-East Queensland Water Grid will make competition a physical and organisational possibility, but the current Queensland Government states that it has no intention of moving to private ownership. We will discuss some of the political and economic tensions and instabilities in the Queensland arrangements and whether a shift to private ownership can be precluded, given pressures from business and orthodox liberal opinion and the experience of eventual privatisation in the electricity sectors of Queensland and New South Wales.

Keywords: Water, water grid, water market, privatisation, corporatisation, public water management, Queensland, political economy

INTRODUCTION

> Water is a renewable natural resource, but its delivery as safe drinking water makes it also a commodity. (Hukka and Katko, 2003, p. 8)

Solutions to the turn of the century drought have been sought in prayer, engineering and the market among others. The Queensland Premier called for a week of prayer[1] while the engineering solution is dams, desalination and recycling plants and the call to the market is for competition, private enterprise and water prices set by supply and demand. We are interested in the confluence between engineering and the market in the South East Queensland Water Grid due for completion on or before 2010. This is a very substantial construction including new dams, desalination, water recycling and a set of mains which will allow water to be transported across the region. It serves an area north from the Tweed to Noosa (about 200 km) and west from the coast to Toowoomba (about 140 km) and a population of 2.8 million which is predicted to rise to 3.96 million by 2026 (Department of Local Government, Sport and Recreation, 2007).

The physical grid is designed to move water anywhere in the region from competing suppliers to competing retail distributors. The management and operating structure of the Grid separates the control of bulk supply of water from wholesaling and retailing and envisages competition in bulk supply and in retail. Water users will be able to choose their retailer and, indirectly, their bulk supplier and source of supply. These arrangements will facilitate an urban water market and the number of bulk suppliers and retailers; the trading rules and the institutional arrangements are to be finalised by 2010. There are some very close analogies with the common transmission systems and market rules that have been established for electricity in Australia and many other countries. However, the South-East Queensland arrangements seem to be leading the developments in urban water. According to the Productivity Commission, a fully competitive market for urban water has not been achieved anywhere in the world[2] (Productivity Commission, 2008, p. xxii).

The promotion of the South East Queensland Water Grid by the Queensland Labor Government has emphasised how the engineering works will 'drought-proof' the region (Osborne, 2006). Publicly at least, political contention has been confined to three issues: the transfer of water assets from local government to state government; opposition to the proposed new Traveston dam on the Mary River; and the use of recycled sewage in potable supplies. The proposal to structure the grid as a competitive water market has had little public discussion, although a commitment has been given to the Queensland Council of Unions that public ownership of water assets will be maintained (Queensland Council of Unions, 2007, p. 6). Yet over the same period, some prominent voices including the Business Council of

---

[1] In October 2006, Queensland Premier Peter Beattie and Queensland clergy made a joint call for a week of prayer for rain (News Ltd 2006). News Limited, 2006 http://www.news.com.au/story/0,23599,20654467-1702,00.html , Retrieved 30 May, 2007.

[2] We take a fully competitive market to be one where consumers can chose between suppliers and purchase the quantity they wish at prevailing prices, which, subject to any overall regulation, are set by interaction of supply and demand. New South Wales has established procedures for 'third party access' in the Sydney and Hunter Water Corporation areas. These "make it possible for private sector access seekers to access certain water infrastructure services and sewerage infrastructure services (services) and so compete with the service providers who control those services" (IPART, 2008, p. 1). We follow the Productivity Commission in treating

Australia, the former Coalition Government and the Productivity Commission have urged that urban water shortages can be solved by water markets and greater private sector involvement. They argue that these will overcome shortages, ensure adequate investment and remove the need for quantitative water restrictions. We note that 'private sector involvement' is an imprecise term and can be applied to a range of situations from partial to full private ownership and/or operation of the water system. In particular, "Privatization can be partial, leading to so-called public/private partnerships, or complete, leading to the total elimination of government responsibility for water systems" (Gleick, Wolff, Chalecki, and Reyes, 2002, p. 21).

Our central question is whether the arrangements being made in Queensland in 2008-2009 will lead to the privatisation of water supply and distribution and sewage collection in the South East. By the term privatisation, we mean that the entire infrastructure for the collection, distribution and supply of water and sewage including dams, pipes and treatment plants moves from public to private ownership and control. Also, and subject only to any water resource plans and other regulations, natural water will become private property at the point that it enters or is retained in the infrastructure[3]. Using South East Queensland examples, this will be when rain is retained in a dam such as Wivenhoe, when sea-water is fed into a desalination plant such as that at Tugun, or groundwater is taken into a bore such as on North Stradbroke Island.

This is the most extensive form of privatisation described by numerous writers (see Gleick et al., 2002, pp. 25-28; Godden, 2008, p. 160; Hukka and Katko, 2003). It is also the 'divestiture' option raised as one of four possible private sector arrangements by the Coalition Federal Government in 2006 (Australian Government, Department of the Prime Minister and Cabinet, 2006). It is similar in scope to when the water systems of England and Wales were privatised in 1989, when ten regional water authorities were sold via a share float. The new owners operate and own all aspects of the water business in these catchments (Lobina and Hall, 2001). To date, this is the only extensive privatisation of water to have occurred in a developed country. We acknowledge that there are other models of private sector involvement such as contracting out, public-private partnerships and franchising and that these are described sometimes by friends and foes as 'privatisation'. We intend to restrict the term privatisation to what we have described as full private ownership of water and sewage and of urban water and sewerage systems[4]. We also note that 'third party access' regimes are principally about introducing competition into monopoly systems. While they may provide some private sector involvement, they do not amount to full privatisation of the form we have defined[5].

---

such arrangements as introducing competition but not private ownership of the system. See also Gray and Gardner, 2008.

[3] Sewage will become privately owned when it enters the sewerage system. However, this would have an exchange value, if it were to be treated and recycled. Conceivably, people might be paid for their sewage and/or the system operator might on-sell it to recyclers.

[4] Given that the lead author has been very closely involved in union and community opposition to 'public sector reform' since the 1980s, we note that the concentration on full privatisation does not indicate any acceptance of public private partnerships or franchising arrangements.

[5] Indeed, the competitor admitted under third party and common carriage arrangements could be a public or social provider; for example, the publicly owned Queensland Rail Corporation operates trains in New South Wales and South Australia. There is also the question of third party access to private infrastructure such as the long running National Competition Council case over access to the rail lines in the Pilbara operated by BHP-Billiton and Rio-Tinto (Business Spectator, 27 Oct, 2008).

We know that that our claim is very contentious and we recognise that the Queensland Government[6] in power at the time of writing has categorically stated that the water grid will not be privatised. However, we will document that the physical and organisation arrangements of the Grid entail a fully competitive market for water. In its present form, this is a competitive market system with public corporations but without private ownership. We will argue that experience in the electricity industry shows that this could be unstable and open to recurrent calls to complete the process of deregulation by combining private ownership with market arrangements. We also see a paradox between the Queensland situation and that in England and Wales. In the latter, an expressly neo-liberal government privatised water, but created private monopolies. The companies were given an exclusive licence to their river basin area for 25 years and protected against takeover for five years. Except for some existing small private companies, the Government effectively created ten private monopolies with complete control of water and sewage in their respective areas (Lobina and Hall, 2001). In contrast, in Queensland, a Labor Government has introduced a competitive water system with public ownership. In our view, public ownership will be more difficult to uphold politically than will private monopoly. As such, and short of radical changes in prevailing economic ideas, private ownership without competition will be more durable than a public system subject to market competition.

We consider that the South East Queensland Water Grid will become a case study to examine the appropriateness and adequacy of market arrangements in urban water supply and sewerage provision. Our subject is unfettered market operations for urban water supply or what economic regulators term 'competition in the market' as against competition 'for the market'. This distinction is used by Swier who is a member of the Australian Energy Regulator and the Australian Competition and Consumer Council. He writes that "The urban water industry internationally has various models for competition for the market but until recently there has been little interest in the scope for competition in the market" (Swier, 2008, p. 2). The model of 'competition for the market' involves the state selling off operating rights for a monopoly service – for example, a contest to gain sole rights to garner and supply water in a region usually under the close purview of a regulator or price controller. In contrast, 'competition in the market' means having competing suppliers of water and sewage services and prices set by supply and demand. In practice, this would mean that corporate and domestic consumers would either contract to get their water and sewage requirements for a stated period with a particular supplier and/or buy them on a spot market[7]. In essence, this would be the same as entering a long turn phone contract with a company such as Telstra or Optus or Internode, or buying some electricity from Origin or Macquarie Power and some short-term on the National Electricity Market[8].

[6] The Australian Labor Party (ALP) is a left-of-centre party which has held power in Queensland from 1998 to 2009; first under Premier Peter Beattie (1998-2007) and then Premier Anna Bligh (2007- to date of writing, 2009). A State election was held in March 2009 and Labor was retained.

[7] A spot market is essentially an auction of short-term supply contracts. These are usually additional to long-term contracts and operate for a wide range of goods, services and financial products including electricity, irrigation water, bulk shipping and even hotel rooms. For example, "wholesale trading in electricity is conducted as a spot market where supply and demand are instantaneously matched in real time through a centrally-coordinated dispatch process" (Energy Supply Association of Australia, 2009).

[8] These are illustrative – in the electricity industry in Australia, there are separate generating and retailing/distribution companies; whereas in the telecommunications industy, Telstra and Optus provide the networks and the retailing, allowing a plethora of retailing companies or resellers who purchase 'line space' on a wholesale basis from Telstra and Optus. Either arrangement is conceivable for urban water and sewerage.

We believe that the reorganisation of urban water supply in South East Queensland is internationally significant and deserves to be documented and analysed. In addition, we want to ask whether the changes will diminish the public realm. The plan for discussing this issue in the chapter is straightforward. The advocacy for a move to market arrangements will be discussed and evaluated in the next section of the chapter. The model for competition 'in the market' proposed for the South East Queensland Water Grid is then described. Our claim about the likelihood of privatisation of urban water supply in South East Queensland will be covered next and in the final part of the chapter, we briefly relate our account of South East Queensland to a wider canvass. This is relevant because privatisation of water in a considerable number of developing countries of the Global South has produced a very considerable counter reaction built around objections to the commodification[9] of water and claims that clean water is either a community and/or a human right.

PULLING THE PUBLIC PLUG

The administration of urban water supply has been transformed since the late 1980s through a combination of National Competition Policy (Australian Government, Department of the Environment, Water, Heritage and the Arts, n.d.), specific reforms under the 1994 Water Reform Framework (Council of Australian Governments, 1994[10]) and the 2004 National Water Initiative (National Water Commission, 2004). The main changes are pricing of water to ensure full cost recovery including commercial return on capital, separating regulation, and delivery and institutional reform via commercialisation and corporatisation[11] of State and Local Government water agencies. The standard model of corporatisation involves the separation of the trading business from day-to-day ministerial control, setting commercial objectives and the appointment of an independent board of directors responsible for pricing and investment decisions. Such arrangements have been applied to all Commonwealth and State government trading agencies, whether they be rail, ports, electricity or lotteries. The 'new' corporations remained in public ownership, at least in the initial rounds of change made in the late 1980s under the Greiner Coalition Government in NSW and the Hawke and then Keating Federal ALP Governments. In contrast, the Kennett Liberal Government in Victoria, in the 1990s, favoured privatisation instead of corporatisation, as did the Liberal Governments of South Australia and Western Australia (Quiggin, 1996). The Federal Labor Government also turned to privatisation including Qantas and the Commonwealth Bank. More generally, according to the Reserve Bank, the shift of public entities and institutions to private ownership in Australia of former was the most extensive in the OECD and greater than in the United Kingdom (Reserve Bank of Australia, 1997, p. 8). A range of government owned corporatisations has been privatised under the Queensland ALP

Perhaps one of the 'psychic' benefits of a market in sewerage services will be choosing whether or not to favour Haliburton which is one of the world's top four water companies as well as a logistics supplier in Iraq.

[9] Commodification – turning an item into a traded good or service.

[10] See 2004 reference in list.

[11] Institutional reform was the final principle of the Water Reform Framework. This applies the general pattern of public sector reform to the separate management and regulatory responsibilities and corporatisation of public water enterprises.

Government since 1998 including the Totaliser Agency Board (TAB) in 1999, Golden Casket in 2004 and the retail power firms of Energex and Ergon Retail in 2006.

The State and Territory Governments have been more reticent about privatising urban water, although some private sector involvement has been arranged. The Liberal Government of South Australia outsourced the management of the sewerage and water services outside Adelaide to Riverland Water in 1995. Riverland Water is a consortium of United Utilities Australia, AMP Investments, and Bechtel Enterprises. This consortium was also awarded a 27-year contract to finance, design, build and operate 10 new water treatment plants (Woodward, Parkin, and Summers, 2002, p. 425). In 1999, United Water, a subsidiary of consortium of Veolia Water, was awarded a 15.5 year contract to manage and operate the metropolitan Adelaide water and waste water systems (SAWater, 2004) (United Water, 2008). These are both what are termed "concessional" arrangements[12] whereby a private operator acquires the exclusive operating rights, but not the ownership. In 1998, the ACT Liberal Government tried to privatise the electricity, water and sewage services, but was stopped by the Territory Legislative Assembly[13] (Quiggin, Saddler, Neutze, Hamilton, and Turton, 1998, Woodward et al., 2002, p. 425). In another arrangement in New South Wales, the French company Veolia has a 25 year contract which started in 1994 to construct, finance and operate two urban water plants located in Illawara and Woronora near Sydney (Veolia, 2007). In 2000, the Victorian Labor Government has also contracted a Veolia owned company, Bendigo Water Services, for 25 years to treat and distribute drinking water to residents of the greater Bendigo area and construct three advanced water treatment plants (Birnbauer, 2007, p. 14). In 2007, the Victorian Government also announced that a desalination plant to be built at Wonthaggi would be constructed as a public private partnership (Wilson, 2007).

Some neo-liberals believe that these changes have not gone far enough. Accordingly, there has been a sustained call for private, rather than public, ownership and for more deliberate attempts to introduce competition from private operators through either third party access regimes or competitive tenders for new sources of supply. For example, Hargreaves writing in the journal of the Centre for Independent Studies which is a free-market think-tank, advocates "a national regime for third-party access: a system by which the owners of water and waste water pipelines carry water at regulated prices for all comers"[14] (Hargreaves, 2008, p. 26). One of the earliest statements in Australasia calling for full privatisation of water was made by the Executive Director of the New Zealand Business Roundtable in 1998. He stated that "Privatisation locks in the gains from corporatisation which are always at risk with political control and adds to them" (Kerr, 1998). Kerr cites the World Bank for support stating that "As the World Bank has put it, ownership matters, and private ownership is a significant determinant of economic performance in established market economies"(Kerr, 1998, p. 8). This argument that corporatisation is inferior to privatisation rests on the idea that

---

12 The term 'concession' is redolent of various empires and the grants of territory and trade in China, Africa, the Middle East and other parts of the globe. These arrangements meant that the writ of the colonial power operated within the national territory of the colonised.

13 According to Quiggin et al. (1998), the ACT Government proposed to privatise ACTEW through the sale of electricity and to issue a single 50 year lease for the operation and management of water and sewerage assets. The arrangement for ACT Water seems very similar to that adopted in South Australia.

14 To date, the key case on third party access is that of Services Sydney which obtained access rights to the Sydney Water Corporation system in order to extract and treat sewage. See Gray and Gardner, 2008 and Marsden Jacob Associates, 2005.

share trading and the financial markets will provide a discipline over private corporations that cannot be applied to government corporations. Put plainly, private business is said to face three risks that do not apply to governments. One is going bankrupt, a second is a falling share price and the third is a takeover. These financial disciplines apply irrespective of the pattern and level of legislative controls and economic regulators.

In 2006, the Business Council of Australia called for "Governments to let water markets work like they do for all other products, including for other essentials like electricity and food. Let prices settle where they will once consumers decide how much they wish to consume at prices that bring forward various increments of new supply" (Business Council of Australia, 2006, p. 21). As summarised by a journalist, "No water crisis if price is right" (Wahlquist, 2006). The same theme was used by the Hon. Malcolm Turnbull, at the time Commonwealth Minister for Water, who said the report "supported his view that cities could have as much water as they were prepared to pay for it"[15] (Turnbull, 2006a, Wahlquist, 2006). These are strong views arguing for market solutions, albeit within regional water resource plans and the rules governing water trading. They want private ownership as well as water markets. Thus the Business Council of Australia (2006) proposes to "Introduce competition into water supply, and develop effective access regimes to water pipes and other relevant monopoly infrastructure. There is no logical reason why urban water supply should remain a monopoly, just as there was no justification for an electricity supply monopoly" (p. vi). The same point was made by Turnbull. In a February 2006 speech at the Centre for Independent Studies, he said that he intended examining "to what extent private sector involvement in Australia's water industry can be expanded" (Turnbull, 2006b).

The Commonwealth Government published a discussion paper in August 2006 on private sector involvement in water supply, stating "the objective of this discussion paper is to explore the role of private sector in supplying safe and efficient water and waste water services in Australia" (Australian Government, Department of Prime Minister and Cabinet 2006, p. iv). The Productivity Commission reported on urban water reform in 2008 and concluded that "a well-functioning urban water market could provide more timely investment signals, a wider range of innovative supply options, greater choice of products and options for water users and more efficient use of nationwide water resources (Productivity Commission, 2008, p. xiv). Similarly the Allen Consulting Group advocated that creating a water market, facilitating water trading and maximising competition was the best means of overcoming reliance on water restrictions (Allen Consulting Group, 2007, pp. iv, 9-10, 22-25). Other major bodies advocating competition and private sector involvement include the Economic Regulation Authority of Western Australia which proposes open competition in providing bulk water (Economic Regulation Authority, 2007).

In what we think is a curious choice of phrase, the Productivity Commission suggests that market reform will mean that "water use would be 'democratised'" (Productivity Commission, 2008, p. xxviii) as users would make their own decisions and negotiate with

---

[15] Note this seemed to be about how a higher price driven by consumer demand would make new water sources such as desalination viable. Although it could easily be construed as buying water from distant sources delivered by pipeline or even bulk tankers. One of the associated issues is the extent to which urban water systems should be able to participate in and buy water from rural irrigation schemes. On the latter point, see Productivity Commission (2008, pp. 59-84). Also Marsden Jacobs (2006) makes the point "that purchase of irrigation entitlements may be among the few alternative water supplies available" for major inland cities (ES iii).

suppliers about price and water security. Phrasing aside, there is little new in the evocation of markets and seemingly little need to explain or justify how the market will weave its magic, to use President Reagan's well-known expression. Perhaps the epitome of this is Robinson's bald advice on what should happen to water in Scotland. He says there are two elements to efficiency - first, the market for the industry's products should be competitive and second, companies in that industry should be held privately (Robinson, 2006, p. 201). This chapter is in a collection by the Sustainable Development Network whose credo is that "most of the problems of the poor world result not from the actions of those in the rich world but from the adoption of unsustainable policies by third-world and first-world governments" (Sustainable Development Network, 2006).

THE SOUTH EAST QUEENSLAND WATER GRID

There was a water crisis in Brisbane (capital of Queensland) for the whole of 2006 and 2007 and for some years previously in certain South-East Queensland (SEQ) cities and towns. During the most severe drought for 150 years (Department of Natural Resources and Water, 2007, p. 1), combined major dam levels were below 20% from April to December 2007 and very tight restrictions were imposed on domestic and industrial water use (SEQWater, 2007). The State Government responded with an extensive programme of new dams, desalination, waste water recycling and a 'water grid' in order to 'drought-proof South-East Queensland' (Osborne, 2006)[16] or in more prosaic terms "to secure the essential water supply needs of the region" (Office of the Queensland Parliamentary Counsel, 2007, p. 2). The existing and new water storages and desalination and recycling plants for completion between 2008 and 2010 will be linked by a network of new pipelines, an undertaking that was called the SEQ Water Grid (Queensland Water Commission, 2007b, p. 2).

The government response is, on one hand, an engineering one reflecting the traditional approach taken since colonisation. However, there is a second strand of what the Queensland Water Commission called "structural and regulatory reforms" including improved, simplified business structures and enhanced economic regulation and pricing (Queensland Water Commission, 2007a, pp. 36, 84, 102). In the words of the then Premier, Peter Beattie, "The present system of water delivery is unwieldy, cumbersome and long overdue for reform. There is no point having a world-class water grid without matching world-class management arrangements" (Queensland Parliament, 2007, p. 1694). The Queensland Water Commission was asked by the State Government to report on possible reforms to the SEQ water industry (Queensland Water Commission, 2007a, p. 16). In its report, which drew on advice from PricewaterhouseCoopers, the Water Commission argued that the current system had problems of financing infrastructure and confusion was evident between the roles and responsibilities between local and state government (Queensland Water Commission, 2007a, pp. 1, 13). These issues were to be met by transferring control of all bulk water and the operation of the system to new corporations owned by the state government-owned.

---

[16] The phrase 'drought-proof' has been used in numerous media releases and reports; perhaps the seminal one was the announcement on 8 August, 2006, of an agreement between the then Premier (Mr Beattie) and the Mayors of the 18 South East Queensland Councils to declare the construction works as emergency measures under Water Regulations (Osbourne, 2006).

As well as these issues which are essentially administrative, the Commission also urged institutional reforms to promote competition and to make the production and price of water subject to market forces (Queensland Water Commission, 2007a, pp. ix, 36). The critical features of the 'Enhanced Water Grid Manager Model' promoted by the Commission are the 'vertical' separation of bulk supply, mains transmission, local reticulation and retailing into distinct business entities. A stand-alone state government corporation will operate the bulk transport of water and a single company owned by all South-East Queensland councils will operate local reticulation throughout the region. An independent public agency termed the Water Grid Manager will oversee the sharing of water across the region and the making of contracts between the bulk supply, transport and the retail businesses (Queensland Water Commission, 2007a, p. x).

The Water Commission proposed competition in bulk supply and at retail level. Three competing bulk water corporations were proposed; two of these would each have half of the natural catchments in the region, with the third operating desalination and water recycling (Queensland Water Commission, 2007a. pp. x, xiii). Each of the corporations would be able to use the grid to transport water anywhere across the region. The Commission proposed three retail companies and strongly recommended the introduction of a contestable market with these retailers competing with each other across the entire SEQ region. It also proposed the entry of other utility providers such as electricity or communications retailers (Queensland Water Commission, 2007a, pp. 30-31)[17]. The Commission envisaged that one of the easiest ways of introducing competition was to allow retailers to supply water across the region. They might also offer different 'products' - for example, pricing water according to the level of security of supply or different prices for dam, desalinated and recycled water.

The final model adopted by the State Government in September 2007 followed the recommendations of the Water Commission with two changes. There are two, instead of three, bulk entities with one business operating all the natural water storages and sources across the region and the other operating the 'manufactured' water from desalination and recycling. There are ten retail businesses each of which will operate exclusively within a local Council area[18] for a transition period (Council of Mayors, 2007). The eventual extent of competition in the water grid will be determined in 2009-2011 and, in the meantime, the Water Grid Manager will determine water prices. After the transition period, water prices will be set by and in market competition, subject to an overall regulator (Queensland Parliament, 2007, pp. 1693-1694). However, the Government has expressly and repeatedly stated that the system would remain under the management and ownership of the State Government and would not be privatised[19] (Bligh, 2007; Roberts, 2007, p. 12; The Courier Mail, 2007, p. 8).

Although final decisions are to be made, the SEQ Water Grid creates a system in which third party access and full competition would be relatively straightforward given the necessary administrative and political will. The reticulation system allows water to be moved

---

[17] The Commission specifically recommended that over the first two years of the operation of the grid (2009-2011), there should a detailed evaluation of retail competition and contestability (Queensland Water Commission, 2007a, p. xiii).

[18] This coincides with the changes in the number and boundaries of Councils due to Council amalgamations being implemented by the State Government at the same time.

[19] For example, in a Ministerial Statement of 24 May 2007, (then) Deputy Premier Bligh said "All assets will continue to be owned by government. The Queensland Water Commission report does not recommend, nor is the government considering, any privatisation of any part of the water asset system" (Queensland Parliament, 2007, p. 1695).

throughout the region and the separate operation of the mains network is very similar to the transmission authority in electricity. A schema for competition in urban water using common carriage was described by Webb and Ehrhardt in 1999 in a World Bank Forum (Webb and Ehrhardt, 1998, pp. 1-3). This involves “several water utilities that compete for customers using a single set of pipes” (Webb and Ehrhardt, 1998, p. 3). The proposal is strikingly similar to the Water Grid proposed by the Water Commission and both have competition between bulk suppliers and a wholesale market between competing retailers and bulk suppliers[20]. Moreover, the Webb and Ehrhardt schema appears to be an influential source[21]. It is referenced by many of the advocates of water privatisation including Schier and in the Federal Government Discussion Paper on Private Sector Involvement.

Webb and Ehrhardt (1998) state that common carriage “has been successfully used to introduce competition in gas and electricity in many countries” (p. 3). The same approach is supported by Cowan who writes that “in the absence of national or even regional water grids there is little prospect for competition via common carriage” (Cowan, 2005, p. 356). Webb and Ehrhardt (1998) conclude that while the conditions for creating competition in water are ‘rare’, “it is a good idea to build in the potential for competition rather than locking in monopolies for thirty years or so” (p. 7). Whether advertently or inadvertently, the Queensland Government has built in the potential for competition. They are congratulated for this by Hargreaves in the Centre of Independent Studies Journal, writing that “The Queensland government is to be commended for implementing these reforms, which put the cautious, if not pedestrian, pace of reforms in other jurisdictions into the shade” (Hargreaves, 2008, p. 6).

IS THERE A PUBLIC FUTURE?

The question of the privatisation of the South East Queensland Water Grid is covered in this section. We commence by again acknowledging that the current Queensland Government has undertaken to retain the Grid in public ownership. Nonetheless, there will be at least three pressures for full or part privatisation. The first is about ideas and ideology. Market advocates hold, as a matter of faith, that public ownership is inferior because organisations are protected from the 'ultimate' disciplines of bankruptcy and takeover. Robinson asserts that “competition in capital markets will mean that the company’s shareholders will press its managers to perform efficiently, reinforcing the pressures from competitive product markets” (Robinson, 2006, p. 201). Arguments that corporatisation is inferior to privatisation and that efficiency gains remain to be made will continue from those who believe in the superior power of share-markets. This combination of business organisations, private sector think-tanks and government economic policy agencies has been highly successful in developing and maintaining the momentum for privatisation of other corporatised utilities especially electricity (Cahill and Beder, 2005).

[20] There is no reference to Webb and Ehrhardt in the Queensland Water Commission Report nor, as far as can be ascertained, in any supporting documents by their consultants.

[21] Ehrhardt appears to be extremely influential in the promulgation of water reforms. He is Chief Executive of the New Zealand-French ‘strategic advisers’ Castalia, which is evidenced as having had a very large involvement in introducing competition to water systems in many countries and as joint authors of the World Bank Approaches to Private Participation in Water Services (Castalia Strategic Advisors, 2007;World Bank, 2006).

The second pressure for privatisation is likely to come from the very complexity of any water market, especially if futures trading, options and hedging and the like are introduced to order to deal with the uncertainty of variable rainfall. These are usual practices in commodity training and are being adopted in rural water markets[22]. However, complex market dealing involves losses as well as gains. Public corporations or public agencies involved in water trading must be expected, at some time, to incur 'on-market' losses from futures and hedging. Yet, while these are a structural necessity of the market, governments are politically and electorally vulnerable to accusations of 'wasting' public money. At some point, government will decide that the electoral liability of defending actual or potential losses from water market dealings is too high and will make themselves immune to criticism by shifting the entire water business to the private sector.

Pragmatic or opportunistic politics will be the third pressure for privatisation. Put plainly, at some point, a government will be tempted to decide that selling some, or all, of the water system constitutes a sensible method of revenue raising for other public expenditures. This was happening in New South Wales at the time of writing (2008). The (then) Iemma-led ALP Government justified privatising electricity assets by stating "We have to act now to keep the lights on in NSW. That left the Government with two choices. Spend billions of dollars on new baseload power stations or encourage greater private sector involvement in the electricity market and use those billions of dollars on other infrastructure" (New South Wales Government, 2008). The Queensland Government used this logic in 2006 as one of its reasons for selling the retail arms of the electricity corporations. The proceeds from privatisation would fund essential infrastructure including the priorities of two new dams for the South East Queensland water grid. In addition, privatisation would ensure that the public retail arms were not "at a competitive disadvantage against larger, more experienced retailers" in an expanded electricity market (Beattie, 2006; Beattie and Mickel, 2006). This shows how the Queensland Government justified the particular privatisation in terms of spending priorities and ensuring that public corporations should not be in 'risky' markets. These bring together the pressures for privatisation we discussed above and we suggest that it is relatively easy to envisage similar circumstances arising about the Water Grid.

WATER RIGHTS AND PROPERTY

Water privatisation has attracted considerable hostility and criticism and we briefly review some of the concerns. The UN Human Development Report 2006 on global water while not totally opposed to private participation concludes that the overall record of privatisation is "not encouraging" and urges the "need for greater caution, regulation and a commitment to equity in public-private partnerships" (United Nations, 2006, p.10). The cost and quality of water is one concern. Whelan and White (2005) conclude a detailed review of instances of privatisation[23] by commenting "privatisation raises significant questions about democratic accountability, good health, and the transparency of decision-making, regardless

[22] In 2006, the Australian Stock Exchange published a discussion paper on trading in water futures (Chapman, 2006). A comprehensive report on rural water trading was prepared by PriceWaterhouseCoopers for the Federal Government in 2006 (PriceWaterhouseCoopers, 2006).

[23] They use a 'wide' definition of privatisation including public private partnerships and management franchises.

of location" (p. 140). They document how recent incidents including the occurrence of cryptosporidium and giardia in Sydney water, the 'big pong' in Adelaide, outbreaks of cryptosporidiosis in Wisconsin and cholera in South Africa are all consequences of profit seeking by private water operators (Whelan and White, 2005). Similarly, there is a claim that privatisation of water in some parts of England and Wales led to "deterioration of the quality and quantity of drinking water and environmental problems" (Martin, 1996). Increasing water charges are another major issue with, for example, claims that "high charges, poor quality and inefficiency are common daily features with privatised water companies in Britain" (Morgan and Chapman, 1997, in Hukka and Katko, 2003, p. 45). More generally, economic modelling shows that if there was an open market for water in Sydney, residential water prices would need to increase between 62 per cent and 143 per cent to reduce usage by 19 per cent to that of Level 3 water restrictions (O'Dea and Cooper, 2008).

Water privatisation is also attacked on more conceptual grounds. One approach is that privatisation infringes on the notion of water as a human right. Another approach is that privatisation is incompatible with water being a communal resource and part of the 'commons'. Gleick and the Pacific Institute are major proponents of water as a human right. He reviews the development of human rights instruments and concludes that "international human rights laws must be interpreted to include the right to sufficient water, at appropriate quality, to satisfy the explicit right to life and the broader rights to health and well-being"[24] (Gleick, 2007, p. 2). Put another way, water is a necessity to which all are entitled and lack of adequate water means social exclusion and material deprivation as measured by the levels relative to the particular society. Moreover, the right to water comes from being a citizen not a 'customer'.

Gleick argues that "meeting a basic water requirement for all humans .....should take precedence over other water management and investment decisions" (Gleick, 2007, p. 3). On one reading, this suggests that human rights could provide an alternative logic for the organisation of urban water supply. This is supported in an earlier statement that "the headlong rush toward private markets has failed to address some of the most important issues and concerns about water. Water has vital social, cultural and ecological roles to play that cannot be protected by purely market forces" (Gleick et al., 2002, cited by McIntosh, 2003, p. 91). However, the problems are not necessarily failures of privatisation and the market. Gleick (2007) recognises "that water should be paid for" (p. 4) and says that subsidies and regulations can ensure that everyone gets water, irrespective of whether the supplier is public or private. In a somewhat similar vein, Whelan and White (2005) conclude that "The provision of a safe water supply requires a transparent and accountable, public process: regardless of whether it is provided by a private company or a public agency"[25] (p. 142).

The second strand of criticism emphasises water as a communal resource. Bakker (2007) finds the human rights argument inadequate because it "does not foreclose private sector management of water supply systems" since "private sector provision is compatible with human rights in most countries around the world" (p. 438). She describes three main features

[24] Gleick (2007) indicates that the amount of water needed to meet basic rights and whether users pay for this varies between jurisdictions, and that this is no different from how other social, cultural and political rights are met.

[25] Against this, Whelan and White say that private ownership means that water is "first and foremost treated as a source of profit" and most of their analysis is about this requires leads to efforts to maximise revenue, minimise costs and maximise surplus value. In turn, these are incompatible with the public interest in a clean and affordable water system. By and large, such political economy is absent from human rights advocacy.

that make water part of the commons (pp. 441-443). One feature is that water cannot be kept private and exclusive because it is a natural 'flow resource' that must be shared with the environment. A second feature is that critical management issues are not easily converted into prices[26] which is the only metric of market allocation. Third, issues of ecology, heritage, landscape and public health entail a high degree of regulation of water and water use that restrict the rights of property owners. Taking the three features together, Bakker concludes that water should be treated as part of the 'commons' rather than a commodity for private trade and ownership. This is a more fundamental critique of privatisation than the human rights approach. The key demand is for the ownership of water to be treated as a "collective property right" which has a different basis to both private and state (public) ownership. As such, opposition to privatisation needs to extend to "reforming rather than abolishing state governance while fostering and sharing alternative local models of resource management" (Bakker, 2007, p. 446). There are elements of this thinking in the approach being taken by the European advocacy group Reclaiming Public Water Network' who seek to combine opposition to privatisation with "an exploration of which models of public water management are desirable alternatives to private or commercialised public models" (Hachfeld, Terhorst and Hoedeman, 2009, p. 2).

Earlier we cited the Productivity Commission's statement that turning to the market will mean that "water use would be 'democratised" (Productivity Commission, 2008, p. xxviii). This seems to mean no more than that customers in a market are 'free to choose' how much water to use. This rests on the primacy of private ownership and individual choice. The approach is self-evidently diametrically different from treating water as collective property. Another view of where democracy fits with water reform comes from the Swiss country report to an OECD forum on competition. This says that for Switzerland, "liberalisation of the water sector is generally believed to be contrary to the constitutional principle of sustainable development" (2004, p. 157). Moreover, because water is recognised in the constitution as a natural resource, "strategic decisions in water supply and regulation should be kept under public and democratic control" (2004, p. 157). Indeed, according to the chair of the conference, "Switzerland saw democracy at stake in the process of corporatisation of water providers" (Heimler, in DAFFE/OECD, 2004, p. 189). Very similar views are made about local control and management of water in Finland; "The private organisation does not guarantee democratic decision making, control, transparency and legal protection of citizens in the same way as the local administration would" (Rasinmäki, 1997, p. 391 cited and translated by Hukka and Katko, 2003). We think that these are two examples of widespread awareness that privatisation reduces public and political controls over water (and indeed other services and utilities). Perhaps more than anything else, this is the central objection to turning water into a private commodity.

[26] For example, water prices can effectively allocate water between different uses, such as types of irrigation, manufacturing or households. Such water has, or can be assumed to have, an exchange value. However, water for environmental flows for maintaining wetlands, vulnerable ecosystems or scenic water features is much more difficult to price, since it has a use and/or symbolic value, but not an exchange one.

Conclusion

The South East Queensland Water Grid is an engineering solution to water shortages. The proposals for new dams and for recycling sewage into potable supplies have attracted considerable public debate and dissent. We believe that we have shown that the "world-class management arrangements", promised by the Premier and set up for eventual privatisation, make the system one of the first fully competitive markets for urban water in the world. These two features, unlike recycling and dams, have hardly been debated. In consequence, we think that there could be substantial opposition, if or when a government proposes to sell off the system. The opposition will be especially strong, if it is based around the idea of water as collective property. Rather than leading the world, we conclude that a supposedly progressive government has missed a very real opportunity to demonstrate how to move beyond old style engineering and old style economics.

References

Allen Consulting Group. (2007). *Saying goodbye to water restrictions in Australia's cities: Key priorities for achieving water security.* Retrieved July 2, 2008 from http://www.infrastructure.org.au/research/pdf/reports/IPA_Water_Restrictions_2007.pdf

Australian Government, Department of Prime Minister and Cabinet. (2006). *A Discussion Paper on the Role of the Private Sector in the Supply of Water and Wastewater Services.* Retrieved September 2 2007 from http://www.environment.gov.au/water/publications/urban/pubs/water-discussion-paper.pdf.

Australian Government, Department of the Environment, Water, Heritage and the Arts (n.d.).The Council of Australian Governments' Water Reform Framework: Extracts from Council of Australian Governments: Hobart, 25 February 1994 Communiqué 1994. Retrieved September 20, 2008 from http://www.environment.gov.au/water/publications/action/policyframework.html

Bakker, K. (2007). The "Commons" Versus the "Commodity": Alter-globalization, Anti-privatization and the Human Right to Water in the Global South, *Antipode 39,* 3, 430-55.

Beattie, P. (2006). *Billion Dollar Fund to Tackle Climate Change*: Media Statement by Premier of Queensland 26th April 2006. Retrieved from http://statements.cabinet.qld.gov.au/MMS/StatementDisplaySingle.aspx?id=45787

Beattie, P. and Mickel, J. (2006). *Sale to Help Drive Energy Market Reform Process*: Media Statement by Premier of Queensland and Minister for Energy and Aboriginal and Torres Strait Islander Policy, 26th April 2006. Retrieved from http://statements.cabinet.qld.gov.au/MMS/StatementDisplaySingle.aspx?id=45786

Birnbauer, W. (2007, 4 November). Overseas companies line up to give Locals a service. *The Age,* p. 14. Retrieved from http://www.theage.com.au/news/national/overseas-companies-line-up-to-give-locals-a-service/2007/11/03/1193619205929.html

Bligh, A. (2007). Ministerial Statement: Water Grid. Hansard May 24, 2007. Retrieved 26 June, 2008 from http://parlinfo.parliament.qld.gov.au/isysquery/efa7fbbc-cf7b-479f-b66f-fba93a67ab57/4/doc/Anna%20Bligh%20spk%20South%20Brisbane%202007_05_24_32

.pdf#xml=http://parlinfo.parliament.qld.gov.au/isysquery/efa7fbbc-cf7b-479f-b66f-fba93a67ab57/4/hilite/

Business Council of Australia. (2006). *Water under pressure: Australia's man-made water scarcity and how to fix it.* Retrieved June 2, 2007 from http://www.bca.com.au/Content.aspx?ContentID=100665.

Cahill, D. and Beder, S. (2005). Regulating the power shift: the state, capital and electricity privatisation in Australia, *Journal of Australian Political Economy 55*, 5-22.

Castalia Strategic Advisors. (2007). *Home.* Retrieved July 8, 2007 from http://www.castalia.fr/.

Chapman, K. (2006). *Establishing a futures market in water availability risk, Australian Stock Exchange Limited.* Retrieved from http://www.sfe.com.au/site/forms/discussionpaper_waterfutures.pdf

Council of Australian Governments. (2004). *Intergovernmental agreement on a national water initiative between the Commonwealth of Australia and the governments of New South Wales, Victoria, Queensland, South Australia, the Australian capital territory and the Northern Territory.* Retrieved 20 September 2008 from www.coag.gov.au/meetings/250604/iga_national_water_initiative.pdf

Council of Australian Governments (1994). *Communiqué*, 25 February. Retrieved 20 September, 2008 from www.coag.gov.au/meetings/250294/index.html

Council of Mayors. (2007). *Council feedback shapes new water system.* Retrieved September 19, 2007 from http://www.councilofmayorsseq.qld.gov.au/content/files/Deputy%20Premier%20- %20Final%20Reform%20Model%20-%204%20Sept%2007.pdf.

Cowan, S. (2005). Competition in Regulated Industries: Some Reflections. *Agenda*, *12*, (4), 351 362.

Department of Infrastructure and Planning. (2008). Population projections: Queensland's future population 2008. Retrieved 12 September, 2008 from http://www.dip.qld.gov.au/population-forecasting/population-projections.html

Department of Natural Resources and Water. (2007). *The South-East Queensland drought to 2007.* Retrieved June 6, 2007 from, http://www.climatechange.qld.gov.au/forecasts/pdfs/seq_drought_2007.pdf.

DFFEA (Directorate for Financial, Fiscal and Enterprise Affairs), OECD. (2004). Competition and Regulation in the Water Sector: Policy Round table: Paris, OECD

Economic Regulation Authority. (2007). *Inquiry on Competition in the Water and Wastewater Services Sector Draft Report Dec 2007.* Retrieved 6 January, 2008 from, http://www.era.wa.gov.au/cproot/6232/2/20071203%20Draft%20Report%20-%20Inquiry%20into%20Competition%20in%20the%20Water%20and%20Wastewater%20Services%20Sector.pdf

Energy Supply Association of Australia. (2009). *Australia's Electricity Markets* http://www.esaa.com.au/electricity_markets.html, Retrieved 21 Feb 2009

Gleick, P.H. (2007).The Human Right to Water. Retrieved 12 June, 2008 from The Pacific Institute www.pacinst.org/reports/human_right_may_07.pdf

Gleick, P. H., Wolff, G., Chalecki, E.L., and Reyes, R. (2002). *The New Economy of Water: The Risks and Benefits of Globalization and Privatization of Fresh Water.* Retrieved August 4 2007 from http://www.pacinst.org/reports/new_economy_of_water/new_economy_of_water.pdf

Godden, L. (2008). Property in urban water: Private rights and public governance. In Troy, P. (Ed.) (2008). *Troubled Waters: Confronting the Water Crisis in Australia's Cities* (Chapter 8). Canberra: ANU E-Press.

Gray, J. and Gardener, A. (2008). Exploiting the unspeakable: Third-party access to sewage and public-sector sewerage infrastructure. In Troy, P. (Ed.). (2008). *Troubled Waters: Confronting the Water Crisis in Australia's Cities* (Chapter 7). Canberra: ANU E-Press.

Hachfeld, D., Terhorst, P. and Hoedeman, O. (2009). *Progressive Public Water Management in Europe: In search of exemplary cases*. Reclaiming Public Water discussion paper: Transnational Institute and Corporate Europe. Retrieved 3 March, 2009 from http://www.waterjustice.org/uploads/attachments/Progressive%20public%20water%20management%20in%20Europe.pdf

Hargreaves, S. (2008). One Good Big Idea for Water Reform. *Policy,* 24, No.1, 26-31. Retrieved June 16, 2008, from http://www.cis.org.au/POLICY/winter08/links/hargreaves_winter08.pdf

Hukka, J. J and Katko, T. S. (2003). Water Privatisation Revisited; Panacea or Pancake? *IRC Occasional Paper Series, Series No. 33-E.* Retrieved June 3, 2007, from http://www.servicesforall.org/html/WaterPolicy/Hukka%20and%20KatkoPanacea%20or%20Pancake.pdf.

Independent Pricing and Regulatory Tribunal of NSW (2008). *Water Industry Competition Act 2006; The access regime – an overview*, WICA Access Fact Sheet No 1. Retrieved 3 March 2009 from www.ipart.nsw.gov.au

Kerr, R. (1998). Corporatisation and Privatisation of Water Supply', paper presented at AIC Conferences 6th Annual NZ Water Summit, 26 February, 1998. Retrieved 22 August, 2007 from www.nzbr.org.nz/documents/speeches/speeches 98/corporatisation-privatisation- of-water-supply.pdf

Lobina, E. and Hall, D. (2001). *UK Water privatisation - a briefing*, Public Services International Research Unit, University of Greenwich. Retrieved 11 October, 2007 from http://www.psiru.org/publicationsindex.asp

McIntosh, A.C. (2003). *Asian Water Supplies: Reaching the Urban Poor. A Guide and Sourcebook on Urban Water Supplies in Asia for Governments, Utilities, Consultants, Development Agencies, and Nongovernment Organizations*: Bangkok, Asian Development Bank and International Water Association.

Marsden Jacob Associates. (2005). *Third Party Access in Water and Sewerage Infrastructure: Implications for Australia*, Canberra, Department of Agriculture, Fisheries and Forestry.

Marsden Jacob Associates. (2006). *Securing Australia's Urban Water Supplies: Opportunities and Impediments. A discussion paper prepared for the Department of the Prime Minister and Cabinet*, Canberra, Department of Prime Minister and Cabinet.

Martin B. (1996). From the many to the few: Privatisation and globalisation. *The Ecologist 26*, 4, 145-155.

Morgan S.P. and Chapman J.I. (1997). *Issues surrounding the privatisation of public service.* University of Southern California, School of Public Administration.

National Water Commission. (2004). Intergovernmental agreement on a National Water Initiative. Retrieved May 4, 2007, from http://www.nwc.gov.au/nwi/docs/iga_national_water_initiative.pdf.

New South Wales Government. (2008). *NSW Future Electricity Supplies.* Retrieved 28 June, 2008 from http://www.nsw.gov.au/energy/

O'Dea, G. and Cooper, J. (2008). *Water scarcity: Does it exist and can price help solve the problem?* Water -Working paper January 2008, Independent Pricing and Regulatory Tribunal of New South Wales retrieved 3 March, 2009 from http://www.ipart.nsw.gov.au/files/Water%20scarcity%20-%20Does%20it%20exist%20and%20can%20price%20help%20solve%20the%20problem%20-%20Working%20Paper%20January% 202008.PDF

Office of the Queensland Parliamentary Counsel. (2007). *Water Amendment Regulation (No. 6), 2006*. Retrieved August 6, 2007, from http://www.legislation.qld.gov.au/LEGISLTN/SLS/2006/06SL202.pdf.

OECD (Organisation for Economic Co-operation and Development). (2004). *Competition and Regulation in the Water Sector*. Retrieved September 16, 2007, from http://www.oecd.org/dataoecd/44/1/33691325.pdf

Osborne, P. (August 8, 2006). Beattie and Mayors sign deal. *Courier-Mail*. Retrieved 18 July, 2007, from http://www.news.com.au/couriermail/story/0,23739,20048904-3102,00.html.

PriceWaterhouse Coopers. (2006). National Water Initiative Water trading study Final Report, Australian Government: Department of the Prime Minister and Cabinet. Retrieved 14 July, 2008 from http://www.environment.gov.au/water/publications/action/nwi-wts-report.html

Productivity Commission. (2008). *Towards Urban Water Reform: a discussion paper*. Retrieved March 2 2008 from http://www.pc.gov.au/data/assets/pdf_file/0003/78150/urbanwaterreform.pdf

Public, Private Industry Advisory Facility (PPIAF) and the World Bank (WB). (2006). *Approaches to private sector participation in water services – a toolkit:* Washington DC. Retrieved 21 June, 2008 from http://rru.worldbank.org/Documents/Toolkits/Water/Water_Full.pdf/

Queensland Council of Unions. (2007). SEQ Urban Water Arrangements Reform: Workforce Framework 2007. Retrieved December 12 2007, from http://www.qcu.asn.au/water_reform_framework_12dec07_final_final__3_.pdf.

Queensland Parliament. (May 24, 2007). *Hansard*. 1692-1694. Retrieved 2 June, 2007, from http://parlinfo.parliament.qld.gov.au/isysquery/99a9cfc0-f8ea-4656-addd82330bd423ab/2/doc/2007_05_24_WEEKLY.pdf#xml=http://parlinfo.parliament.qld.gov.au/isysquer/99a9cfc0-f8ea-4656-addd-82330bd423ab/2/hilite/

Queensland Water Commission. (2007a). *Our water: Urban water supply arrangements in South East Queensland, final report*. Retrieved June 24, 2007, from http://www.qwc.qld.gov.au/myfiles/uploads/institutional%20arrangements/Urban_Water_Supply_Arrangements_in_SEQ.pdf.

Queensland Water Commission. (2007b). *Fact sheet 4- SEQ Water Supply Projects*. Retrieved August 14 2007 from http://www.qwc.qld.gov.au/myfiles/uploads/fact sheets/SEQ Water Supply Projects.pdf.

Queensland Water Commission. (2007c). *Target 140*. Retrieved October 12, 2007 from, http://www.qwc.qld.gov.au/Water+use+measurement.

Quiggin, J. (1996). *Great expectations: microeconomic reform and Australia*. St. Leonards, N.S.W: Allen and Unwin.

Quiggin, J., Saddler, H., Neutze, M., Hamilton, C., and Turton, H., (1998). The Privatisation of ACTEW: The fiscal, efficiency and service quality implications of the proposed sale of

ACT Electricity and Water. The Australia Institute Discussion Paper No 20, Canberra: The Australia Institute

Rasinmäki J. (1997). Yksityistäminen kunnallishallinnossa. Hallinto-oikeudellinen tutkimus yksityistämisen oikeudellisista taustoista, edellytyksistä, rajoituksista, muodoista ja vaikutuksista. The privatisation of the administration of local government: A judicial study of the background, provisions, restrictions, forms and effects of privatisation. Kauppakaari Oy: Helsinki. (Original in Finnish, summary in English)

Reserve Bank of Australia. (1997). Privatisation in Australia. *Reserve Bank of Australia Bulletin,* December 1997.

Roberts, G. (2007, January 17). Warning on heavy cost rises for water. *The Australian,* p. 12.

Robinson, C. (2006). How not to reorganise an industry: Privatisation, liberalisation and Scottish water, In K. Okonski (Ed.), *The Water Revolution: Practical Solutions to Water Scarcity* (pp. 201-226). London: International Policy Press.

SAWater. (2004). *Our Partners, Government of South Australia.* Retrieved August 11 2007 from http://www.sawater.com.au/SAWater/AboutUs/AboutSAWater/Our+Partners.htm.

SEQWater. (2007). *Current Dam Levels.* Retrieved January 2 2008 from http://www.seqwater.com.au/content/standard.asp?name=DamOperationsAndMaintenance.

Sustainable Development Network. (2006). About Us. Retrieved June 20, 2008 from www http://www.sdnetwork.net/main/page.php?page_id=2.

Swier, G. (2007). Is there a role for Urban Water Trading? *Network: a journal of the Utility Regulators' Forum,* 26, 2-9.

The Council of Australian Governments. (1998*). National Competition Policy Agreements: (Second Edition),* Retrieved August 11 2007, from, http://www.ncc.gov.au/pdf/PIAg-001.pdf.

The Council of Australian Governments. (2004). *Water Reform Framework.* Retrieved May 2, 2007, from http://www.environment.gov.au/water/publications/action/pubs/policyframework.pdf.

The Courier Mail. (2007, January 16). State denies privatisation, p. 8. Retrieved January 22, 2007 from, http://www.news.com.au/couriermail/

Turnbull, M. (2006a). *Why should our cities be short of water? 'Is urban water a finite resource?* Speech to the Committee for the Economic Development of Australia. Retrieved June 8 2007, from www.malcolmturnbull.com.au/Pages/Article.aspx?ID=495.

Turnbull, M. (2006b). *Malcolm Turnbull discusses Water Policy at CIS.* Retrieved March 4, 2007, from www.malcolmturnbull.com.au/Pages/Article.aspx?ID=379.

United Nations. (2006). *Human Development Report 2006: 'Beyond scarcity: Power, poverty and the global water crisis.* Retrieved October 18, 2007, from http://hdr.undp.org/en/reports/global/hdr2006/chapters/

United Water. (2008). *Veolia Water in Australia and New Zealand*, Website http://www.veoliawater.com.au/en/veoliawater/uwi/ Retrieved Dec 12, 2008.

Veolia. (2007). *Veolia Water in Australia.* Retrieved 26 July, 2007, from http://www.veoliawater.com/veoliawater/keyfigures/locations/asia- oceania/australia.htm

Wahlquist, A. (2006, September 18). No water crisis if price is right, *The Australian*, Retrieved 22 September, 2007, from http://www.theaustralian.news.com.au/story/0,25197,20429015-2702,00.html.

Webb, M, and Ehrhardt, D. (1998). Improving Water Services through competition. *Public Policy Journal*, *164*, 1-8. Retrieved June 2, 2007, from http://rru.worldbank.org/documents/publicpolicyjournal/164webb.pdf.

Whelan, J. and White, R. (2005). Does privatising water make us sick? *Health Sociology Review* 14, 2. Retrieved 3 March 2009 from Health Reference Center Academic database.

Wilson, K. (2007, September 19). $3b private sector cash for Vic desal plant. *ABC News*, Retrieved 30 September, 2007 from http://www.abc.net.au/news/stories/2007/09/19/2037393.htm

Woodward. D., Parkin. A., and Summers. J. (Eds.). (2002). *Government Politics Power and Policy.* Melbourne: Addison Wesley Longman.

Contributed by Geoffrey Frost, Tasmania.

Photo: Civilization and Water Enjoyment.

PART 3:

CAN WE AFFORD TO TAKE CHANCES WITH OUR FUTURE?

Contributed by David Chant, Tasmania.

Photo: Little Creatures are Depending on Us.

In: Meltdown: Climate Change, Natural Disasters…
Editor: Kathryn Gow
ISBN 978-1-60876-153-1

Chapter 18

GLOBALISATION AND CLIMATE CHANGE

Mark J. King
Queensland University of Technology, Australia

ABSTRACT

We have become accustomed to hearing that we live in a globalised world, even though the term "globalisation" has only become popular in recent years. The process of economic integration it describes, however, has been underway for much longer. What is more, the economic underpinnings of globalisation mean that it has both contributed to climate change, and will continue to do so. Alongside the "macro" phenomena of globalisation and climate change, we also find the "micro" phenomena of their actual (or rather anticipated) effects on us as individuals. In this chapter, I explain what globalisation is, in terms of both economics and the exercise of power by transnational corporations and nation states, and attempt to show how the individual experience of potential climate change relates to globalisation. I also make some comments on how developing countries are likely to be affected, and reflect upon the current financial crisis and what it tells us about how we prioritise our globalised economy relative to the global phenomenon of climate change.

Keywords: globalisation, development, climate change, third world, economic effects

INTRODUCTION

I'm looking out my window on a cool clear winter's day at trees blowing in the wind. It's a beautiful day, cold overnight but becoming warm enough to let me go round in shirtsleeves in the middle of the day. In the summer it gets hot and sticky, but that's the way it's been here for as long as records go back. Every now and then we have a year which is exceptionally hot, or wet, or dry, or cold, but most of the time the weather records show that things were more extreme in 1946, say, or some time in the 1890s. So we get used to the pattern of the weather according to the seasons, the variations in our climate that we think of as natural.

> But things are changing. Last year we had a lot of trouble here in Queensland, Australia with water supplies; rainfall had been low for so long that the dams were emptying out. True, our population had been increasing rapidly in the past few decades; true, our land is notorious for its periods of drought and flooding; but now we had the spectre of permanent change, the possibility that this kind of weather might become more common because of global climate change. As usual, after much discussion and debate, the State government developed a strategy and put a series of programs into place. We are building new dams and a desalination plant, establishing more water recycling plants, and constructing a network of pipelines to shift water between dams over thousands of kilometres. This year we have had ample rain, but the ball has been set rolling and concern about future rainfall is still high.
>
> The government wants to meet the needs of industry, keep the tourists coming in, and maintain a good quality of life. On a broader scale the State government is concerned about the future of farming, and has gotten together with other States and our national government to anticipate climate change and put measures in place to deal with it. This is typical of a management-style response to a possible change in the "environment": assessing the current situation, making projections, considering options which should exploit the positive things expected in the future, and blunt or avoid the negative things. This is logical and sensible, but it doesn't really get to the heart of people's fears. What is going to happen when the temperature increases a couple of degrees? Will it mean an end to cool winters and a shift to longer, hotter summers? If the rainfall doesn't increase, can the vegetation cope? Perhaps the trees I can see from my window will all die. Maybe everybody's lawns and gardens will dry up and blow away. With all the extra evaporation, will the dams meet our needs? What does it mean for our quality of life? Will I have to pay more for water, and for food because it costs more to produce? Perhaps industry will find it too difficult to operate here. Perhaps tourists will find it less attractive to come here, especially if oil prices increase the cost of travel. Will some of my neighbours move away, to better prospects elsewhere? What about the people whose houses will be affected by rising water levels? Will the value of my property decrease? Should I move somewhere else too? But where will the good jobs be, where will the best quality of life be? And can I get there, and do I really want to leave anyway?

These are questions which link climate change to the way we make a living, how we think about our security and quality of life, and our ability to adapt our lives when these things are challenged. They are not the only questions we have about climate change and how it affects us, but they are the questions which link climate change to economic issues and hence to our globalised economy. More importantly, the very fact of globalisation means that the economic effects of climate change on individuals will be experienced everywhere, from the poorest of countries through to the richest. The recent global financial crisis illustrates perfectly how global economic ties mean that everyone bears the pain.

What this chapter attempts to do is to show how climate change and globalisation are linked, and to explore how this affects people's concerns for the future. In doing so, it will show how globalisation is linked to power, politics and development, and what this means for people in developing countries who are confronting climate change.

What Is Globalisation?

If you looked at newspapers and magazines twenty years ago, you would rarely come across the word "globalisation". Then, in the 1990s, "globalisation" began to be talked about, though often people didn't really know what it meant. It was as if someone used the term "worldisation" – it obviously had something to do with "getting things together" on a world scale, but what things, and with what significance?

The year 2000 brought an explosion of books, articles and official pronouncements about globalisation which explored a range of ideas about what globalisation was (see, for example, Beck, 2000; Bonefeld and Psychopedis, 2000; Burnham, 2000; Coyle, 2000; Derluguian and Goldfrank, 2000; Giddens, 2000; Grunberg, 2000; Micklethwait and Wooldridge, 2000; Mittelman, 2000; Rupert, 2000; Scholte, 2000; Suter, 2000). Some considered the concept as a narrow economic term referring to the structure of nations' economies (perhaps reflecting the establishment in 1995 of the World Trade Organization to oversee global trade), while others saw it as either affecting, or indeed constituting, almost every social change occurring around the world. Some construed it in negative terms as a hostile bid by rich countries to exploit developing countries as markets and sources of cheap production, and yet others glimpsed in it the promise of a better future, both economically and socially.

The elements of these divergent ideas about globalisation remain (e.g. Goldin, 2006; Spaargaren and Mol, 2008; Sklair, 2006), although the discourse has crystallised around some key notions. Sklair (2006), while pointing out the lack of an agreed definition and the multiple ways the prodigious literature can be classified, settled on four approaches which I consider really constitute only three that are intimately interconnected: two concern the economic integration of the world (with the underlying ideology this entails), one is about cultural homogenisation (and can be viewed as a consequence of economic integration), and the last addresses movements to integrate civil society globally (which can in part be understood as a reaction to economic integration).

As the parenthetical comments in the previous sentence imply, and the more detailed discussion below elaborates, I conceptualise globalisation as basically an economic phenomenon. This accords with academic approaches (e.g., Spaargaren and Mol, 2008) and with the evolving popular understanding of globalisation in terms of the expansion or creation of markets where any country that wants to buy and sell can be involved. These "markets" are not just for goods but for labour, production, technology and services, such that a manufacturer can choose to locate production in a country where the labour is cheap, or where there are technological synergies, etcetera. This idea of global markets has become so familiar to us that it is hard to remember the days when Australia automatically sold all its export butter to Britain, guaranteeing a closed market with no competition, but where the buyer had a lot of power. At the same time, this kind of closed market still exists in many forms, as will be discussed below.

Although there are both positive and negative popular attitudes to globalisation, economists, politicians and captains of industry have generally endorsed it as a desirable evolution of our economies. The philosophy underpinning this endorsement is the same as that which provides the rhetoric for free trade agreements and reductions in tariffs: an economic approach which draws on the basic arguments of Adam Smith (1723-1790, author of *The Wealth of Nations*), the tenets of neoclassical economics and the ideology of neoliberalism. Adam Smith believed that economic development would involve ever more detailed division of labour and consequent specialisation (Nolan, 2006); while he saw this as an inevitable process within countries, the same argument applies across nations. Neoclassical economics provided the simple analytical tools which claim to demonstrate how such specialisation leads to more economically efficient allocation of resources and hence greater overall utility (Cohen, 2006).

The general thrust of the argument is that it is more efficient for a country to produce only those goods and services which it can produce more cheaply than other countries, and then to use the income to purchase the other goods and services needed from countries which can produce them more cheaply than anyone else. The same analytical approach can be used to demonstrate that interventions in economic processes (government actions such as taxation, tariffs and the like) interfere with the ability of these processes to provide benefits to economic entities, be they individuals, companies or nations. One of the deceptively simple implications is that overall benefits are maximised by minimising these interventions, which supports the doctrine of free trade and the operation of "the market" as the most rational way of setting prices for goods and services based on supply and demand.

There is a paradox in the way governments approach this doctrine, however. Anyone who has observed the debates and deliberations on free trade agreements will have observed that the nations which call most stridently for free trade are also among those most likely to intervene to protect particular industries of their own, in direct contradiction of free trade principles (Goldin, 2006; Bolton, 2007; Ponting, 2007). This may mean satisfying interest groups who want to keep certain industries going, maintaining industries which are inefficient, but have a role in securing the continued existence of the government (e.g. defence industries), and/or strategically maintaining certain industries to avoid loss of bargaining power or potential disaster, if the most efficient producer cannot meet demand. Al Gore (2007) gives specific examples of US Government interventions of this nature to protect industries considered to be directly responsible for climate change (through greenhouse gas emissions) from decreases in profits which would ensue, if they were forced to decrease their emissions. I will return to this issue of the inconsistency between government rhetoric and action on climate change later.

Neoliberalism is an ideology which combines neoclassical faith in the operation of the market with an agenda of political and civil change (Beck, 2000; Stein, 2006). The point has been made that the neoliberal view of economics has become so embedded in the institutions, discussions and politics of our lives that it has become invisible; that is, most people believe there is no other natural or sensible way for economies to function successfully (Coyle, 2000). International agencies such as the World Bank and the International Monetary Fund have been encouraging (or forcing: Bolton, 2007) poorer countries to participate in globalisation by tying aid and loans to progress towards the neoliberal recipe, with results that often benefit richer countries more than poorer ones (Bolton, 2007). It is widely agreed that the dominant conception of economics today, and therefore the conceptual base of globalisation, is neoliberalism (Spaargaren and Mol, 2008).

Adam Smith's views of economics, and even the perspective of neoclassical economics, did not directly encompass the global; they had more to do with how nations can prosper. Smith's work came at a time when the "nation state" was on the rise: the linking of geographic boundaries, demographics and government in a way that highlighted the sovereignty of the state, rather than the sovereign. Nation states are the main political players on the world stage, and have a big influence on economics, but transnational corporations have become larger, wealthier and correspondingly more powerful. By 2000, the Institute for Policy Studies reported that just over half the largest economic entities in the world were corporations rather than countries (cited by the Corporate Accountability Project, http://www.corporations.org/system/top100.html). These corporations have a number of advantages that nation states do not. They are not bound by geographic limits, but can operate anywhere; their responsibilities to their employees are much less than the responsibilities of governments to their citizens; and they have a simple profit-making objective, while governments must pursue multiple goals. Corporations can also become so large that governments feel obliged to provide support for them when they are suffering financially. This has been well-illustrated by recent events in the US, where both the Bush and Obama governments spent large amounts of money to prevent the collapse of banks and automobile manufacturing companies. Carson and Vaitheeswaran (2008) document a series of similar government interventions in the US over the past few decades, all aimed at keeping large multinational corporations afloat. In 2008-2009, the bailing out of large corporations was repeated around the world, as countries faced financial recession and feared a financial depression.

Corporations do not owe the same such loyalty to governments. They can switch their activities from country to country on the basis of profitability, influence governments to provide better deals for them, and manipulate their accounting to maximise after-tax profits. For example, different countries have different tax rates, and transnational corporations can transfer money internally by having one part of the company buying from another, with the price set so that the profits occur in the lowest-taxing country. Countries also compete with each other for transnational investment by offering special benefits (subsidies, tax breaks, infrastructure support) (Huber, 2008). It has been argued that globalisation has led to the weakening of nations because political power has been replaced by market forces, often to the detriment of the majority (Habermas, 2001; Whitfield, 2001), and that this fact has implications for major environmental issues like climate change (Huber, 2008; Spaargaren and Mol, 2008).

In today's world, globalisation is driven by the activities of transnational corporations, seeking the cheapest ways of producing goods and services for the best profits. Although the recent world financial crisis in 2008-2009 has shaken their confidence somewhat, many developing countries see themselves as gaining from globalisation because it could mean more jobs for their citizens, with correspondingly more taxation revenue, more money in people's pockets to consume goods, and faster economic development. Transnational companies benefit because more countries are competing to become suppliers of raw materials, human resources and skills, and they can seek the best deal. For consumers, especially in Western countries, globalisation delivers more products than ever before, from all over the world (Spaargaren and Mol, 2008), and at prices that are often decreasing in real terms. In my house, our dining room table and chairs come from Vietnam; our washing machine is a prestigious European brand that used to be expensive, but is now manufactured in Thailand and competes easily with the local products; when I seek technical support for problems with my landline (rented from an Australian company), computer hardware and software (usually American) I speak to technicians in India or the Philippines whose skills and knowledge are supplied much more cheaply than Australian or American expertise; and appliances and technology which were once luxury items, or a major household outlay (such as when the first televisions and personal computers came out), have become commonplace with many houses boasting several such items which are now cheaper to replace than to repair.

It is easy to see globalisation in terms of what it delivers for us, in the West, today. In practice, globalisation has historically been a far less benign process than outlined above, and there are many ways in which it does not have a good fit with the ideal of minimal government involvement and the pre-eminent role of the market. Further, as will be explained below, globalisation has contributed to climate change as a by-product of its operation, and can be expected to throw up obstacles to our attempts to address climate change.

Globalisation and Power

While the term "globalisation" is of recent provenance, the phenomena it describes are not. Globalisation is a simple process on a grand scale, and has been taking place for centuries at least (Huber, 2008). One of its features is the exchange of goods and services across political boundaries, and this has been happening for tens of thousands of years. Archaeological evidence shows that obsidian for stone tool-making, stone tools themselves, certain styles of pottery, sea shells and so on have travelled hundreds and even thousands of kilometres across the range of human habitation, at a time when humans themselves did not move beyond fairly stable boundaries (Gross, 1992). Many of the products of Southeast Asia and China were available in the medieval markets of Europe at a time when these parts of the world had a mythical status at best, being shown on biblically correct maps as being located on the other side of the Garden of Eden, populated by fabulous beasts, and near the realm of the legendary Christian monarch Prester John (Boorstin, 1985).

Globalisation is also characterised by the exercise of power to promote and profit from these exchanges. The power of transnational corporations is mentioned above, but a more significant historical influence on globalisation has been the exercise of power by one state over another, reaching its height during the period of European colonisation of Africa, Asia and South America (Howe, 2006; Ponting, 2007).

A theoretical approach which pre-dates the development of "globalisation" as a construct, but which puts its history into perspective (albeit a very particular kind of perspective) is Immanuel Wallerstein's World System Theory (Chase-Dunn, 1989; Hobden and Wyn Jones, 1997; Hopkins and Wallerstein, 1996). At base, World System Theory is an application of Marxist concepts of power, labour and control of production to the world stage, expounding the exercise of economic and political power by countries or blocs in relation to each other, whereas traditional Marxism deals with the exercise of economic and political power within nation states, by those who control the means of production. Sklair's "global systems theory" (Sklair, 2002) takes a similar approach. The main argument of World System Theory is that economic integration has tended to occur, whenever there have been dominant states (or blocs of states) which seek to extend political and economic hegemony beyond their borders. In so doing, they create a larger system of political power and economic distribution. In classical times, the "world" was a smaller place, and "world systems" were correspondingly constrained.

The current world system is considered to have its beginnings in the 16th century, with European exploration, market development and the colonisation of Africa, Asia and the Americas. In the present day, almost every part of the globe is part of the world system, with very few countries (notably North Korea) attempting to maintain a separate existence. The process of achieving this world system, and the processes which maintain interdependency between its component countries, are what we now call globalisation, and it is not uncommon for theory-based discussions of globalisation to allude to it as the more recent expression of the development of the "world system" (Huber, 2008).

An important feature of World System Theory is its classification of countries as being in the core of the system, or on its periphery (Hobden and Wyn Jones, 1997). Core countries are those that dominate political and economic relationships, and there are flows of capital from peripheral to core countries. Where countries are located, in terms of the core and the periphery, can change; Portugal is a notable example of a country which moved from the core (during the 16th and 17th century, when it dominated trade through its naval expeditions and conquests) to the periphery (as one of the poorer countries in Europe, until it became part of the EU). The 20th and 21st centuries have witnessed the rise of peripheral countries such as South Korea, India and China to the point where it has become widely expected that China and India will eventually be the new economic powerhouses of the world. It is worth recalling that both India and China used to be the core countries in their own regions in times past.

In addition to the historical perspective offered by World System Theory, it provides an insight into how power operates in globalisation. As mentioned above, the liberal economic approach taken by Adam Smith and neoclassical economists entailed the diminution of government involvement in the economy on the basis that all government intervention creates inefficiencies in the economy by interfering with the market. This assumption lies behind the rhetoric of right wing (and many centre) politicians around the world, and has been the impetus behind Thatcherism in Great Britain, Reaganomics in the United States, and microeconomic reform and deregulation in other parts of the world. As mentioned, it forms the basis of the rhetoric behind the free trade agenda, which is a good place to focus on how political power affects globalisation.

Australia is a primary producer, obtaining most of its foreign income from mining and agriculture. Although Australia is not a developing country, it shares with many of them the capability to produce agricultural output at lower cost than it can be produced in Europe and the United States, even when shipping is taken into account. Yet both the European Union and the United States have imposed several kinds of barriers (including tariffs and quotas) in order to protect their own agricultural producers. This is by definition an intervention in the market which creates inefficiencies, and it is done for political reasons (Bolton, 2007; Ponting, 2007). Ironically, the United States is at the forefront of efforts by the World Trade Organisation to reduce trade barriers. There are many who interpret this as a hypocritical attempt to "have their cake and eat it too", by maintaining their own barriers against cheaper produce from other countries, while forcing these countries to remove the barriers which have stopped the US from competing in those markets where it is competitive. Sometimes, this is done in a disingenuous way; for example, by removing tariffs which discriminate against foreign producers, but then providing subsidies to local producers so that they can compete (Ponting, 2007). The US is a prominent example, but by no means the only one: the European Union behaves in a similar way for many agricultural products, Australia maintains barriers to protect its sugar industry; Japan spends more on subsidising domestic rice production than the rice produced is worth, and so on (Ponting, 2007).

These are examples of political power operating mostly at the borders, regulating the goods and services that enter the country, but there are many examples of its operation directly on other countries, sometimes in a quite coercive way. Conquest and colonisation are the most obvious of these, but other means were used to achieve the same economic ends. The Opium Wars of the 19th century in China were essentially about Britain using its military might to force China to open its markets in order to facilitate the lucrative opium trade (Fay, 1975). In the 20th century, both Britain and Australia used their political power to force the inhabitants of Nauru and Ocean Island to allow these islands to be rendered almost uninhabitable through the mining of phosphate, with only a belated and thoroughly inadequate gesture at compensation (Ponting, 2007). There are many other examples over the past few centuries, encompassing at a minimum all European countries, North America, Australia, Japan and China. Over the longer spans of history, it seems that almost any country in a position of power has used it to force unfair economic terms on other countries.

What I wanted to emphasise above is that not only have globalising processes been operating for centuries, they have never been confined to simple economic principles; the exercise of political power has always featured in them. As an additional note, it is worth mentioning that transnational trading organizations, such as the Dutch East Indies Company, have been integral to globalisation over the past few hundred years, and their operations have been intimately bound up with the exercise of political power (Ponting, 2007). Time and again, in various parts of the world, military forces have been deployed to protect the operations of transnational companies where the government (sometimes colonial, sometimes not) benefits from their operations, but the populace does not.

How Is Globalisation Relevant to Climate Change?

The discussion of globalisation above seems a long way from climate change. However, seeing "globalisation" as a recent label for a longstanding process is more important than it might seem at first glance. While climate change is a very recent concern for most people and their nations, we have some centuries of experience in the way that our globalised (or rather, globalising) world works. It is quite likely that the nations of the world will respond to climate change using the same patterns of behaviour that we have witnessed over the centuries, and with much the same effects. Furthermore, globalisation has contributed to climate change both directly and indirectly, and the inertia that has been established by the widespread acceptance of its neoliberal underpinnings means that it is likely to provide barriers to the ways that climate change is addressed.

It was mentioned above that the operation of the market is a central part of neoliberal thinking and hence of globalisation. Unfortunately, the market approach involves a limited conceptualisation of what is important – goods are produced using inputs (including labour and processing) which must be paid for, and sold for a price which should cover the cost of the inputs and provide a profit. This approach considers those things which have a cost, not those things which are considered to be "public goods", or are simply taken for granted. At different times, in a number of parts of the world, an enormous effort has gone into the building of wooden ships; Elizabethan England and 15th century China are examples. Timber was of course the major input, and its cost was the expense of cutting it down, transporting it and dressing it. What was not factored in was the value of the years spent growing it, or the habitat it provided for many animals, and hence the living it provided for people who collected or hunted food in the forests. Vast forests were chopped down in a brief period of time and never replaced (Ponting, 2007). These uncosted items are called "externalities", and we are now more aware of their significance than ever before, especially where they concern collective or common "goods" such as air and climate (Spaargaren and Mol, 2008).

Globalisation has enabled economic exploitation of resources to spread across the world. This exploitation has always been at the expense of externalities, including the health, freedom and quality of life of different peoples. Ponting (2007) lists numerous examples, of which I will mention just one. From the 15th to the 20th centuries, there was widespread introduction of plantation agriculture across Africa, Asia and the Americas. Typically this involved clearing forest, exhausting the soil and clearing more forest. In search of economies of scale, plantations became larger, slaves and indentured labourers were widely used, and competition meant that further efficiencies were sought until it became uneconomic to continue (Ponting, 2007). The brutal power that was needed to make slavery, indentured labour and the acquisition and destruction of other people's land possible points again to the way that power underlies globalisation in practice. The large scale loss of forest cover has contributed to the climate change that we are experiencing today, and globalisation has contributed to stimulating the speed at which it has happened across the globe (Sonnenfeld, 2008; Spaargaren and Mol, 2008).

Because climate change appears to be a result of the unfactored external effects of economic activity, it might be possible to argue that it is not a direct effect of globalisation, but just an unfortunate side effect. Similarly, those with faith in the market argue that the market will solve our climate change problems by supplying the solutions that people demand, at a price that they are prepared to pay. Given the apparent invisibility of these important externalities to date, it is not surprising that there is some skepticism about this. Typically the operation of the market has been a process of improving efficiency and extracting profits, and rebelling against too much regulation, and it has been argued that it is unwise to place too much faith in the ability of consumers to make decisions which benefit the environment (Huber, 2008; Roberts, 2008).

The only certain way of ensuring that externalities are incorporated in the costs of goods and services is for governments to ensure that it happens. Goldin (2006) calls for national governments to take into account global externalities including climate change when developing policies. Huber (2008) takes another tack, arguing that governments still have a role in fostering environmental technologies in a globalised world, through the regulation of markets to encourage the development of solutions to environmental problems. There are examples of this in different parts of the world, such as the need to include the cost of recycling in the price of a computer, along with actually supplying the recycling service (see Huber, 2008, for further examples). However, the ability of transnationals to switch activity from one country to another undermines the efforts of governments acting on their own. Developing countries in particular are left with limited options, either at a government or at an individual level. Governments do not want to risk losing profitable activity by raising costs and driving transnationals elsewhere, but on the other hand, risk the large scale environmental damage that will further exacerbate climate change. Individuals in those countries have limited ability to move around, and must choose between work and wages now and an uncertain environmental future, or lack of employment and the same uncertain environmental future.

This mention of individual choice brings us to another issue. What has been outlined above is an analysis of globalisation and its link to climate change. What has been neglected until the last paragraph is a consideration of how people react to globalisation and climate change, especially in developing countries.

What People Think about Globalisation and Climate Change

Globalisation by itself offers promises to people about jobs, income and the availability of goods; it also offers threats of different kinds. In Western countries, many trade unions and workers rightly fear that jobs will move to countries where labour is cheaper. Those most concerned about developing countries can be seen protesting at G8 summits and meetings of the World Trade Organisation, because they believe globalisation will be just another form of exploitation which will ultimately further disadvantage developing countries (Coyle, 2000). These concerns are very much about now and the immediate future, and centre around economics, personal security and politics. Certainly the discourse on globalisation has not led naturally to discussions of climate change, which is illustrated by responses to the current global financial crisis. In 2008, the price of oil soared: from a climate change point of view, expensive oil is desirable as it dampens demand for emission-producing activities; however, pressure was brought to bear on oil-producing countries to produce more oil because of concerns for the economy. Then the price of oil plummeted because demand dropped (but now seems to be on the rise again), which should be positive news for climate change, but in economic terms, can be seen as another marker of a global recession. Clearly the priority of governments is immediate global economic health, rather than climate change, and similarly many of their citizens respond more readily to the immediacy of economic pain than to the long term threat of climate change.

Unlike globalisation, with its mix of positive and negative outcomes for people, climate change has very little of a positive nature to offer. Climate change will potentially affect much of the world: rising sea levels will flood many islands and coastal areas, causing death, population displacement and economic problems; rainfall will diminish in some areas and increase in others, affecting the viability of the land and crops, as will changes in temperature; tropical diseases will increase their range; and there will be mass extinctions among some species and migration among others. Many of these changes are relevant to globalisation, because they affect where crops may be produced, which countries are likely to be economically successful, and where cheap labour is likely to be found. People's feelings about climate change involve a combination of diffuse fears about the medium to long term future, and more focused fears about the meaning of current events.

An added level of uncertainty is introduced by the lack of a scientific consensus, at least in the public's mind. After the Intergovernmental Panel on Climate Change released its reports stating that climate change was real and caused by human activity, CNN (2007) reported that a mass study of blogs had found that 38% of contributors disputed the report, while 31% supported it and 31% were neutral or "unknown". Many of the comments were based on people's immediate experiences; for example, among posts disputing climate change were comments about current unseasonably cold weather, while among those supporting it were comments attributing the recently (at time of study) experienced Hurricane Katrina to climate change.

Blogs are not just places where people can post their own opinions and reactions; they can either legitimise a blogger's opinion if it accords with the view taken in the blog, or challenge it. Climate Feedback[1] presents news with a scientific flavour on which people can comment, and also provide links to a range of related blogs with divergent stances. World Climate Report[2] claims to be "the world's longest running climate change blog" (although the evidence appears to be against this) and is devoted purely to discrediting global warming with a mixture of science, insinuation and personal opinion. Climate Audit[3] treads an interesting line - that of the scientific skeptic: the aim is to seek scientific substantiation of climate change claims, by making assumptions and analyses transparent, and by reanalysing data that has been used to support climate change claims. To a committed climate change activist, the site would appear to be devoted to discrediting climate change, but from a scientific perspective it represents the ideal of the scientific process: making sure that the methods are correct, the findings are valid, and the interpretations justified. The New Scientist magazine[4] is not dedicated to climate change, but publishes science news stories which sometimes feature research and commentary on climate change, and runs blogs on a variety of topics. People can post comments on news items, and very quickly the threads are filled with sometimes heated exchanges between proponents and discreditors of the reality of climate change running the same well-worn arguments. Among these posts, however, can be found those from people genuinely concerned about current weather events, and what climate change will mean for them and their families.

This suggests that, even though the links between climate change and globalisation could well bring major economic and social dislocation globally, these larger considerations tend not to impinge directly upon people's concerns. Raising the possibility of climate change triggers uncertainty, and if globalisation is alluded to at all, it is as a context which could potentially exacerbate the effects of climate change, or offer solutions through the operation of the market. All it takes is one or more unusual climatic events, and the blogs feature suggestions that perhaps these events are signs of climate change, repudiations, affirmations, and further exchanges that often fly off at tangents into discussions about religion, science, politics and economics. Underlying these discussions are fears about the future and the pace of change.

At the personal level, it is this fear – or denial of the problem – that forms the most basic response to climate change. For some, the fear stems from the discomfort of knowing that a threat may exist, that it is potentially catastrophic, but that the timing of its arrival and its scale remain uncertain. All the imponderables create a situation where action is difficult to take – it is difficult to justify significant changes to lifestyle and behaviour, just in case there is little to fear, but it is also hard to justify inaction because the consequences might be so great. People are faced with choices like this every day, and the patterns of our responses tend to become fixed. We know that the chances of our children being harmed, if we let them walk to school, are low, but we don't dare allow them to do this because the consequences of the occasional abduction and assault are so great. We also know that our lifestyle is typically unsustainable, in terms of the resources we consume and the negative impact of our ecological footprint, but the effort required to change our lives is too great.

---

[1] http://blogs.nature.com/climatefeedback/.
[2] http://www.worldclimatereport.com/.
[3] http://www.climateaudit.org/.
[4] www.newscientist.com.

Another aspect of our response to climate change is the essentially social nature of human beings – we live with and among other people and we constantly interact with them in a variety of ways. For all that, the West values independence in thought and action; but as individuals we generally don't take action without having the thoughts, opinions and potential reactions of others in mind. The fact that people express their opinions in discussion and blogs, view and read the opinions of others, and joke about current topics, has something to say about climate change even in passing; all activities point to the operation of social processes in influencing what we think about climate change and what we will do about it. It is this social process which is both a strength and a weakness when we confront the larger questions of human existence. The strength lies in the "working through" of issues - all the things that need to be weighed, balanced, considered; the weakness comes from the time it takes to decide upon action, the possibility that the action which emerges from discussion may be a suboptimal compromise, and the risk that no action may ensue. We witness these situations occurring all the time in politics and government, and in our own lives.

It is possible to steer out of this morass of uncertainty, and once again we are back to a discussion of how power operates. If a sufficiently powerful entity cuts the Gordian knot and makes a decision to act swiftly on climate change according to a particular set of assumptions, then the way becomes clear – although of course, over time, it may prove to be the wrong way. The decision to refuse to accept climate change is one such decisive position, which has a lot of impetus behind it, because it justifies "business as usual", and therefore carries the power of inertia with it. The position that some governments have taken reflects this stance, and it is bolstered with a range of arguments which are designed to counter criticism rather than reflecting considered positions. Gore (2007) provides examples of this from the US, although the possibility of a partisan bias cannot be denied. Ironically, the term "business as usual" is peculiarly appropriate to our globalised world, as the recent global financial crisis has shown. Roberts (2008), in an overview of four academic papers linking climate change, environmental issues and globalisation, showed remarkable prescience in predicting that "If we have one major economic crisis, I fear that all voluntary emissions reductions and labeling programs will be utterly forgotten luxuries of a bygone day" (Roberts, 2008, p. 377). The subsequent global crisis (which has had its scale and impact attributed to globalisation of financial markets (Schwenninger, 2009) has seen many governments rein in their response to climate change because that would constitute an added negative economic impact.

A second irony is that, even though we have a globalised economy and climate change is a global issue, there is a lack of global governance. The United Nations cannot offer leadership unless its strongest members (and in particular the US) do so, and some of these members have vested interests in keeping the UN relatively weak (Traub, 2006). Scientific leadership is even more problematic. It could be argued that the lack of a decisive government response to emerging change, or slow and inadequate responses, would be addressed by placing scientists in a more powerful role. This argument portrays scientists as objective discoverers of truth, who will also know the best way to proceed. This is couched in a mechanistic way, with scientists being seen as collecting and processing data which reveals what is happening and points to what should be done. The irony here is that science itself is characterised by the same debates and uncertainty about questions that are seen in popular discourse, and the contested nature of science can be seen readily in the discourse on climate change.

These kinds of uncertainty are as familiar in the developing world as in the West, but many developing countries have additional concerns because they already experience periodic problems with existing weather patterns. Much of Asia relies on the monsoon which is sometimes late, and every few years does not come at all, leading to crop failures and hunger. In Bangladesh, flooding is almost an annual event, taking many lives, while at the same time providing the silt on which their farming livelihood depends. In Ethiopia and other parts of Africa, there are frequent droughts. These "routine" climate problems have dramatic effects, because these countries do not have the resources to deal with them effectively, or at least not yet (Bolton, 2007). A recent retrospective study of the impact of temperature rises on the economies of rich and poor countries shows that rich countries were relatively unaffected, while poor countries lost about 1% of GDP when the average temperature increased by about 1°C (Dell, Jones and Olken, 2009). A more dramatic change in climate is therefore potentially catastrophic for developing countries (Roberts, 2008).[5] On the other hand, developing countries have a strong motivation to make the climate change situation worse, in ways that are related to globalisation. These countries need to develop industries, infrastructure and institutions to help them eliminate poverty and disease and to improve standards of living, elements which are now considered internationally to be universal rights and are enshrined in the Millennium Development Goals (Bolton, 2007). They cannot achieve this without increasing their emissions, and indeed the levels of emissions from China and India have been increasing greatly (Ponting, 2007). However, their emissions per head of population remain much lower than those of the US and other industrialised nations, and the developing countries argue that placing constraints on their emissions is unfair and possibly racist. The Indian government argues that there is a qualitative difference between the emissions of developing countries, which it calls "survival emissions", and the "lifestyle emissions" of the developed countries (Sehgal, 2008). It is in the interests of developing countries to use the machine of globalisation to burn more fossil fuels in pursuit of industrial expansion, increased motorised transport and other infrastructure; they can access customers in markets around the world to increase the incomes of their citizens. In this "prisoner's dilemma" scenario, they would prefer to share the potentially worse climate change scenario on a more equal footing with the developed countries, rather than entrench economic inequality and be forced to deal with climate change as impoverished and powerless countries.

CONCLUSION

There are a number of striking parallels between globalisation and climate change as phenomena, the most obvious of which is their effect on virtually every human being on the planet. The all-embracing nature of both phenomena inevitably means that they impact on each other, but not necessarily in obvious ways. Globalisation is purported to be no more than a process of economic integration, but history and experience demonstrate the deep involvement of politics and power. The interpretation of climate change has been appropriated at the political level as well, with scientific arguments being used on either side,

[5] [Editor's note: Both negative and positive effects will be felt in agriculture in Africa for instance, and these ideas are taken up by Kelbessa in his chapter in this book.].

although the weight of evidence in support of climate change has become increasingly more persuasive.

The process of globalisation, for all its political content, shows the contrast between the mobility of economic entities across countries and the constraints of citizens of nation states. The citizens of developing countries are already at a disadvantage in a globalised world and the impact of climate change is unlikely to change this. Against these analytical considerations is the actual experience of climate change, which is a vague threat that may materialise sooner rather than later, and with unfathomable consequences – except it seems likely that poorer countries will fare much worse than richer countries.[6]

> So, to return to my ruminations at the beginning of this chapter, I (as a citizen of a developed country) am in a much better position to cope with the effects of climate change in a globalised world. Although I am not employed by a transnational company, and am subject to controls on my ability to move from country to country, I can be confident that the pain of climate change will be shared globally, and that I will probably be protected from some of its more extreme effects on people's lives because I live in a developed country. Somehow, this is only a partial comfort. Perhaps, we in the developed nations should be looking at ways of assisting less developed nations with the impact of climate change because, to a large extent, our own development has taken place through globalisation that favoured our countries at the expense of theirs, and helped to produce the climate change which now threatens their countries more than ours.

REFERENCES

Beck, U. (2000). *What is Globalization?* Cambridge: Polity Press.

Bolton, G. (2007). *Poor Story: An Insider Uncovers How Globalisation and Good Intentions Have Failed the World's Poor*. UK: Ebury Press.

Bonefeld, W. and Psychopedis, K. (2000). The politics of change: Ideology and critique. In W. Bonefeld and K. Psychopedis (Eds.). *The Politics of Change: Globalization, Ideology and Critique* (pp. 1-6). Hampshire: Palgrave.

Boorstin, D.J. (1985). *The Discoverers*. New York: Vintage.

Burnham, P. (2000). Globalization, depoliticization and "modern" economic management. In W. Bonefeld and K. Psychopedis (Eds.). *The Politics of Change: Globalization, Ideology and Critique* (pp. 9-30). Hampshire: Palgrave.

Carson, I. and Vaitheeswaran, V. (2008). *Zoom: The Global Race to Fuel the Car of the Future*. Camberwell: Penguin.

Chase-Dunn, C. (1989). *Global Formation: Structures of the World Economy*. Cambridge MA: Blackwell.

CNN. (2007). What's the buzz on ... global warming? CNN exchange (website), 12th Feb 2007, http://edition.cnn.com/exchange/blogs/umbria/2007/global.warming.html

Cohen, A. (2006). Cambridge controversies in growth theory. In D.A. Clark (Ed.) *The Elgar Companion to Development Studies* (pp. 27-31). Cheltenham: Edward Elgar Publishing.

Coyle, D. (2000). *Governing the World Economy*. Cambridge: Polity Press.

[6] [Editor's note: The issues surrounding these impacts on developing countries are taken up and further elaborated by Workineh Kelbessa and Anna Lee Mraz Batra, in their respective chapters within this book.].

Dell, M., Jones, B., and Olken, B. (2009). Climate shocks and economic growth: Evidence from the last half century. Paper (dated November 2008) presented at the 2009 American Economic Association Conference (part of the Allied Social Sciences Association Conference), January 3-5, 2009, San Francisco. Paper available at http://www.aeaweb.org/annual_mtg_papers/2009/.

Derluguian, G.M. and Goldfrank, W.L. (2000). Introduction: Repetition, variation, and transmutation as scenarios for the twenty-first century. In G.M. Derluguian and S.L. Greer (Eds.) *Questioning Geopolitics: Political Projects in a Changing World-System* (pp. 1-12) Westport: Prager Publishers,

Fay, P.W. (1975). *The Opium War, 1840-42: Barbarians in the Celestial Empire in the Early Part of the Nineteenth Century and the War by which they Forced her Gates Ajar.* Chapel Hill, North Carolina: University of North Carolina Press.

Giddens, A. (2000). *Runaway World: How Globalization is Reshaping our Lives.* New York: Routledge.

Goldin, I. (2006). Globalisation and development policy. In D.A. Clark (Ed.) *The Elgar Companion to Development Studies* (pp. 204-207). Cheltenham: Edward Elgar Publishing.

Gore, A. (2007). *The Assault on Reason.* London: Bloomsbury.

Gross, D.A. (1992). *Discovering Anthropology.* Mountain View: Mayfield.

Grunberg, I. (2000). Globalization, governance and the role of the United Nations in economic and social affairs. In I. Grunberg and S. Khan (Eds.) *Globalization: The United Nations Development Dialogue* (pp. 1-12). Tokyo: United Nations University Press,

Habermas, J. (2001). The postnational constellation and the future of democracy. In M. Pensky (Ed.). *The Postnational Constellation: Political Essays: Jurgen Habermas* (pp. 58-112) Cambridge: Polity Press.

Hobden, S. and Wyn Jones, R. (1997). World-system theory. In J. Baylis and S. Smith (Eds.) *The Globalization of World Politics: An Introduction to International Relations* (pp. 125-145). Oxford: Oxford University Press.

Hopkins, T.K. and Wallerstein, I. (1996). The world-system: Is there a crisis? In T.K. Hopkins and I. Wallerstein (Eds.) *The Age of Transition: Trajectory of the World-System 1945-2025* (pp. 1-10). London and New Jersey: Zed Books; Leichhardt NSW: Pluto Press.

Howe, S. (2006). Colonialism. In D.A. Clark (Ed.) *The Elgar Companion to Development Studies (*pp. 72-76). Cheltenham: Edward Elgar Publishing.

Huber, J. (2008). Pioneer countries and the global diffusion of environmental innovations: Theses from the viewpoint of ecological modernisation theory. *Global Environmental Change,* 18(3), 360–367.

Micklethwait, J. and Wooldridge, A. (2000). *A Future Perfect: The Challenge and Hidden Promise of Globalisation*. London: William Heinemann.

Mittelman, J.H. (2000). *The Globalization Syndrome: Transformation and Resistance.* New Jersey: Princeton University Press.

Nolan, P. (2006). Adam Smith (1723-1790). In D.A. Clark (Ed.) *The Elgar Companion to Development Studies* (pp. 554-559). Cheltenham: Edward Elgar Publishing.

Ponting, C. (2007). *A New Green History of the World: The Environment and the Collapse of Great Civilizations*. London: Vintage Books.

Roberts, J.T. (2008). Challenges and opportunities for global environmental governance in the 21st century, *Global Environmental Change*, 18(3), 375–379.

Rupert, M. (2000). *Ideologies of Globalization: Contending Visions of a New World Order.* London: Routledge.

Scholte, J.A. (2000). *Globalization: A Critical Introduction.* New York: St Martin's Press.

Schwenninger, S.R. (2009). Redoing globalization. *Nation,* 288(2), 30-32.

Sehgal, R. (2008). Shyam Saran: "Can lifestyle emissions be pitted against survival emissions?" InfoChange News and Features, May 2008, www.infochangeindia.org

Sklair, L. (2002). *Globalization: Capitalism and its Alternatives.* Oxford: Oxford University Press.

Sklair, L. (2006). Globalisation and development. In D.A. Clark (Ed.) *The Elgar Companion to Development Studies* (pp. 200-203). Cheltenham: Edward Elgar Publishing.

Sonnenfeld, D.A. (2008). Globalisation and environmental governance: Is another world possible? *Global Environmental Change*, 18(3), 341–342.

Spaargaren, G. and Mol, A.P.J. (2008). Greening global consumption: redefining politics and authority. *Global Environmental Change*, 18(3), 350–359.

Stein, H. (2006). Structural adjustment. In D.A. Clark (Ed.) *The Elgar Companion to Development Studies* (pp. 596-601). Cheltenham: Edward Elgar Publishing.

Suter, K. (2000). *In Defence of Globalisation.* Sydney: University of New South Wales Press Ltd.

Traub, J. (2006). *The Best Intentions: Kofi Annan and the UN in the Era of American Power.* London: Bloomsbury Publishing.

Whitfield, D. (2001). *Public Services or Corporate Welfare: Rethinking the Nation State in the Global Economy.* London: Pluto Press.

Contributed by Barbara Crawford.

Photo: Global products being sold in Ba Ria, Vietnam rural market.

In: Meltdown: Climate Change, Natural Disasters…
Editor: Kathryn Gow
ISBN 978-1-60876-153-1

Chapter 19

SERIOUS FEARS ACROSS CULTURES ABOUT COLLAPSE OF THE WORLD ORDER

Elizabeth Tindle and Waveney Croft
Queensland University of Technology, Australia

ABSTRACT

This chapter addresses the consequences and implications of pleonexia or desire of man on a finite planet. While all creatures including Homo sapiens strive for survival and a replication of their own DNA into future generations, it is man, the authors argue, who has done so at the expense of his fellow man, all other fauna and flora and the natural environment. The juggernaut of climate change is a looming threat in the 21st century. The authors argue that how we respond to it will determine the survival or demise of life on earth as we have known it.

Keywords: Malthus, collapse world order, climate change, survival, fear of future

INTRODUCTION

What's the point of doing anything? We're just going to die"
"Why make new friends, we'll just leave them."
"Life in the next world has got to be better than this!"
(Comments of young men, 2008)

CIVILIZATION AND MALTHUS

We are increasingly uneasy about many things, one of which is the state of the planet earth (Lavelle, 2008). According to researchers from the US-based Pew Global Attitudes Project (Pew Research Center, 2007) and the Australian Fear Survey August 2007 (cited in Lavelle, 2008), many of us are concerned about environmental degradation, climate change, war and drought, but perhaps even more so since the great financial meltdown in the latter part of 2008.

Indeed, we might well ask if the Malthusian collapse of our civilization has begun. Thomas Robert Malthus (1798), in his essay on population, proposed that the "two principal hungers of man", food and sex, can never be "quelled or controlled" and because of the exponential growth in population, man will come up against a situation where resources will be "spoiled or limited". At the Deakin University Debate on Environment, "That climate change is the only issue", one of the debaters, Don Henry (cited in Australian Broadcasting Corporation (ABC), 2008a), Executive Director of the Australian Conservation Foundation, claimed that "Malthus is back on the agenda". The affirmative team that won the debate argued vociferously that "we are living in fear… there is a societal malaise and many look to the future with trepidation" (Henry cited in ABC, 2008a).

In this "culture of fear" (Williams, 2008), we will need money to save our skins when the full effects of the unfolding global financial crisis sets in. Only those with great wealth, many believe, can buy everything they need to compete with a swarming and increasing humanity to keep themselves and their families, safe and viable. This extends to safety from the catastrophe of global warming, climate change and an impending environmental disaster such as coastal inundation displacing up to 145 million people by the end of the century, if there is the predicted one metre sea rise (Bun, 2008). Tragic stories and graphic images of destructive cyclones and hurricanes such as Gustav, Katrina and Ike in the Gulf of Mexico with wind forces of over 200 kilometres per hour (Casey, 2008), tsunamis and earthquakes, floods, fire and drought have filled our airways and TV screens in recent times. All of these disasters have left death tolls of many thousands. In addition, millions of people have been left injured, homeless, destitute and hungry to the point of starvation. Those in the Irrawaddy delta of Myanmar (formerly Burma) were in this situation in May 2008 and at least 84,000 people are said to have lost their lives ("The Mekong's Fatal Floods", 2008, p. 10). Many survivors of these disasters join the millions of political or economic refugees in our world today. Malthus (1798) was of the view that only vice and misery (in the form of war, pestilence, famine, crime and disease) limit the world's population and that there was nothing that man could do to prevent this. Leakey and Lewin (1995) argued in their book "The Sixth Extinction" that we are now on the verge of a man-made disaster caused by the relentless expansion and limitless appetites of human beings around the world.

How much fear has the message of potential environmental collapse, natural disasters and even the threat of extinction brought to us in the twenty-first century? Does it influence our behaviour? Scarcely a day goes by that we are not reminded through the media, of the impact that humans have had on our planet (Bun, 2008). David Attenborough, an elected Fellow of the Australian Academy of Sciences, also believes that man's activities are implicated. In his interview with Andrew Denton (2008), in the programme "Enough Rope" on ABC television, he propounded a pessimistic view of the planet's future, citing over-population as one of his major concerns.

We have heard about the shortage of essential grain foods, especially rice (Ryan, 2008) and wheat and the impact this will have on many countries. As in past famines, people in many parts of the world may have to spend long hours queuing for basic food such as rice. Even in the USA, we were informed (Ryan, 2008) that long before the world financial crash, people were afraid of going short and had been stocking up on these commodities. In Paul Roberts' recent book "The End of Food" (reviewed by Timmer, 2008, p. 722), he claims that the "global food system is heading for a cataclysm" and "A collapse in any [one] part of the global system will have extraordinary ramifications for everyone" (p. 722).

Many people have ignored the pessimistic messages and gone on with life as usual. They were unaware of anything in their life changing and continued to assume that food and energy would be provided from somewhere; they concerned themselves with trivial inconveniences such as a bus or train arriving late. Some young people have not heard of any of the world's problem, as they are engrossed in the lives of their teenage friends (personal communication, June 2008). Even after the global financial crisis and the loss or employment being aired regularly on television, those in work appeared to change little of their behaviours.

Moreover, over the past decade, if one "scratches the surface" of many university students and more aware members of the public, one witnesses a frenetic competition and drive for survival, safety and security by means of increased wealth, in an increasingly unstable world.

Environmental Concerns

Historically, there have been many such scares about damaging the environment, and concern for environmental protection is not new. A number of papers addressed concerns with water, air and soil contamination during the agricultural revolution in the Fertile Crescent of the Middle East some centuries ago (see Geist, 2005). Most of the Mesopotamian and Sumerian civilizations eventually collapsed (Lawler, 2008).

Edward the 1st of England is said to have banned the burning of sea coal in 1272, because it produced a thick soot and smoke pollution over the cities (Environmental Graffiti, 2008). London was famous for its 'pea-soup' smog and even Charles Dickens (1859) set the first scene of his novel "A Tale of Two Cities" in a thick London fog which resembled the phenomenon afflicting Beijing in 2008, as it attempted to clean up the air pollution before the Olympic Games. Since the 18th century, concerned people have called for the removal of many toxic factories, especially tanneries and paper mills, sited along our formerly pristine rivers and coast lines. Domestic coal fires in Britain were banned by Parliament in 1952 after 400 Londoners died as a result of four consecutive days of thick smog (Urbinato, 1994).

In America, the poet Henry Thoreau (1854) argued for a simpler life and the protection of nature, while Scottish-born American émigré, John Muir (1838-1914) strongly influenced the formation of the modern environmental movement. Muir was instrumental in establishing National Parks in North America and was the Founder and first President of the now famous Sierra Club (Sierra Club, 2008); they both believed that nature had inherent rights. The international eco-village movement today embraces a similar philosophy of living with nature.

Mahatma Ghandi, in India, inspired the practice of peaceful resistance which was taken up later by Indian women villagers in the Uttaranchal region in 1973. These women prevented the cutting down of trees by forestry contractors by encircling the trees with their bodies. This practice, known as "Chipko" (meaning tree-hugging), inspired other environmental groups in Australia and America (Mishra and Tripathi, 1978) in to activism. These days 'tree-hugging' is often a pejorative label for those who are against industrial progress, the equivalent of modern-day Luddites. However, deforestation continues to be one of the major contributors to climate change, and environmentalists in Australia have tried to stop 'Gunns', the logging company in Tasmania, from felling ancient forest for pulp milling, wood chips and paper making (Gettler, 2003). Similarly, the Amazon rainforests, said to be the 'lungs of the planet', are reported to be "pillaged and raped" for valuable hardwoods by Brazilian loggers with false permits (Negus, 2008). The loggers are described as being "clever and corrupt" and they clear felled timber under the cover of darkness, often protected by the police (Negus, 2008).

In the twentieth century, serious environmental threats such as aerial spraying of pesticides and DDT, which were brought to our attention by Rachel Carson (1962) in her book 'Silent Spring', environmentally destructive oil spills such as that of the Exxon Valdes in Alaska (Exxon Valdes Oil Spill Trustee Council, n.d.), as well as the destruction of the largest inland sea in the world, the Aral (Welsh, 2000), have raised the consciousness of many. The latter was the result of irrigating cotton and the international squabbles over the water after the collapse of the Soviet Union. Today the soil is so toxic and the water so polluted that birth defects have grown exponentially (Welsh, 2000). The irrigating of cotton crops in dry areas can mean the demise of large river systems downstream. Cubbie Station in the head waters of the Murray- Darling basin in Queensland is an example of such a threat (Guest, 2008). The owners had planned to raise money to expand their irrigation to enable the growing of vegetables and to launch their enterprise on the stock-market, but they have since decided to sell the whole enterprise to the Federal Government.[1]

Since the 1960s, James Lovelock, a British scientist, has been proclaiming that the earth is a totally self-regulating living system with a homeostatic balance which man has upset with overpopulation and the modern way of life (Allaby and Lovelock, 1983). This is the Gaia hypothesis, Gaia being the Greek goddess of the planet earth. Many consider Lovelock to be a doomsday pessimist to be "taken with a pinch of salt"; however, is he actually forecasting an Apocalypse for us and the planet earth?

When he was interviewed by Phillip Adams on Radio National, Lovelock (2007a) stated that 1200 of the best scientists on climate change predicted that Europe and many other parts of the world would become scrub and desert by the year 2040. Today we see the frantic fight by the Chinese to stop the sand dunes of the Gobi Desert from spreading westwards and inundating heavily populated areas with sand. Already sandstorms are so thick that they reduce visibility to a few metres in Beijing. By 2040, Lovelock (2005) claims, temperatures will be approximately five degrees hotter than they are today. He explained that the same thing happened 55 million years ago and the world population could use only 20% of the earth's surface, and that was only in the polar regions. Lovelock (2007b) believes that there will not be many survivors when it happens again, as there will be insufficient food to feed more than six and a half billion people. Already at various spots around the world, there are

[1] Australian Pastoral Co Ltd (1888 -). http://www.gabr.net.au/biogs/ABE0022b.htm.

those who have secreted themselves in "safe" hideouts or bunkers to avoid the devastation that they believe is imminent (Fortuna, 2007). Some even put a date on this cataclysmic event; one group claimed that it was to be May 14th 2008 (Fortuna, 2007).

The "Greenland catastrophists" (Witze, 2008) speculated that the melting of the Arctic's biggest ice sheet, Greenland, which is occurring today, would result in global sea levels rising by seven metres, thereby inundating many islands and large tracts of coastal land throughout the world. This, Witze says, has not been factored into the IPCC's (Intergovernmental Panel on Climate Change) estimates of sea level rise given in their earlier reports. Witze (2008, p. 798) states that "the collapse of the Greenland ice sheet is in the front rank of potential climate catastrophes". The ABC World News ("North Pole Ice Cap", 2008) reports that unlike all previous years in recorded history, there was no ice left on the Arctic seas during the northern hemisphere's summer of 2007.

Lovelock (2007b) maintains that even if all the fossil fuel burning stops now, the greenhouse heat cannot be prevented. He proposes that the only technological way it can be stopped is to put sunshades in space and to use aerosols in the short term. This, he says, is like putting the earth on a dialysis machine. Lovelock explains that 70 percent of the earth is ocean and it has turned into an ocean desert, hence the disappearance of many of the major fishing grounds of the 1950s and 1960s (Dogger Bank, North Sea, Newfoundland). His theory is that if the colder water can be raised from below the surface by means of gigantic pipes ten metres in diameter and 100 to 200 metres long, it would help the earth to save itself. In 2007, Lovelock and colleague Chris Rapley (2007) proposed the construction of the ocean pipes solution. He believes that humans cannot save the earth. Lovelock (2007a) emphasises that this is not a debate: "Science", he says, "is fact". He has compared some of the smaller solutions suggested to "moving the deck chairs on the Titanic!" However, Dawson (2007) challenges this expression in his article titled "Shut up about deck chairs!" and reports on peak moments of inspiration when humans realise a paradigm-altering truth focussing on the "ice-berg [of resource depletion] towering over the ship" and not the ship itself. Those who embrace the resulting eco-village solution are leaving cities and establishing sustainable settlements, such as Findhorn in the highlands of Scotland, to survive the impending calamity.

Many countries are aware of the resource depletion and are attempting in different ways to address it. China has one of the largest programmes for weather modification in the world (Qiu and Cressey, 2008, p. 970). Their 'armoury' includes 7000 anti-aircraft cannons, 5000 rocket launchers, radar, satellite and weather balloons as well as over 32,000 employees engaged on the problem. Weather modification, one of the key projects in their next five year plan, will attempt to seed suitable clouds to avert the water crisis that is being felt by farmers in many parts of the country. In a similar way, Israeli researchers at the Hebrew University of Jerusalem are working with cutting edge technologies to try to replenish the water reserves in the Sea of Galilee in Northern Israel (Qiu and Cressey, 2008, p. 970).

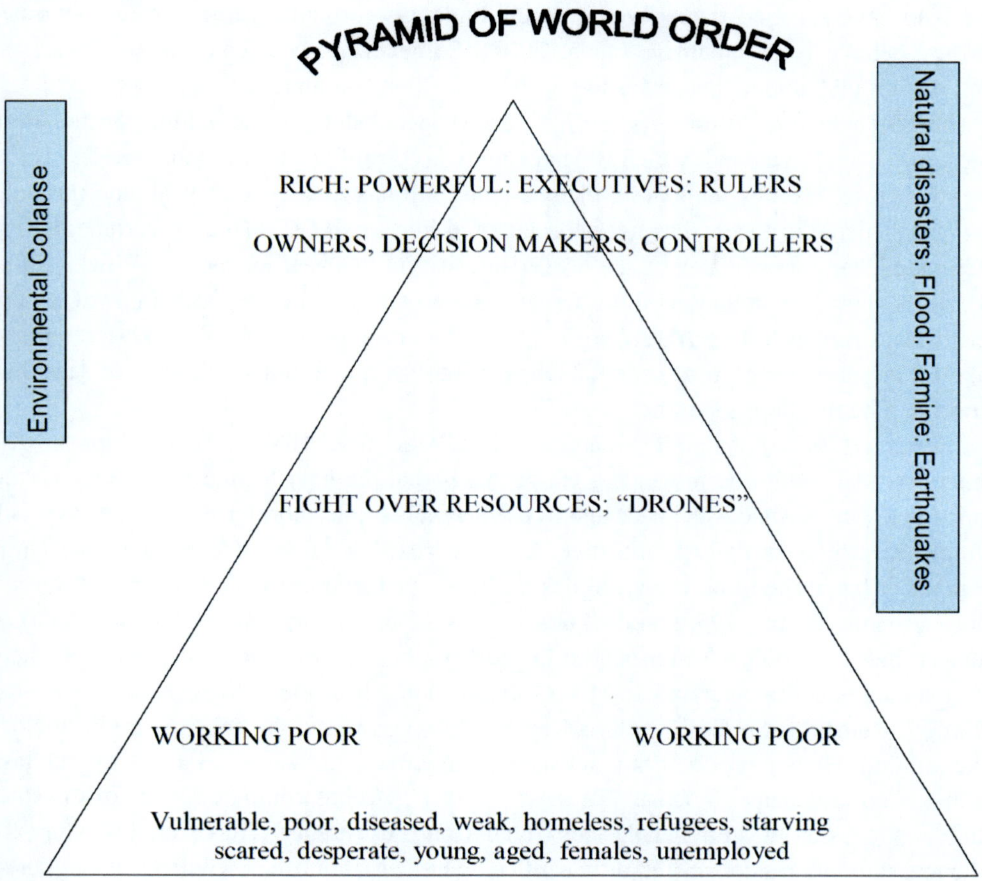

Figure 19.1. Who's where in the hierarchy at time of collapse.

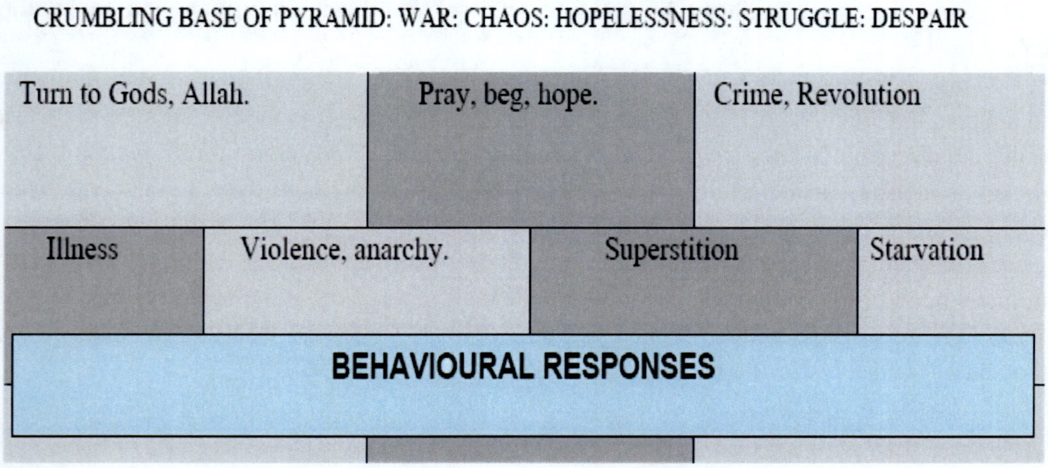

Figure 19.2. Behavioural Responses at Base of Pyramid during and after collapse.

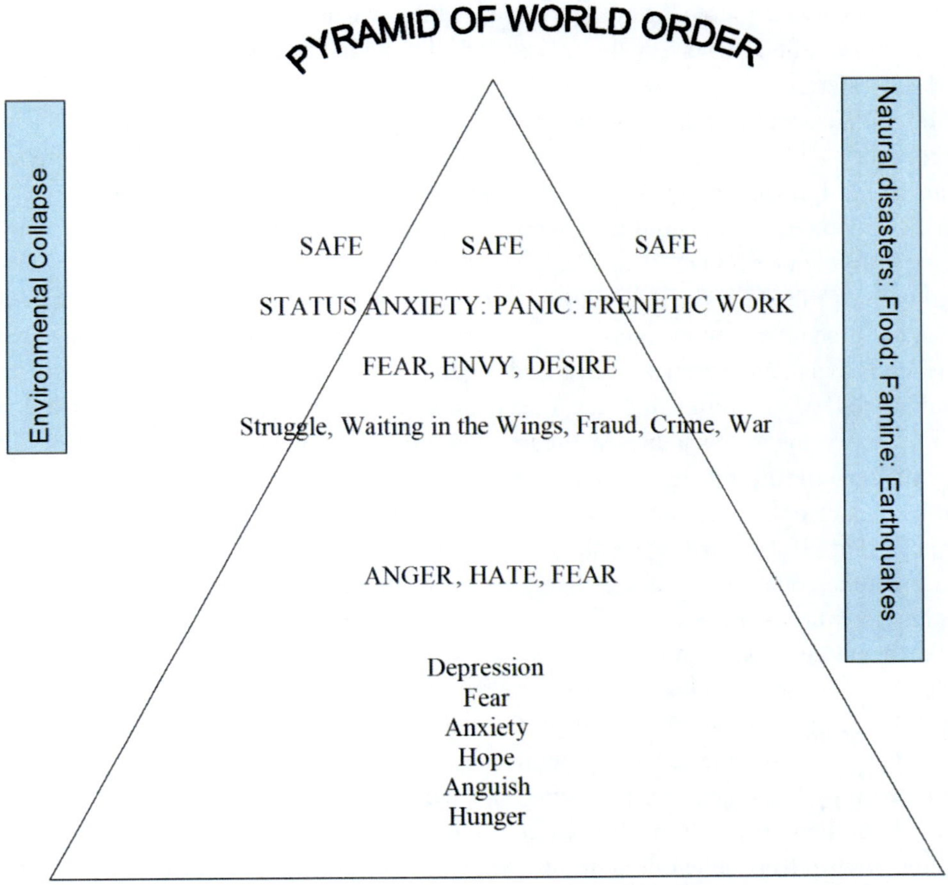

Figure 19.3. Emotions and behavioural responses at top, middle and base of Pyramid during and after collapse.

Fears about Societal Collapse

The next discussion refers to a set of figures (Figures 19.1, 19.2, and 19.3 that make up the whole picture of Figure 19.4); these figures relate to the pyramid of world order, the place of the role takers in that order and the emotions and behaviours that accompany one's position in the hierarchy, during and after the collapse of the world order, either through an economic, environmental, social, or political collapse. Figure 19.1 is the hierarchy at the time of the collapse; Figure 19.2 outlines the behavioural responses of people at the base of the pyramid during and after collapse; Figure 19.3 lists the emotions and behavioural responses of people at the top, middle and near the base of the pyramid during and after collapse; and Figure 19.4 shows the whole picture with all the players' reactions to collapse of world order.

The societal collapse line of thinking has been continued in various ways with the teachings and writings of a number of scientists including Paul Ehrlich (1978; Ehrlich and Ehrlich, 1975; 1981; 1990), Jared Diamond (1997; 2005), David Suzuki (2006), Tim Flannery (2006), Joseph Tainter (1990) and many others. In Jared Diamond's (2005) book "Collapse: How societies choose to fail or succeed", he reminds us of the previous collapses of civilisations and why they occurred. It is Diamond's theory of hierarchical societies that applies to our world today. At the apex of this hierarchical pyramid are the super-wealthy, powerful, aristocratic owners of the means and distribution of the world's production. They have been referred to by Diamond as "The owners of the world" and include CEOs and corporation owners; they have status, are opulent and (theoretically) safe. In the global markets which are said to be driven by greed and fear, those at the apex make superabundant wealth. In his famous speech "Greed is good" from the Oliver Stone (1987) movie "Wall Street", Gordon Gecko argued that "Greed is good; greed is right and greed works". He went on to try to convince his "Teldar" shareholders that greed captures the essence of the evolutionary spirit. He argued that greed for money is as normal as greed for life, a greed for love and a greed for knowledge. In a convoluted way he seemed to be arguing that this behaviour is what biological survival and safety are all about. Were we to apply this to Zimbabwe today, we would find that President Mugabe, who has the 'power of the pointy end' of the pyramid of wealth (see Figure 19.1) lives in a mansion resembling the French Palace of Versailles, protected by his army and security staff (Sithole, 2008). He flies in and out of his palace on his personal helicopter. From lowly beginnings, it has been noted that, he ensconced himself in luxury while the country's economy collapsed. There are some famous instances of the collapse of societies being due to the uprising or anarchy of the lower classes or proletariat who overthrew and assassinated the wealthy, ruling elite; such was the case in the French and the Russian revolutions.

The fraud, that was being uncovered by the Federal Bureau of Investigation (FBI) in the Prime Mortgage meltdown in the USA, seemed to provide some evidence of the financial world being operated under a 'Gecko philosophy'. Was this another example of what Malthus (1798) meant by "vice and misery" being inevitable? Over 400 bankers, lawyers, real estate agents and brokers, many from Wall Street Banks, were charged with conspiracy to defraud by being involved in malicious mortgage fraud. Adam Smith (1776), a moral philosopher who lectured at Edinburgh University and who wrote "The Wealth of Nations" in the 18th century, proposed that self-interest was basic to man's survival and flourishing. This underpins the race to gain personal wealth by means of international trading and industrial productivity. Friedman and Friedman's (1990) book "Free to Choose" further builds on the concept of making as much profit as possible in an open, unfettered, preferably growing, global market. Donald Trump, a multi-billion dollar businessman at the apex of the pyramid, who had speaking engagements planned in Australia in 2008 (but then cancelled) titled "Think like a billionaire", would have earned $4-$5million for his seminars (Trump, 2008).

Population increases translate into more buyers and more profits for owners of company shares and executive officers. They also provide more opportunity to find deviant and devious means to acquire riches and reach the top of the pyramid of success (National Nigerian Fraud Symposium, 2007). Australia's fourth richest man, also a multi-billionaire, was fined $36m by the Federal Court in 2007 for price fixing (Business Spectator, 2008). The Conservative British Government under Margaret Thatcher, elected in 1979, and the United States Government led by Ronald Reagan, took up this monetarism in a big way, condoning financial greed and thereby enabling the competitiveness and the survival of the financially fittest.

Michael Shermer (2007) refers to this as *homo economicus* and argues for evolutionary economics. Misha Glenny (2008), a former BBC World Service reporter, describes in his latest book how in 1989, with the collapse of Communism in eastern Europe, the powerful state employees, many of them former KGP officials, have extricated the wealth of the USSR industrial machine and that of other nations, for themselves. They have organised major crime syndicates and set up corporate enterprises protected by former strong-men, wrestlers, body-builders, weight-lifters and bouncers, and they trade anything that makes a profit, including drugs, oil, women and children. Much of the proceeds were invested in the lavish lifestyles and the property boom in places such as Dubai and Abu Dhabi (Glenny, 2008).

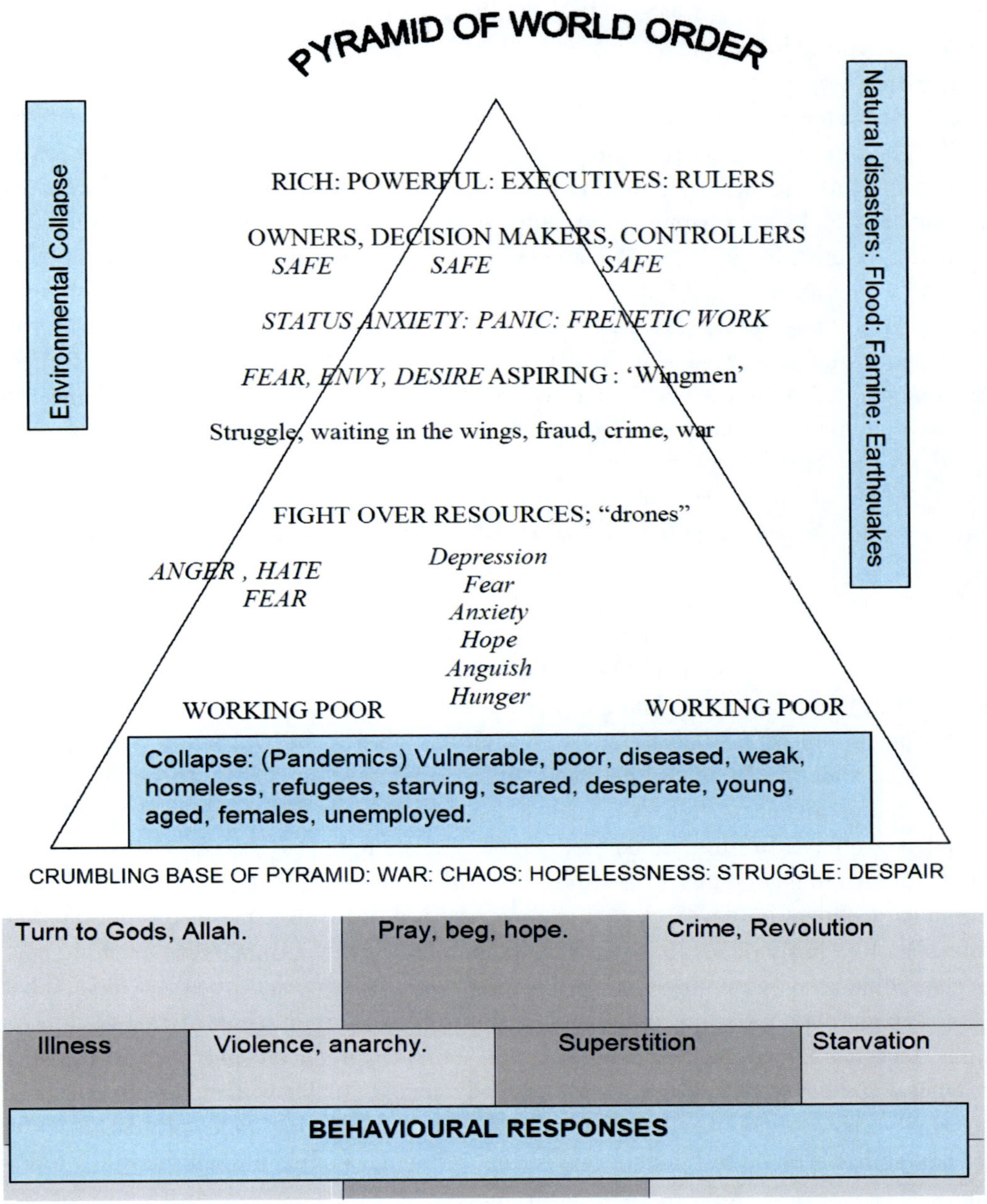

Figure 19.4. The whole picture: reactions to collapse of world order: emotions and behaviours.

The capitalist model of commerce has been adopted throughout most of the world. We are conditioned to believe that anyone can become super-rich, and this message is the motivating force for many. All it takes to become a billionaire is hard work, smart risk-taking and business know-how. It can be everyone's dream. The message is that poverty is the fault of the individual and is due to stupidity or laziness. The neo-conservative theory in modern-day business has been described as 'get in, get rich and get out', and then theoretically one is secure and safe. There is ample money to invest in one's self or one's family and to ensure that the latter have a big slice of the world's resources on offer in the way of education, luxury goods, lavish lifestyle and expensive possessions. Certainly, they will have more materially than their parents had; at least that was the projection before the global financial meltdown occurred in 2008/2009.

These messages are internalised by the young. "The only thing that motivates me is money!" says a teenage youth talking to his friends (personal observation). When counselling, we see children whose parents have retired in their mid-thirties, having realised their ambitions of financial security, and the family often lives in their dream home. The clients, predominantly male, usually present with little interest in anything. They report that they have no passion for anything in particular and are unmotivated to study or work. Throughout their short lives, they have wanted for nothing and have not had to take any risks; everything that money can buy has been provided for them. They have usually attended a 'good' private school and achieved a reasonable academic score to be accepted into university. But once they commence, they often lack the motivation or are too socially inept to do anything other than play computer games; some have said that they don't know how to feel happy. Sometimes they remain well into their twenties, in their physical 'comfort zone' within the family edifice which can resemble a modern mansion. They are safe there, but can feel depressed and socially anxious. Some are scared to face a real world of people and challenges.

On the economic and sociological pyramid, below the super-wealthy minority, are those aspiring to join the increasing numbers of wealthy and privileged people at the apex. They have been described by Diamond (2005) as "waiting in the wings". They are the "wingmen", the "hopefuls" desiring to step into the shoes of those above them on the hierarchy. Many of the aspiring group have "status anxiety" which has been identified in the writings of the philosopher Alain de Botton (2004) and others. This position in the hierarchy demands constant competitiveness, vigilance, energy and performance. They frequently have a full-time job as well as a full-time study load. Some work more than 70 hours a week for success in life. One example of this is a young man who presented in a hot sweat over losing thousands of dollars on the stock market as a result of the 2007 American prime mortgage fallout. He, like many others. believes that the purpose of life is to make as much money as one can. Often a favourite movie is "Wall Street" with the famous "Greed is Good" advice. Clients explain that they are members of Generation Y who are in a hurry to get the 'big job' on the stock market, in an investment bank or similar, and get out into the world to make money. They are examples of the one-child family creating males with an appetite for wealth and some with an insatiable appetite for bought sex. Some young males present to services after having had a blood test due to risky sexual behaviour on one of their many sex tours to Asia.

Other young men present to services throughout Australia with extreme anxiety or panic. They have an excellent academic record, but are terrified that it might drop and their goals of obtaining high-paying jobs in the stock market or in top investment banks would be scuttled. Some have a burning desire to get rich and be safe (from poverty). They cannot, or are unwilling, to lower their expectations even though their mental health is seriously affected by the worry of maybe not making it to the top. They self-medicate their anxiety with pot or alcohol and continue to burn the midnight oil. Similar stories are appearing from other sources; one report from New Delhi in India reported that "It is suicide season when students buckle under parental pressure to get high marks… and a chance of a high paying job" ("India exam pressure", 2008). The newspaper report went on to say that in 2006, "5857 students - or 16 a day - killed themselves".

Situated below those with status anxiety and a fear that they will not become rich and move up the pyramid to safety, are the millions of workers. They fill the bulk of the pyramid and are described as the "working bees" or "drones". They may fit the category of the middle class, lower middle class or working poor.

If we scan back in history to the time of the sculptor Rodin, we are reminded of his strong argument immortalised in his famous gigantic sculpture "The Gates of Hell". He tried to depict man's energy and libido which, he believed, would create "Hell on earth"; in it, he created a giant portal squirming with bodies in a state of eternal motion as a portrayal of male libidinal and restless energy. Earth is that Hell for many people today.

Those who are unable to continue the daily struggle to climb the ladder or pyramid of work often feel burnt out, suffer from a severely depressed mood and frequently start experiencing panic attacks and various sleep disorders. For some in the first world (Australia, North America, UK and Europe), life is considered boring and pointless. People, especially the young, suffer an ennui and anomie in the overcrowded and impersonal cities. Anomie has been described as a lack of purpose, identity or ethical value in a person or society, which results in personal rootlessness. Many citizens cannot find their place in a fast, complex, competitive society. Identity eludes them; some are alone and suffer in their isolation and disconnection. The increasing homelessness in Australia, especially amongst young people, exemplifies this. Many people experience ennui, a weariness and dissatisfaction with life resulting from a lack of interest and boredom. A lead article in Time Magazine (Mayer, 2008) describes British youth as "unhappy, unloved and out of control". The western affluent society is clearly not benefiting all. The 'trickle down' effect which is sometimes presented as a justification for continuing growth for growth's sake may never reach those on the bottom rung of society. The British National Union of Teachers debated why so many pupils are so "unhappy and anxious" (Andalo, 2008).

The aspiring position, just below the wealthy apex layer, brings with it a fear that these 'wingers' may not make it to the safety at the top (see Figure 19.2). By failing, many are terrified that they may join the hoards at the base who are the losers in the world: the weak, poor, destitute, homeless, jobless, stateless, landless, hungry, sick, dirty, old, illiterate, crazy and powerless (see Figure 19.1). It will be the demise of this group through war, famine, floods, disease and other climate-related disasters that will occur first, if our civilization collapse follows the usual pattern. We hear and see daily on our TV screens the imminent devastation of climate related effects. The Catalyst (2008) programme on the "Dying River Murray" brought the potential disaster close to home for Australians; large tracts of inland Australia have soils that are as acid as a car battery, brought about by inappropriate land management and overuse of the whole River Murray basin system. The mouth of the river is dead. Farmland that once produced a superabundance of food for the world is no longer able to do so. Adelaide and other towns that rely on the river for drinking water are in a very vulnerable position. Professor Richard Kingsford (ABC, 2008b) from the University of New South Wales, in his discussion with Dr. Michael Young of the Wentworth Group, spoke of disaster in the lower lakes of Alexandrina and Albert. Many dependent on the Murray and its tributaries are scared and angry. They have a fear of the impending changes that have to be made in an effort to survive. Tim Flannery (cited in Davis, 2004) at a presentation in Sydney predicted that Perth is likely to become the first "ghost metropolis" because of the significant decrease in rainfall in Western Australia and the major water problems the city has experienced.

Diamond (2005) argued that the 'rot' will move up the crumbling pyramid as it has in the collapse of other civilizations such as the Mayan, Sumerian, and Easter Island. To Diamond's collapsed civilization examples, we could add others such as the Nile and the Indus (Lawler, 2008). All previous societies have frantically appealed to their gods to save them as their looming demise stares them in the face (see Figure 19.3). We saw pictures of Burmese people praying for help in their environmental plight after the cyclone hit the Irrawaddy Delta (Casey, 2008). Sydney hosted thousands of young Roman Catholics (Conway, 2008) from around the world and there were multiple celebrations of mass, sermons and prayer with thousands of people taking part. Praying for the planet and life on it has been part of these ceremonies. "In ancient times the Chinese believed that dragons controlled the weather and elaborate rituals were performed to bring about sufficient rainfall and good harvests" (Qiu and Cressey, 2008, p. 970). The same rites occurred in other cultures, such as the American Indians who performed their rain dances around the totems. The Maya of Central America sacrificed young children to their god, the Plumed Serpent, Quetzalcoatl, in order to bring rain in times of drought or famine (Sharer and Morley, 1994). Blood was believed to be the supreme food of the Gods; the smell of blood, it is recorded, was in the air for great distances from the sacrificial spots on the huge Mayan pyramids, the largest of which is at Cholula, near the town of Puebla in Mexico. Rain eventually came confirming the veracity of the practice.

In the 21st century, religious fundamentalism has expanded. If modern society follows the same pattern as previous failing societies, fundamentalism will continue to grow. Threatened societies will pray, beg and appeal to their specific higher authority, Allah, Vishnu, the Blue Peacock, the Plumed Serpent, Yahweh, Christ, Mohammed and others to save them from starvation, disease or attack. The inhabitants of Easter Island built more and more Moui, enormous stone statues, to their god as they, the islanders, slowly removed every tree from a once thickly forested island. They had no timber to make any more fishing canoes, but had not the wherewithal to change their practices when the looming demise stared them in the face. The population of the island increased until the inhabitants survived by cannibalism. Rival groups exterminated each other in the battle for survival. Easter Island is symbolic of the planet earth. Is it any wonder that people are universally scared?

Paul Krugman (2007), Professor of Economics at Princeton University, postulated that the post-World War II middle class, which appeared as a result of the welfare states and socialism when the fighting had ended in 1945, is fast disappearing. The New Deal society in America "has been dismantled" he argues. The decline in the power of the union movement to negotiate for those at the bottom of the pyramid contributed to the widening of inequality. The gap between the obscenely rich and the direly impoverished has widened to the chasm it is today. The lowest strata of world society, including American society, is thickening. In many parts of the world, it is in chaos (see Zimbabwe, Ghana, Somalia, Ethiopia, Sudan, Ghana and Kenya - post 2008 election where over one thousand people were killed) and people are churning in desperation for their own survival and the survival of their families.

The Western form of democracy does not appear to be working in Africa, and at such a level of desperation, countries are vulnerable to military dictatorships. Such countries sink into social unrest and civil war or genocide, tribe against tribe, powerful against powerless (Uganda), settled farmers against nomadic pastoralist (Sudan); and Muslims against Christians. Further afield, there are Hindus set against Muslims, Turks against Kurds, Haredi Jews against secular Jews (Israel), and human strife goes on ad infinitum. Conflict is the norm; it is part of the human condition to fight when they believe that their livelihood is threatened. There are said to be over thirty countries that are vulnerable to uprising, if rice is not available and food prices rise ("UN Sets Up Task Force", 2008).

Theoretically, only those who are safe at the top of the pyramid are *not* afraid, although even they build their gated communities and hire private security guards and bouncers to protect themselves, their interests and their property. The aspiring group are afraid of not making it to this refuge of opulence where they believe that money can provide anything they desire. Some resort to criminal behaviour to get there (Glenny, 2008). The middle and working classes continue to beaver away giving a good day's work for their salary and trying to do their small part for the environment by recycling waste or replacing their light bulbs. Meanwhile, those in the thickening bottom layer are worried about not having the basic food, medical and shelter needed to feed themselves and their families, in order to survive. Fear is a part of their daily existence. (See Figure 19.4 for the whole picture). Thirty-seven million of these are refugees dependent on the altruism of the world for humanitarian help (Rodriguez, 2008). At the G8 meeting of world leaders in Hokkaido, Japan, during July 2008, all countries agreed to address world poverty. Resolving differences on climate change and the environmental impact of carbon and other aspects of global desecration were on the agenda for the G8 meeting held in Copenhagen in 2009. In the meantime, all countries will resort to 'moving their own deck chairs', as they deem appropriate.

CONCLUSION

In 2009, the bicentenary of Charles Darwin, who first introduced us to the radical concept of evolution and adaptation, we find that we are witnessing a speeded up planetary evolution and an exponential growth of human population. Some questions touched on in this chapter challenged the belief that man and the environment can adapt in order to keep pace with the rapid climatic and environmental changes to planet earth. Our main concern is whether or not human beings are able to change sufficiently in their utilitarian approach to the world in order to arrest a foreboding global disaster.

REFERENCES

Allaby, M., and Lovelock, J. (1983). *Great extinction*. New York: Doubleday.

Andalo, D. (2008, March 14). Teacher unions take aim at spiralling workload. *Education Guardian*. Retrieved May 14, 2008 from http://www.guardian.co.uk/education/2008/mar/14/schools.news

Australian Broadcasting Corporation. (2008a, June 20). Climate alert: Top priority or obscurantist twaddle? *[Radio broadcast]. Sydney: ABC Radio National.* http://www.abc.net.au/rn/nationalinterest/stories/2008/2280736.htm

Australian Broadcasting Corporation. (2008b, March 27). *Murray and Darling Rivers still dying* [Radio broadcast]. Sydney: ABC Radio National. http://www.abc.net.au/rn/breakfast/stories/2008/2200292.htm

Bun, M. (2008). *Ockham's razor.* Presentation on climate and sea level rise. ABC, Radio National, 31st August, 2008. http://www.abc.net.au/rn/ockhamsrazor/stories/2008/2349127.htm

Business Spectator. (2008, June 20). *Pratt to defend charges.* Retrieved from http://www.businessspectator.com.au/bs.nsf/Article/Pratt-to-defend-charges-FSCNB?OpenDocumentandsrc=srch

Carson, R. (1962). *Silent spring.* London: Hamilton.

Casey, M. (2008, May 10-11). Huge winds a deadly lesson. *Courier-Mail*, p. 48.

Catalyst. (2008, May 1). *Acid mud* [Television broadcast]. Sydney: Australian Broadcasting Corporation. http://www.abc.net.au/catalyst/stories/2232992.htm

Conway, D. (2008, July 15). Faith overflows as cross arrives. *Courier-Mail*, pp. 8-9.

Davis, A. (2004, May). Sydney's future eaten: The Flannery prophecy. *Sydney Morning Herald* May 19th 2004. http://www.smh.com.au/articles/2004/05/18/10847835177132.html.

Dawson, J. (2007, Sept. 27). Shut up about the deckchairs! *New Statesman.* Retrieved from http://www.newstatesman.com/blogs/life-at-findhorn/2007/09/peak-moment-collective-energy

de Botton, A. (2004). *Status anxiety.* New York: Pantheon.

Denton, A. (2008, June 16). *Enough rope* [Television broadcast]. Sydney: Australian Broadcasting Corporation.http://www.abc.net.au/tv/enoughrope/transcripts/s2276104.htm

Diamond, J. (1997). *Guns, germs and steel: The fates of human societies.* New York: W. W. Norton and Co.

Diamond, J. (2005). *Collapse: How societies choose to fail or succeed.* New York: Viking.

Dickens, C. (1859). *A tale of two cities.* London: Chapman and Hall.

Ehrlich, P. R. (1978). *The population bomb.* New York: Ballantine Books.

Ehrlich, P. R. (1998). *Recent developments in environmental sciences.* Retrieved August 9, 2008 from http://dieoff.org/page157.htm

Ehrlich, P. R., and Ehrlich, A. H. (1975). *The end of affluence: A blueprint for your future.* New York: Random House.

Ehrlich, P. R., and Ehrlich, A. H. (1981). *Extinction: The causes and consequences of the disappearance of species.* New York: Random House.

Ehrlich, P. R., and Ehrlich, A. H. (1990). *The population explosion.* New York: Simon and Schuster.

Environmental Graffiti. (2008). *Environmentalism in 1306.* Retrieved September 19, 2008 from http://environmentalgraffiti.com/offbeat-news/environmentalism-in-1306

Exxon Valdes Oil Spill Trustee Council. (n.d.). Retrieved October 5, 2008 from http://www.evostc.state.ak.us/

Flannery, T. (2006). *The weather makers: The history and future impact of climate change.* London: Allen Lane.

Fortuna, Q. (2007). *Armageddon cocktail hour: Russian cult predicts May 2008 apocalypse.* Retrieved July 31, 2008 from http://armageddoncocktailhour.wordpress.com/2007/11/14/russian-cult-predicts-may

Friedman, M., and Friedman, R. (1990). *Free to choose: A personal statement.* San Diego, CA: Harcourt, Brace Jovanovich.

Geist, H. (2005). *The causes and progression of desertification.* Aldershot, UK: Ashgate.

Gettler, L. (2003, Nov 24). Greenpeace takes on Gunns logging. *Sydney Morning Herald.* Retrieved August 9, 2008 from http://www.smh.com.au/articles/2003/11/23/1069522472555.html

Glenny, M. (2008). *McMafia: Crime without frontiers.* London: Bodley Head.

Guest, A. (2008, June 28). Cubbie expansion sparks Murray-Darling fears. *ABC News.* Retrieved June 30, 2008 from http://www.abc.net.au/news/stories/2008/06/28/2288704.htm

India exam pressure leads to wave of student suicides. (2008, May 19). *The Straits Times.* Retrieved October 5, 2008 from http://www.straitstimes.com/Latest%2BNews/Asia/STIStory_238926.html

Krugman, P. (2007, November 25). "Where is the middle class?" *Background Briefing* [Radio broadcast]. Sydney: ABC Radio National. Retrieved August 3, 2008 from http://www.abc.net.au/rn/backgroundbriefing/stories/2008/2160260.htm

Lavelle, P. (2008). *Global anxiety.* Retrieved October 17, 2008 from http://nationalfearsurvey.net.au/feature_globalanxiety.aspx

Lawler, A. (2008, June). Indus collapse: The end of the beginning of an Asian culture? *Science, 320,* 1281-1283.

Leakey, R., and Lewin, R. (1995). *The sixth extinction.* New York: Doubleday.

Lovelock, J. E. (2005). *Gaia: And the theory of the living planet.* Stroud, UK: Gaia Books.

Lovelock, J. E. (2007a, July 12). 2007 Adelaide festival forum. *Late Night Live* [Radio broadcast]. ABC Radio National. Retrieved from http://www.abc.net.au/rn/latenightlive/stories/2008/2120512.htm

Lovelock, J. E. (2007b). *The revenge of Gaia: Why the earth is fighting back- and how we can still save humanity.* London: Penguin.

Lovelock, J. E., and Rapley, J. C. G. (2007). Ocean pipes could help the earth to cure itself. *Nature, 449,* 403. Retrieved September 16, 2008 from http://www.nature.com/nature/journal/v449/n7161/full/449403a.html

Malthus, T. R. (1798). *An essay on the principle of population, as it affects the future improvement of society with remarks on the speculations of Mr. Godwin, M. Condorcet and other writers.* Retrieved July 4, 2008 from http://www.ac.wwu.edu/~stephan/malthus/malthus.0.html

Mayer, C. (2008, April 7). Mean streets. *Time Magazine*, pp. 40-46.

Mishra, A., and Tripathi, S. (1978). *Chipco movement: Uttaranchal women's bid to save forest wealth.* New Delhi: People's Action/Ghandi Book House.

National Nigerian Fraud Symposium (2007). *Queensland Police Fraud Squad Annual Conference, State Crime Operations Command.* Holiday Inn, Brisbane, 7-9 October 2007. http://www.police.qld.gov.au/News+and+Alerts/campaigns/synergy/testimonials/natNigFrdSymp2007Testmls.htm

Negus, G. (2008, July 16). Report from the Brazilian Amazon [Television broadcast]. *Dateline.* Sydney: SBS.

North Pole ice cap 'melting faster than ever'. (2008, August 28). *ABC News.* Retrieved September 19, 2008 from http://www.abc.net.au/news/stories/2008/08/28/2349041.htm

Pew Research Center. (2007). *Global unease with major world powers: Rising environmental concern in 47-nation survey.* Retrieved October 11, 2008 from http://pewglobal.org/reports/display.php?ReportID=256

Qiu, J., and Cressey, D. (2008, June 19). Taming the sky. *Nature*, *453*(7198), 970-974.

Rodriguez, J. (2008, June 20). 37 million refugees in spotlight this week. Retrieved October 15, 2008, from http://www.commondreams.org/archive/2008/06/20/9761

Ryan, P. (2008, April 28). Rice ration sparks revival of survivalists [Radio broadcast]. *Radio* National. Retrieved October 3, 2008 from http://www.abc.net.au/am/content/2008/s2228717.htm

Sharer, R., and Morley, S. (1994). *The ancient Maya* (5th ed.). Palo Alto, CA: Stanford University Press.

Shermer, M. (2007). *The mind of the market: How biology and psychology shape our economic lives*. New York: Henry Holt and Co.

Sithole, C. (2008, May 24). Mugabe's £5 million palace complete. *The New Zimbabwe.* Retrieved September 15, 2008 from http://www.newzimbabwe.com/pages/palace13.12328.html

Sierra Club. (2008). *Who was John Muir?* Retrieved June 30, 2008 from http://www.sierraclub.org/john_muir_exhibit

Smith A. (1776). *An inquiry into the nature and causes of the wealth of nations*. Retrieved July 9, 2008 from http://www.adamsmith.org/smith/won-index.htm

Stone, O. (Director) (1987). *Wall street* [Motion picture]. United States: Amercent.

Suzuki, D. (2006). *The sacred balance: Rediscovering our place in nature.* Toronto: Greystone Books.

Tainter, J.A. (1990). *The collapse of complex societies.* Cambridge: Cambridge University Press.

The Mekong's fatal floods. (2008, Sept. 1). *Time Magazine*, p. 10.

Thoreau, H. D. (1854). *Walden*. Retrieved August 18, 2008 from http://thoreau.eserver.org/walden00.html

Timmer, C. P. (2008, June 5). Staving off the global food crisis [Review of the book *The end of food* by P. Roberts]. *Nature, 453*, 722-723.

Trump, D. (2008). *Think like a billionaire*. Seminars presented in Melbourne, Sydney and Gold Coast, November 12-14, 2008.

UN sets up task force to tackle food crisis. (2008, April 29). *ABC Radio National.* Retrieved June 16, 2008 from http://www.abc.net.au/news/stories/2008/04/29/2230926.htm

Urbinato, D. (1994, Summer). London's historic "pea-soupers". *Journal of the U.S. Environmental Protection Authority*. Retrieved July 30, 2008 from http://www.epa.gov/history/topics/perspect/london.htm

Welsh, P. (2000, March 16). The Aral sea tragedy. *BBC News*. Retrieved August 17, 2008 from http://news.bbc.co.uk/2/hi/asia-pacific/678898.stm

Williams, A. (2008, June 20). That climate change is the only issue. *Big ideas debate* [Radio broadcast]. Sydney: ABC Radio National.

Witze, A. (2008, April 17). Climate change: Losing Greenland. *Nature, 798-802.* Retrieved October 12, 2008 from http://www.nature.com/news/2008/080416/full/452798a.html

Contributed by Geoffrey Frost, Tasmania.

Photo: Humans and Nature co-existing.

In: Meltdown: Climate Change, Natural Disasters...
Editor: Kathryn Gow
ISBN 978-1-60876-153-1

Chapter 20

THREE ECONOMIC TSUNAMIS: SHOULD WE HAVE SEEN THEM COMING? A BOUNDARY RIDER'S PERSPECTIVE

Paul Wildman
Economist and Community Development Consultant, Australia

ABSTRACT

Right now, we in the West are being immersed in three economic tsunamis. Although we should have known better, we walked onto the beach as the water went out. "Great", we thought, "now we can gather more shells and dig some more pipis and still get away at the last moment when the tide comes in". Yet at this very moment, the emergent economic catastrophe that we see engulfing us today, was on the horizon in the whispers of the wind announcing its arrival. Yet these whispers have been there for the past generation as an increasingly faint and evanescent memory in our panicky minds. This chapter seeks to answer the question: 'How did we - ignoring the roar, looking at the onrushing wave - (now freeze framed by a photographer from an apartment block), ever get here?' In order to achieve this, we consider these economic tsunamis and explore several behind- the- scenes reasons that we should have known the world financial washout was to occur.

Keywords: economic catastrophe, financial crisis, economic collapse, forecasting, tsunami

INTRODUCTION

This chapter presents an overview about the economic catastrophe engulfing us today, from an outsider's perspective. Here I identify three economic tsunamis approaching us and our children. In many ways, we have forgotten the lessons of our parents, ones they learned in the depression. Tragically, we have also left the management of the economy in the hands of the mandarins and now we find that they never really had any intention of managing the system in our favour, but rather in theirs and their mentors. Moreover, having become besotted by debt, we have forgotten the wisdom of the ancients and abandoned the importance of the physical economy for the leveraged levitation of the monetary 'no payments for 48 months' position.

Fears and concerns about our economic future abound; instantly the tsunamis are personal and collective; national and ultimately global. Our political systems just seem to be impotent, other than to offer reactive palliative assurances that more of the same is on the way. For too long, we have given up our right to know and understand the economy. This article argues that now we, as citizens, need to take a more active role in the economics of the world that our children will inherit from us – leaving it all to others, to the mandarins and their masters, is no longer an option.

Further, the chapter argues that our systems of government, based around the Nation State, have clearly been shown to be inadequate and more suited for a stable preschool setting. We should, and could, have known that the washout was coming, yet even then we were wading out towards these oncoming tsunamis.[1,2]

THREE ECONOMIC TSUNAMIS APPROACHING THE WEST

Today in the Western world, we face three tsunamis of economic change. Economics can be defined as the efficacious production and distribution and consumption of goods and services for our everyday life. Clearly this speaks primarily to the physical economy.

Tsunami I: Finance/monetary system. The size of the monetary economy[3] is currently many times, even up to 100 times, the size of the global physical economy of goods and

[1] Tsunami (ts*oo*-nah-mee). A natural disaster resulting from a very large ocean wave that is caused by an underwater earthquake or volcanic eruption and often causes catastrophic destruction of property and loss of life when it strikes land. Tsunamis can have heights of up to 30 m (100 ft) and reach speeds of 1000 kph (600 mph). They are characterized by long wavelengths of up to 200 km (124 ml) and long time periods, usually between 10 and 60 minutes between wave fronts. The speed travelled usually means they arrive unannounced, plus the sea water generally retreats into the ocean for several hundred meters just prior to their arrival. A tragic example herein is the December 2004 Indonesian Tsunami which claimed over 250,000 lives and impacted overall 1.3m people.

[2] For the past decade or so, especially on the web via. YouTube, many pundits have been issuing warnings (I called mine 'Wildman's Warnings'), including a series of 6 DIY manuals for thrival in these coming difficult times (1993) and more recently an interactive CDRom (2003). Early on in the process, I realised that basically no one was listening and, as in entrepreneurial theory, little attention is paid thereto while the 'going's good'. It takes a 'displacement event' such as we are experiencing today to alert people to the need to open their minds and listen afresh - even to heed the warnings.

[3] There are 5 types of Money Supply (MS) in Economics: from M0 (basically coins and notes in circulation). The next narrowest measures of money supply are M1 (MO+ and bank reserves with the central bank) and M2. These add various types of deposits by the private sector with banks and other financial institutions. These are still regarded as narrow measures. They are sometimes referred to as referred to as narrow money. The broad money measures (M3, M4 and further) add other types of money such as bank acceptances, commercial paper and bonds. Some countries (but not the UK) include foreign currency deposits. M4 then is M3 plus financial and debt instruments, with maturities of up to 5 years. The leveraging, derivatisation, collateralisation of debt etc., which we see in the financial economy today, is beyond M4; and in this regard I propose, and use in this chapter, M5 to include these 'intangible' instruments not money, not debt, not physical products - almost a phantom. This chapter uses the term money supply or monetary system in this broader M4 and beyond type manner and beyond, in effect, including all forms of dollar denominated financial instruments whether debt, leverage, hedge, derivative and so forth. As we move from M0 to M4/5, we move from public to private, from public control to private whim and appropriation, from concrete abstract to intangible abstract, from regulation to speculation, from Main Street to Wall Street, from Oikonomia to Chrematistics.

services. We are awash with cash, near-cash, non-cash and just trash cash – all, other than cash itself, effectively internationally generated and yet completely unregulated. Welcome to the global casino. Our concern was that once this speculative bubble burst, so to speak, it would take out the physical economy. The monetary economy has expanded internationally since 1971 to around 100 times the national physical economies, all outside the current controls which were built for a bygone post World War II era. Now the value of the global currency, the US dollar, is outside the US's control and is set by OPEC. That is, until OPEC denominates its petrodollars in another currency; say the Euro or Yuan, or some other potential international currency.[4]

This tsunami has been approaching since 1971, and at an increasing rate since then, when the US broke the link between the two economies by President Nixon revoking the Gold Standard. This tsunami hit as of mid 2008 and now its knock-on effect on the physical economy is being felt, with unemployment increasing, asset deflation occurring especially housing asset bubbles, debt pyramids approaching collapse, more engagement in predatory lending practices and so forth. Many pundits anticipate a three to five year turn around; however, as we will see in the rest of this chapter, a simple return to 'business as usual' will not be an option[5].

Tsunami II: Unemployment. This tsunami has developed in three main arenas, One has occurred in developed nations, as their physical economic basis is exported holus bolus to third world nations and simultaneously their education and training infrastructure has been reoriented towards economic rationalist goals (Wildman, 1998). A second scenario shows intelligent, even sentient, technology will arrive by the mid 2020's and the need for basic green, blue and pink collar workers will rapidly decline and perhaps even disappear in both

---

[4] E.g., China's Premier Wen 29-01-2009 at the Davos conference on World Trade blamed the global financial crisis on 'inappropriate macroeconomic policies of some economies' and 'prolonged low savings and high consumption,' in a lightly veiled attack on the United States. He highlighted the 'excessive expansion of financial institutions in blind pursuit of profit and the lack of self-discipline among financial institutions and ratings agencies' while the 'failure' of regulators had allowed the spread of toxic derivatives. ABC News Report 29-01-2009. Further history shows us that societies that persist over time have social mechanisms for limiting capital growth; i.e., they limit the monetary economy through cultural mores such as Islamic limits to usury, and potlatch systems in North Western Indian tribes in the US at colonisation, to keep it within the bounds of sustainable physical economy (Greer, 2008, p. 237).

[5] For instance, Cordeiro (2008, pp. 6-10 esp. Tables 3 and 4) explores this question of return to business as usual, in relation to monetary management, especially in developing nations, and finds the business as usual approach of floating exchange rates adversely affects developing nations, in terms of growth and stability; e.g. in relation to the 'Lost Decade' of the 1980's in Latin America. Overemphasis on such floating exchange rates and off shore creation of international pseudo-credit can lead to the 'four horsemen of the Economic Apocalypse': exchange rate devaluation, domestic inflation (speculative bubbles), capital flight and ultimately recession. Further, Australia's Prime Minister, Kevin Rudd, has declared a new era of financial reregulation and government intervention in world markets to restore confidence and stop capitalism 'cannibalising itself' as a result of the global financial crisis. Cowboy capitalism must not be allowed to destroy 'real' read 'physical' economy, especially its jobs, he says. [Weekend Australian 31-01-2009 to 01-02-2009 p. 1] Tragically, I suspect most centre-left Governments, such as Australia's, will see the need for regulation as compliance and miss the opportunity to reform the system through transparency, accountability, scrutiny of the monetary economy and re-linking it to the physical economy. Creating money, or what I call 'pseudo credit', on such a massive scale, without creating balancing wealth in the physical economy, is in my opinion, the financial equivalent of an over-unity perpetual motion motor, which is ultimately in defiance of the laws of thermodynamics. I suggest that this indicates, when combined with our related rapid transition to virtuality, and the increasing abstract nature of our civilisation, our seeking for immortality via a simulacrum, wherein we seek to break the bounds from our physical world and our selves; a counterpoint for instance to the Pharaohs who sought immortality through the simulacrum of the physical pyramid and mummification. May I caution strongly against either hubris.

first and third worlds (Wildman, 1999, 2000). Even with present levels of automation, authors, such as Gorz (1980), argue that we can produce everything we need with less than 20% of the current workforce. The third area of concern is in developing nations, where debt repatriations to the West for so-called development loans exceeds their GDP, and corruption in several of the recipient countries largely diverts what aid funds are accorded.

This has been approaching since 1992 via, for instance: (a) the suite of national reports that all call for the 'productivitisation' of training, in line with the needs of international competiveness as demarked by (b) the advent of the World Trade Organisation (WTO) in 1995[6]. This unquestioning embrace of globalisation was rapidly entrenched in industry and training policy as the preparation of Australia's Vocational Education System for full speed globalisation[7]. Since the formation of the WTO, whole countries' physical economies have been decimated and training systems have been wholly 'commodified' and re-oriented as product and productivity. Education, learning and training have been reorganised to be conceptualised within the economic discourse, and as a foundation of the productivity meme.

Tsunami III: Physical Economy/ Manufacturing/ Long Hollowing. Since the collateral explosion of financial instruments post 1971 and the 'enforcement' of globalisation, post the

---

[6] Finn Report (1991), the Carmichael Report (1992), and Mayer Report (1992).

[7] In the preceding decade, the thrust towards Globalisation was proceeding apace. Four key impact areas of an 'economic eugenics' relevant to this chapter, are firstly privatisation primarily under Hawke and Keating, which effectively privatised Australia's physical infrastructure. A second attribute of Globalisation was superannuation, previously used for replenishing such infrastructure, but then directed by Keating into the privatised banking system which ultimately supported the real-estate bubble - the bursting of which has led to the rationale for this chapter. While the third impact was evident by March 1992 when the Employment and Skills Formation Council released its report proposing the Australian vocational certificate training system be highly attuned to international competitiveness and the Globalised needs of capital and its banking; and the fourth impact being the corporatisation and internationalisation of money supply. That council was under the chairmanship of Laurie Carmichael, a former trade union leader. The report recommended a staged strategy for meeting Australia's training needs by the year 2001 through structural reform. The focus of the Carmichael report was on the growing convergence of work and training and the link between general and vocational education.

These reports have had, in my view, a tragic side effect, in that they conflated all training into productivity improvement oriented, structured competency based training, in the service of global competitiveness in line with the introduction of the WTO, and sought to apply Taylorist principles of skill disaggregation into specific and individual behaviourist competencies applied to individual learners. The result was a totally de-contexting and mechanising skill formation; even more tragic was the totalisation of the discourse of learner as a skin encapsulated individual and working class learner who did not need any critical, problem solving, syncretic, creative or citizen formation skills; rather just basic behaviouristic componentised competencies, as Competency Based Training (CBT). The learner was now a commodified part of the training machine and outputs were behaviours - a sort of correct sequence of lever pulling. Such disaggregation has gone largely unrecognised and is, in my view, nothing short of a national tragedy. So keen was the Union movement to identify with the corporate state that it too, gave up its sacred calling to protect the worker and workforce, and rushed head long, and indeed even as the very head/chair of many of these inquires, into these arrangements.

Critically all other forms of training such as ACE (Adult and Community Education), Training Levy sponsored industry training, and community learning processes including learning circles and experientially based learning processes ,such as Action Learning - all foundational to the social fabric and conviviality of our country which forms the basis of economic efficacy, Australia and crucial aware citizen, were, in a period of a few years - has now been almost fully extirpated by the now top down economic rationalist structured, accredited productivity, first discourse, and survives only in isolated informal pockets such as the venerable Great Australian Bush Mechanic. As Deputy Commissioner for Training in Queensland at the time, I well member attending the hearings in Brisbane and seeking, unsuccessfully as it turns out, to make this point face to face with Dr Carmichael. The eventual outcome has been a horrendous 15 years, whereby all but structured formalised training has been eliminated. I posit that such informal education was, only ever at best, half of the overall national training effort and was undertaken in TAFE, ACE, Schools of Art, Mechanics Institutes, Industry, NGOs etc.

formation of the World Trade Organisation post 1995, we have seen the emergence of what the Japanese have been experiencing now for a generation that is 'the long hollowing'. This is the emptying out of the physical economies of developed nations especially the US, Europe, Japan, and Australia whereby their physical economy has basically been exported 'holus bolus' to developing nations with effective labour rates between up to 1/20th that in the developed nation with no oncosts including no superannuation, annual recreation leave, sick pay, with little or no Occupational Health and Safety.[8]

Simultaneously the monetary economy has been growing in staggering leaps and bounds and combined with the above, and largely unnoticed, automation has been underway for several decades. For instance, now we have the majority of vehicle manufacturing being undertaken in third world nations in near fully automated factories. This situation has been in train since 1995 and the formation of the WTO.

The long hollowing. Possibly, the long hollowing is the least known of the above three Tsunamis; in a little over a generation, this 'hollowing out' occurred in the economies of the developed nations. This can lead to what I call 'economic pole' inversion, whereby developed nations enter a process of de-development, and developing nations then locate their factories in ghettos therein (Wildman, 1998). Apparently, BMW is now doing this in the USA. In particular, this applies to the physical economy in such industry categories as manufacturing goods, machine tools, even agriculture. Developed economies then are left with ever contracting niche manufacturing and agriculture, and of course financial services. The one area they can excel in, however, is the monetary economy, that is, until it is outsourced to India. This economy, to our relief, remains essentially outside the purview of the Nation State and with no international regulation.

As the substance of Nation State economies thus becomes 'thinner' or 'hollower', then with perturbations (such as when the three tsunamis, peak oil, global warming etcetera occur), they/we do not have the economic depth to restructure readily, or to adapt laterally for self or national resilience. This, in my view, increased the overall global systems' vulnerability to these types of impacts and thus amplifies the negative outcomes through limiting the developed nation's resilience.

In counterpointing this, the knowledge economy and information economy pundits argue that today, it is the manipulation of symbols and not the manipulation of matter that is of relevance. My response is that if this is ever the case, then in the future we will all be living in the noosphere or will have all left the planet. One can't eat noosphere, biosphere and physiosphere noosphere.[9] Humanity is an organic entity, and economics is a mundane organic science that, I argue, ought to aim at 'providing the goods and services for our everyday life'. The key issues of today relate to our ability to look after one another and our planet – all of these require organic actions, which are certainly 'informed' by knowledge, skill, symbols and other factors of production; however at the end of the day we must eat.

Today's economic superstorm. Today, however, I suggest that we address the immediate concerns at hand, as we face these three key economic tsunamis. Further, I also suggest that

---

8 In Australia, these on-costs are around 30-40% of top of the cost of the actual wage.

9 Here we see the tipoff - the modern epistemology that valorises abstract thought and pejoritises action/doing/organic activity. We see this distinction, for instance, in the separation and naming of Higher and Vocational Education. NB: modern money as per M3-4 and M5 in footnote 3 earlier is such an abstraction; a food van for the homeless is not abstract, nor is a local farmer's market, although they also both have a certain symbolic value.

our present situation, as it is one of a surfeit of monetary aggregates and thus inflationary, is different in this regard from the Great Depression, because it is basically deflationary; that is, there is a shortage of money which leads to deflation. Nowadays, perturbations are instantly transferred worldwide via web based investment and trading algorithms, and we face unstable and weakening Nation States which are increasingly expensive to run, with little efficacious governance nationally or globally.

These can all conflux to generate, if you will, 'an eco-socio-econo superstorm'. To address this superstorm, we need first to look to our own 'personal or household economy' - our own Oikonomia - and then take this to the structural level. Clearly significant changes need to be made to capitalism.[10]

What Has Shaped These Tsunamis?

Now that we have identified the waves, we can ask the question about what gave them their shape. In this section of the chapter, I seek to identify four such shaping forces.

Depression Lessons not Learned

As we face this superstorm over the next three years, are there some lessons we can learn from history? Well....yes there are. In the mid 1970's, I interviewed my parents and their friends (about 10 in all) who had gone through the depression (1929-1934) about economic survival lessons to be applied in the future, based on their depression experience. They are all now long dead. Little did I know that some 35 years later, it would become relevant again. The lessons included the importance of minimising debt, maximising mutual aid, undertaking small scale practical projects such as gardening, DIY (do it yourself) repairs, seamstressing,

[10] We see intimations of this, for instance, from the UK Prime Minister and the President of the USA who have, in the week of 3-03-2009, both said we need to manage our economy upward from our households, not downward from international finance. These changes, however, are beyond the scope of this article, although a few are listed here: (1) usurpation of the physical economy by the monetary economy; (2) legal definition of the corporation as a 'natural' person; (3) Nature's property rights valued at 0; (4) unregulated international finance and banks; (5) international creation and control of capital; (6) commodification, cannibalisation of the social fabric that sustains the economy; (7) commodification of human relations inc. training; (8) immediate obsolescence that means discard, don't recycle, let alone reuse or repair; (9) lack of alternative measures of economic outcomes; (10) an economics which can only value something, if it is within its monetary calculus, such that the ecological has to fit within the social and that within the economic, so the ecological is judged by monetary calculus; (11) resource ecological and economic impacts of population explosion including species extinction; runaway economic footprint; (12) distortions in the progressive taxation system to favour the rich; (13) deeply entrenched structural corruption, - private gain | public pain e.g., CEO's salaries etc.; (14) stock market and derivatives - a case of conflict of interest as the stock market, now privatised and corporatised, relies in a turn -over charge, and hedge funds and derivatives make huge numbers of transitions per day; thus the stock markets profit thereby and become dependent thereon, and beholden thereto; and finally (15) nature of the corporation, as if it were an individual human being.

Clearly many will argue that this does not go far enough and that wholesale and deep changes are needed in governance, markets, property rights and work. This may well be the case; however the author argues for renovation and 'religio' 'relinking' process re-evolution; i.e., building from within capitalism and from the ground up, rather than advocating increased regulation. Further, the momentum built from addressing the three tsunamis and these 15 changes will, I suggest, establish intentionally, determination and possibly community capability, to move into these more challenging arenas. This is still system redesign and not business as usual type incremental adjustment in regulations; however it is, I argue, a 'doable' task.

repairs, remanufacture, reuse, refocusing on alternative uses, "mates rates" labour exchange, and so forth - all what I would call 'bush mechanic' and 'community economic development' type skills (Wildman, 2005). Clearly the theme here was managing uncertainty upwards – a 'big gig'. Perhaps this article could be renamed 'when big systems fail from the top, then what do we do at the bottom?'

After the Great Depression and its drift into the Second World War, it was recognised by political leaders that a system was needed to prevent, as far as practicable, a repeat of the depression in the future. To do this, in 1945, the Bretton Woods agreement, at the close of World War II, provided for the monetary economy to be mechanically linked via gold to the monetary economy through what was called 'the gold standard'. This remained in place basically until President Nixon revoked this agreement on 15th August 1971, and until that date, American dollars were convertible into its equivalent of gold.[11,12] For me, from a

---

[11] Gold as a defacto international currency had been in place, in various countries, for centuries. The issue of the Gold Standard has three dimensions: (1) linking the physical and monetary economies; (2) as a means of international currency to fund global trade; and (3) to regulate internal economies within Nation States as part of monetary policy. The impact of the gold standard viz. (2) and (3) can be seen in, for instance, http://au.youtube.com/watch?v=O7pnjzCuSv8andfeature=related . This chapter uses the Gold Standard in relation primarily to dimension (1) and to an extent (3), while (2) relies on the US$; that is, as an example of one way of linking the physical and monetary economies and one important tool for internal management of a Nation State's market economy.

[12] The Bretton Woods Conference of 1944 recognized the need for a comparable international institution for trade (the later proposed International Trade Organization (ITO) to complement the International Monetary Fund and the World Bank. In this way, the Nation State system sought to (a) gain a global systems reach to help prevent WWIII for instance, and (b) to use this approach to help prevent another depression. This is crucial to our considerations in this chapter, to help prevent what was seen as key flaws in the previous trade arrangements which, in part, contributed to the Great Depression which ended only in mobilisation for WWII. Essentially, these flaws were seen mainly in the fear of protectionism rather than seeing what the strength of the USA economy had achieved through mobilisation - and earlier that included a certain level of protectionism.

Author's note: another key issue was the lack of knowledge of a Keynesian (Government surplus expenditure) solution.

Associated with the Bretton Woods Agreement in the immediate post war years 1944/1945, was the establishment of the International Monetary Fund (IMF) and the Word Bank (WB). The emergence of these global economic management institutions was parallel to that of the UN itself (which came into being in October 1945) with the former growing out of the deliberations of the latter. A third further protocol crucial to our present discussion was established in 1947, the General Agreement on Tariffs and Trade (GATT) which, along with the International Trade Organisation (ITO), was rolled into the WTO on 01-01-1995 (World Trade Organisation). The WTO has 153 members, which represents more than 95% of total world trade. These 'flaws', that WTO is designed to overcome, include trade restriction, economic isolation, tariffs and protectionism etc. There is, to my mind, a strong argument that the WTO agreement violates the UN Charter of Human Rights, even to the point that the statutes of the World Trade Organisation which were drafted in separate closed sessions by technocrats outside the purview of the Governments who subsequently signed it and thus beyond the purview of the people, e.g. us, in the countries now subject to its rules and regulations, and further such a focus on the axis mundi of the economic system is directly opposed to any local economic emergence.

Readers will note please that, although I make these points as an aside in the context of this paper, it is crucial in that (1) the gold standard was extant so the link between the physical and monetary economies was fixed and (2) Government expenditure, for instance on public works and infrastructure, occurs directly in the physical economy. Nowadays, we see Nation State Governments stumbling over themselves to show how they are good Keynesians (yet you can't see this, when the market is firing on all cylinders, so to speak, and Western Governments- e.g. the USA- , are cutting back on welfare expenditure and reducing even further residual stock market and monetary regulation. The USA, Europe including the UK and Australia are exemplars of this 'spend your way of out of recession' approach, which I argue has its roots in economic events some 80 years ago viz. the Great Depression. The big challenge here is that now there is no gold standard, the Nation State Governments can simply print money to obtain this Government expenditure, as indeed the USA is doing at present. Finance corporations use derivatives, hedge funds and collaterialisation of debt etc. to achieve the same results – create money out of nothing, thus running the twin risks of further hollowing out their (host)

theoretical position, the removal of this link between the monetary and physical economies, as represented by the gold standard, is utterly fundamental to understanding the economic tsunamis of today.

The Anatomy of the US 'Mother of All' Ponzi Schemes

The 'inflatability' of the US monetary system is illustrated, for instance, in considering the US Commercial Banks holdings at 30-06-2005. Of Off Balance Sheet Derivatives approximately $100,000b; Assets, approximately $10,000b; Loans and Leases approximately $5,000b; Equity capital approximately $1000b; and Loan loss Reserves $70b. An estimate of the reserve situation, inherent in these figures, is as follows; Liabilities = D+L = $105,000b and, Assets = E+R = $1070b = 1% of liabilities. That results in 2006, for instance, of a Liquid Reserve Ratio of approximately 1% of Liabilities. The article (New Citizen, 2006, p. 2) has the footnote 'highly-risky derivatives dwarf banks' other assets'; and 'indeed derivatives were illegal in both Australia and the US until about a decade ago, because they were not only gambling, but were side bets on gambling'. One could, I submit, legitimately describe the whole derivatives system, even our whole economy, as one giant Ponzi scheme, in which we all live.[13]

One could go a step further and argue that the global monetary/financial system had become, in effect, the *Mother of all Ponzi's* and the best con-game in town (Hsu, 2008, p. 80). It went like this: (a) get paid well then manage one's financial dealings to generate huge amounts of risk/debt from say sub-prime mortgages, with little or no ability to repay basis and as the above example shows few, if any, reserves, then (b) generate enormous income from nothing by securitising or collateralising this risk/debt, which is (c) then on-sold out of the regulated banking system, thus placing it beyond government control which allows you to (d) leverage and hedge these liabilities now appearing as assets on your books on a 3, 4 and even 5:1 basis[14] then (e) invest this loot in even more dodgy leveraged deals, finally (f) when the margin calls[15] are made, file for bankruptcy and then (the next step is the most astounding

---

economies through the international coropor'n'ations and triggering (hyper) inflation through printing money by the nation states.

Interventionist economic management systems based, for instance on Keynesianism or fiscalism (Government expenditure) have been linked sometimes pejoratively, to protectionism, isolationism and pump priming; whereas Friedmanism or monetarism (money supply changes in its price and availability) have been linked, sometimes pejoratively, to unfettered markets, market deification, uncontrollable volatility in the Money Supply and destruction of regional economies. Everywhere in the West today, this approach to monetarism is ascendant. Third sector (NGO based) physical economic development systems such as CED, aren't even on the page for consideration.

[13] And indeed M4-5 as per endnote 3 is up to 100 times the size of M1-2 and even larger in terms of the size of the physical economy. Ponzi – noun- a swindle in which a quick return, made up of money from new investors, on an initial investment which lures the victim into much bigger risks. Further assets (A) are excluded here as they are primarily stocks (capital account – buildings etc.) and not trading flows (current account); further there is an assumption here that these are net assets. Moreover, these Off Balance Sheet (OBS) derivatives and hedges can then be on-sold and securitised, thus generating I suggest a total OBS figure several times that in the above example reducing the Liquid Reserve Ratio (LRR) of 1% even further.

[14] This is, for instance, a leveraged loan whereby a loan of $5 is given for each $1cash lodged. The author is aware of leverage rations several times the 5:1example, which is included here as a conservative estimate.

[15] Margin calls is where the market value of your portfolio, purchased with the leveraged finance falls below the borrowings you have undertaken to establish the 5$ (loans – issued to you):1$ (assets – contributed by you) line of leveraged credit; what is called a margin call is made and repayment of all the lent capital, now several

aspect of this whole debacle), if you're a big enough bank or financial entity you can have (g) Governments fix your problem, at the expense of the taxpayers who now lose their houses, but who never saw a cent of shared wealth to begin with.

Let's Get to the Heart of the Matter – Money

So, hopefully we can now see the centrality of money to our present catastrophe, so let's drill down into this thing we call money - literally the currency of our crisis. Here, we then find ourselves asking two crucial questions (A) well then what is money, and (B) who owns it?

(A). Money comes in various forms: (1) *trust money* – barter and other good will exchanges where 'one's word is one's bond' so to speak; local exchange systems such as LETS are sophisticated forms of barter; (2) *commodity money* - store of actual wealth – gold and precious stones; (3) *representative money, that is*; representative of wealth (paper certificate/dollar that can be exchanged for actual wealth); (4) *fiat money* - in which paper notes are backed only by use of' 'lawful force and legal tender laws' of the Nation State Government) [note well: this is the present situation in all developed nations and clearly, thereby, the link between the physical and monetary economies is broken]; (5) *interface money* – money here is a medium of exchange (such as with a local currency wherein no federal fiat is required); (6) *virtual money* – really virtual fiat money and with the advent of the web has in a sense aided the volatility of international monetary flows, as these are outside any form of control, just as their destination instruments are; (6) *eco money* – here money is a form of earth energy – human energy, carbon trading and the eco dollar (where each eco$ is worth a kilowatt of renewable energy for instance.

(B). Who owns this money? In respect of the above list, money is owned by: (1) *trust money* - by those undertaking the transaction: (2) those with the resources, that is, that can generate the *commodity* that will function as money; (3) those with the power to create the certificates – *representative money* is approaching fiat money, but with a ratio, for example, of 10% of the certificates being held in actual commodity reserves; (4) Governments who can create and enforce the *fiat;* (5) *interface money* - here we have a type of local trust fiat money that would be used to generate a monetarist type local economy; (6) *virtual money* is owned by the computer programs and algorithms that create (7) *eco money*; if undertaken correctly, this is a stewardship style of monetary system so the majority shareholder, so to speak, would be the host of the stewardship itself - that is, Gaia.

Today we find the emerging realisation that in the Great and subsequent Depressions, there was a lot of work to be done and many people with the skills to do it, but yet there was no way that the monetary system would allow the two factors to be linked (Lietaer, 2003). From my perspective, this is for me where a personal↔public economic dilemma emerges; for example, looking for work (a personal dilemma) also becomes a public domain moral question, because the reason the two could not be linked, that is, the person and the work, was that there was no available money; money that is, in the sense of federally produced by Nation State mints. [NB: a local currency would, however, readily link the two]. Note that

times what the portfolio is worth, is required within days. Thus starts the domino effect and the collapse/implosion of the 'House of Ponzi' is underway.

anyone making their own money, to try and link the two, was, and still is, engaged in an act that makes one guilty of a major crime. Such federal creation is, of course, *scarcity based* and regulated by the Governance system, in order to set the price (interest rate) of money and its availability (money supply). As we have seen from the current international monetary fiasco, the price of money is extremely low, but its availability to borrowers can be extremely restricted.

It appears that the institutions, least affected by the present crisis, may well be the Islamic banks which have been somewhat immune to the collapse, possibly because Islamic banking prohibits acquisition of wealth via gambling (or alcohol, tobacco, pornography or stock in armaments companies). Further, Islamic banking apparently forbids the buying and selling of a debt as well as usury; additionally, Shariah banking laws forbid investing in any company with debts that exceed 30% (Hsu, 2008, p. 81).

Wither the Nation State? Its loss of Control of the Vital Economic Factors of Production

The factors of production in economics represent, as it were, the ingredients of any economic activity. These factors are: (i) land, (ii) capital (money in this chapter), (iii) raw materials, (iv) labour, and I would add (v) technology and (vi) process. Clearly the first three and the last two have, post World War II, deserted the Nation State. Today, we witness yet another factor of production - monetary policy - moving outside the Nation States' purview. Consequently, to retain the appearance of legitimacy, the Nation State simply passes more laws on the only factor of production that the Nation State can control currently - that is, labour (see (iv) above; that means us - its citizens - us - as physical taxpaying passport requiring entities. In this sense, the Nation State has become a simulacrum that we collectively construct, so that we can believe that we have some say in what goes on. Consequently, it may be argued that as far as the Nation State is concerned, it has, over the past two generations, given away basically all its controls that allowed it to keep a 'steadying hand' on the tiller of society.[16] Even worse, in the past generation, the defacto international regulators have gone further and passed into law international trade regulations that allow our Nation State Governments to be punished, whenever they try to place an economic hand on the tiller and regulate the monetary and or physical economy via trade etcetera.

Rather, I argue, we have moved to mechanise, then economise, then statisticalise, then computerise, Adam Smith's Invisible hand – and call it 'markets'. The market now is the quintessential expression of our old mechanical Christianity – omnipotent, omniscient and omnipresent, objective, and mathematical and above all 'free', with markets (i.e., consumer votes, cp., purchase), ultimately being considered superior, even to democratic voting, by

[16] For example: even 30 years ago in Australia, for instance, the Nation State Government controlled most of the economic factors of production as well as : (A) the equivalent of superannuation, and used it, at low/concessional interest rates – reducing the need for foreign investment, for infrastructure construction for which nowadays Governments have to pay market rates. In the 1970's, this superannuation rate of return to contributors was about 2% cp. general interest rates of around 10%. Nowadays, infrastructure capital is obtained by Australian Governments from various sources including international capital investment banks; (B) the money supply; (C) interest rates; (D) owned the people's bank - Commonwealth Bank (established as the people's bank in 1921 and privatised in 1991); (E) set statutory reserve deposits of around 20% of deposits (now there are no fixed SRD's).

some economists. Thus we have moved from a 'society as heuristic' viewpoint to a 'society as algorithm' viewpoint[17]. No progressive[18] theologians have believed in this 'theistic' mechanical algorithmic formulation for more than a generation, nor should economists.

Yet we let the instigators, regulators and managers of these systems, both private and public play and even game with our money, with our own and our children's and ultimately the planet's future, without holding them, or the systems they operate, directly accountable for the performance of our economies. Nowadays, most of the top 300 economic entities on earth are Corporations (I call them corporanations) beyond Nation States, and yet uniquely their leaders are not held accountable for their many disastrous and incompetent decisions, with some even ending up with 'golden parachutes' of multiple million dollar payouts, 'when the balloon goes up', or more correctly phrased, when the parachute fails to open. These private and public derelictions of duty have, in no small way, helped produce the three Economic Tsunamis we are discussing in this chapter.

Where to from Here? Riding the Waves to Avoid Sinking

On reflection, to my mind, there are three principle steps consequent to each of the Tsunamis, if we wish to avoid sinking. While a detailed explication is beyond the scope of the chapter, we can outline several key initiatives that synergise with each of these three major Tsunamis (TI, TII, TIII). In all these, we need to see a positive joining of wealth and health, so that our economic success also complements our own and our planet's health (Wildman, 2002).

First, get three wave rider buoys - each set to monitor one each of the particular approaching Tsunamis, then (TI) by 'religo' or relinking the physical and monetary economies and managing the interface, so that the latter does not exceed the former by say more than a factor of 10. Achieving this will require legislation, as well as the establishment of credible and ethical standards, and accountable transnational global financial controls, as well as addressing at least some of the 14 major structural flaws listed earlier in endnote 10; (TII) establishing employment generation and innovation schemes and processes to boost the local employment, through innovation aimed at supporting the physical and socially sustainable economy, while being aware of the onrushing singularity[19] (Wildman, 1998, 2005); (TIII) re-establishing the physical economy within the Nation State and linking sub-national regions in inter and intra regional confederations (Wildman, 1999, 2000).

[17] See for instance the 2007 BBC Documentary Series 3 one hour episodes of the BBC documentary series "The Trap" by Adam Curtis. The series explores the concepts and definitions of freedom, specifically, how a simplistic neo-liberal model of human beings as self-seeking, almost robotic, creatures led to today's idea of market based 'negative freedom', whereby essentially everyone has the freedom not to interfere with the market http://en.wikipedia.org/wiki/The_Trap_(television_documentary_series)

[18] See work of theologians: Armstrong, K. (2000) and Taylor, S. (2005).

[19] There are various meanings for the singularity. In this chapter, I posit two primary types: (1) The Economic S which we see via totalising discourse of Globalisation – the end of economic history type scenario with 'The solution being incandescently, happily and energetically welcomed and applied by all and sundry (sic!), and (2) The Technological S – The technological singularity is a theoretical onrushing future point of unprecedented rapid technological progress, caused in part by the ability of machines to improve themselves using artificial intelligence. This is getting very close; some suggest around 2030, as humans transcend our biology through recombinant forms of life generated through GNR the new black for DNA viz.Genetic Engineering | Nanotechnology | Robotics.

CONCLUSION

This chapter has sought to show that we could, and indeed, should have known at least for the past generation about the Tsunamis of the oncoming economic Dark Age. Further, we should not have so easily turned our backs on the lessons our parents learned in the Great Depression. We, as citizens, have handed our economic sovereignty to the ghosts in the financial machine, over the carcass of the political systems in our effete Nation States. Now we have to take it back. Clearly, we have also failed to hold to account those in power who have directed us towards this catastrophe. Crucially, we must individually and collectively ensure that these structural flaws are addressed.

In stabilisation and recovery mode, our governance process will need to call on all parts of the economy, both formal and informal, and physical and monetary. The answer, however, will be to 'harmonise their diversity, not centralise their conformity'. This will require Governments to legislate governancal prudential provisions, in particular on the monetary economy to keep it in line with the physical economy, not to try to colonise the informal economy, and to allow local currencies and so forth to develop, as it were, as zones of citizen economic innovation. Internationally, as the Nation State continues to eviscerates itself, some sort of third sector civic movement will be required internationally, as a countervailing power, as transnational governance continues to be beyond the capability of our present inept Nation States. So back to the future for business as usual is not an option.

Finally, I have acknowledged that many of us in this post modern world, see such legislation as 'a bridge too far'. We have lost faith in our Governments and politicians at all levels. All we can do is to leave the 'formal' system to a degree and manage our own affairs as best we can, in individual households and intentional communities. If taken to extremes, this can become 'retreatism', 'primitivism' or even 'caveism', where we individually hoard foodstuffs in preparation for the coming apocalypse. We witnessed lots of this type of action in the 1970's, as the hippy experiment turned sour and disappeared into drugs and social welfare. This chapter strongly argues against a repeat of this scenario; we have to stand and make it work this time in our own living areas – house by house, water tank by water tank, and garden by garden. Nor will our fear of flying be assuaged by one hero president or Hollywood celebrity chatting to the Dali Lama, although these may just be a start.

We, as citizens, need to rediscover how to tame the tsunami, even to shape its energy, and even to ride the waves. It is coming our way and is not going away.

The economic system can and must enhance our democracy and be proactive in these instances; yet, as it is presently constructed, our economic systems do not even perform reactively or ethically. Nor does it send us early 'wave rider buoy' type warning signals, and so this time, while it was difficult to *see* the Tsunamis coming, we should have *known* they were heading our way. Let's learn from this and next time see them coming; if not, it may take a decade or more to recover. Our grandchildren deserve nothing less than our best efforts based on our decades of experience.[20]

---

[20] Experience here means practical hands- on experience, not only theoretical. [Editor's note: Stephan Lewandowsky also take up this theme of thinking of our grandchildren – see his Commentary in this book.]

REFERENCES

Arendt, H. (1963). *On Revolution*. London: Penguin.

Armstrong, K. (2000). *The Battle for God: Fundamentalism in Judaism, Christianity and Islam*. London, HarperCollins.

Carmichael, L. (1992). *The Australian Vocational Certificate Training System*. National Board of Employment, Education and Training, Employment and Skills Formation Council. *Canberra:* Australian Government Printing Service. [21]

Cordeiro, J. (2008). *Monetary Systems in Developing Countries: An Unorthodox View*. Chiba, Japan, Institute of Developing Economies (Japan). IDE Discussion Paper 154: 36 pgs.

Finn, B. (Chair), (1991). *Young People's Participation in Post Compulsory Education and Training. Canberra:* Australian Government Printing Service.[22]

Gorz, A. (1980). *Farewell to the Working Class - An essay on Post-Industrial Socialism*. London: Pluto Press.

Greer, J. (2008). *The Long Descent - a User's Guide to the End of the Industrial Age*. British Columbia Canada, New Society Publishers.

Hsu, T. (2008). Death of the American Empire and Economy. *Nexus* Dec 2008 - Jan 2009, NEXUS Magazine Pty Ltd. 16 (1), 23-26, 80-82.

Korten, D. (2009). Beyond the Bailout - Agenda for a New Economy. *YES Magazine* Winter. *www.YesMagazine.org/bailout*

Lietaer, B. (2003). Complementary Currencies for Social Change: Bernard Lietaer interviewed by Ravi Dykema. *Nexus: Colorado*. http://www.nexuspub.com/articles/2003/july2003/interview.htm

Mayer. E. (Chair). (1992). *Putting General Education to Work*. Australian Education Council and Ministers for Vocational Education, Employment and Training. Canberra: Australian Government Publishing Service. [23]

New Citizen. (2006). A vision for the Next Fifty Years. *New Citizen*, p. 2.

Taylor, S. (2005). *The Fall: The evidence for a Golden Age, 6000 years of insanity, and the dawning of a new era*. Winchester: Circle Books.

Wildman, P. (1993). *The Community Economic Development DIY handbook series*. 6 booklets/DIY action learning manuals. Brisbane: Prosperity Press.

Wildman, P. (1997). *Communities Working and Learning: The philosophical, epistemological and methodological underpinnings of a community economic development process that seeks to increase sustainable local employment and training opportunities by linking local economic development and community development through action learning and action research*. Southern Cross University/IMC International Management Centre Australia. Doctorate.[24]

[21] Proposed a new unified entry level training system, the Australian Vocational Certificate Training System (AVCTS), later renamed the Australian Vocational Training System (AVTS) inc. the national registration of training organisations renamed as Registered Training Organisations or RTO's.

[22] Recommended a convergence of general and vocational education and established targets for student participation in education and training following the compulsory years of schooling. Also recommended that 95% of 19 year olds should have completed Year 12 or an initial post school qualification or be participating in formally recognised education or training.

[23] Examined the place of key competencies in education and training.

[24] Three volumes 350pgs. See Appendix 7: After virtues: Two views of economics 'Oikonomics and Chrematistics'. pgs 64-74 – Aristotle 2500yearsBP identified two types of economics – Oikonomia from which we get our

Wildman, P. (1998). Futures of Work: Localism, the Post-Job Economy and the Learning Institution. *Futures Bulletin*, 24(4), 12-13.

Wildman, P. (1999). Blood Sweat and Gears: Some present implications of cloning and other life futures. *Australian Rationalist*, 33-37.

Wildman, P. (2000). Life Futures: An Initial Taxonomy of Terrestrial and Non-Terrestrial Forms of Life. *Journal of Futures Studies, 4*(2), 93-108.

Wildman, P. (2002). Healthy Community Economic Futures. *Journal of Futures Studies*, *7*(1), 135-144.

Wildman, P. (2005). Bush Mechanics: Futuring the Australian way. *Future News*, 2005 (September), p. 1.

Wildman, P. and Schwencke, H. (2003). *Your Community L'earning - action learning circles for learning and earning through community economic development.* Brisbane, Community Learning Initiatives and Prosperity Press: Multi Media CD Rom.

FURTHER READING ON THE INTERNET

Web accessed 10-03-2009

http://www.netbay.com.au/~noelozzy/commbank.htm history of the Commonwealth Bank of Australia.

http://www.finance-broker.com.au/content/australian-bank-home-loan-interest-rate-history history of Home Loan interest rates Australia 1960-2007.

http://au.youtube.com/watch?v=z5MVsm2cpc0andfeature=related Alan Greenspan (in 2007 of all people) in favour of a gold standard. 2mts.

http://au.youtube.com/watch?v=O7pnjzCuSv8andfeature=related Milton Friedman (on the banking credit creation multiplier) (2007). 9mts.

http://au.youtube.com/watch?v=jNrMQazZHDcandfeature=related Milton Friedman (in 2005 on how strong the US economy is/was) 60mts.

https://www.det.nsw.edu.au/reviews/macqt/voca06.htm#Heading30

Australian training reform in the early 1990's.

www.YesMagazine.org/bailout David Korten Beyond the Bailout and After the Meltdown

www.newrules.org New Rules Project – Rules as if Community Mattered

http://model-economy.wikispaces.com/Bernard+Lietaer+Interview Money, Community and Social Change see also http://www.nexuspub.com/articles/2003/july2003/interview.htm

---

English word 'economics' today which stands for prudential household management and 'chrematistics' which stands for speculative love of money. Further they do however strongly point to prudent Oikonomia [German for economics – Ökonomie] i.e., management of the home (often female – I call this nurturing economics – from small seeds large trees grow) cp. Chrematistics i.e. love of money and speculative gain (often male – I call this the economics of big projects big boys' toys). Also the Greeks distinguished between techne - techneque (where the tool enhances hand wisdom) and what we call today as technique where the tool replaces the hand. It is the latter, along with its associated focus on 'international competitiveness' to which most Vocational Education has been directed in the past generation.

In: Meltdown: Climate Change, Natural Disasters... ISBN 978-1-60876-153-1
Editor: Kathryn Gow

Chapter 21

A THIRD WATERSHED: THE CAPACITY TO ADAPT AND SOLVE FUTURE DILEMMAS CONFRONTING OUR PLANET

Ian Plowman
Management and Community Consultant,
Queensland, Australia

ABSTRACT

Characteristics that humans exhibit today are present because they have served an important psychological purpose in aiding survival. In contrast, characteristics that did not aid survival have died out. Through close examination of evolved psychological characteristics, we can make an informed judgement as to the value of these characteristics for individuals, organisations and communities. This chapter draws a chain of inference from niche differentiation in families to human motives to occupational and social role preferences. These preferences lead to social structures, organisations, communities and systems of governance that are naturally conservative and largely incapable of adapting rapidly to external signals. In consequence, conventional social structures and their leadership are proving inadequate to deal with the accelerating changes facing humanity and the globe. With major challenges in finance, climate change and resourcing, new forms of social organisation and leadership will be essential.

Keywords: personality, capacity to adapt, future leadership, leadership styles, psychological characteristics

INTRODUCTION

The task is not so much to see what no one has yet seen; but to think what nobody has yet thought, about that which everybody sees.

(Erwin Schrödinger, 1887-1961)

The First, Second and Third Watersheds

Bob Dick, an Australian academic, psychologist, social consultant and insightful thinker, refers to what he calls the "second watershed" in the evolution of humankind (Dalmau and Dick, 1990). The "first watershed" occurred when the problems faced by society could not be successfully addressed by instinct or pre-wiring alone. Rather, solutions needed to be learned and then passed on from generation to generation. The "second watershed" occurred when the solutions of yesterday no longer worked for the problems of today, and that meant contemporary solutions needed to be found. Therefore the second watershed renders largely irrelevant, in certain contexts, the wisdom of our elders and forebears. This second watershed arrived with a vengeance in the second half of the 20th Century. In a society experiencing rapid change across so many dimensions, old answers were often found to be inadequate.

It might even be argued, given the rapidity of change across every social, technological and environmental facet of the world that we are at the point of a ‘third watershed’, where the problems of today cannot be solved by the solutions of today. New and creative solutions need to be found if we are to adapt to the current financial crisis and the ecological challenge of climate change. Albert Einstein was correct when he reportedly said: "The significant problems we face today cannot be solved at the same level of thinking we were at when we created them."

Yet, as this Chapter will argue, individuals who hold positions of authority and who are therefore best placed to initiate or authorise new and creative solutions are frequently insensitive to the urgency. In fact, as will be argued, they are part of the problem. Large scale change to save the planet will need massive support from governments, institutions, organizations and communities. Yet organized social systems of any sort have the systemic characteristic of maintaining the status quo and an inability to read and respond early to emerging evidence.

This chapter weaves a journey from individual differences to occupational and social role preferences to organisational communication and finally to the consequences for adaptation. The lessons are sobering.

Evolutionary Psychology as A Lens on Adaptability

Evolutionary psychology is based on the Darwinian notion of survival of the fittest. Evolutionary psychology asks the question: "What is the adaptive function of this psychological mechanism?" (Buss, 1995, p. 84). As Pinker (1997) explained, evolutionary psychology is a form of reverse engineering. In ‘forward’ engineering, one designs a machine to do something. In reverse engineering, one examines the machine to see what it was designed to do. Particularly importantly, evolutionary psychology explains why we are the way we are, not at the surface level of presenting behaviour or obvious dispositional characteristics (the stuff of conventional management), but at the deeper unconscious evolutionary level where the entrenched motives are critical for species survival (Wright, 1994).

Human beings are small-tribe animals. For over 100,000 years, we have roamed the earth in small bands of hunter-gatherers and indeed the characteristics that we possess today are the characteristics that equip us best to be effective hunter-gathers. Although humans settled into more agrarian and urban lifestyles in the past 10,000 years, that period of time is far too short for evolution, through natural selection, to have changed any of our fundamental characteristics (Nicholson, 1998).

The secret of natural selection is to avoid extinction. Any ancestral characteristic that did not aid survival was not transferred to offspring; conversely, only those characteristics that favoured survival were retained (Buss, 1995; Miller, 1997).

Two Forms of Generalised Intelligence

Horn and Cattel (1967) suggest that we are all born with two forms of general intelligence; 'fluid intelligence' is the capacity to learn stuff, while 'crystalized intelligence' is the capacity to know stuff. The young have more of the former, while mature adults have more of the latter; so the young have the capacity to learn rapidly, a characteristic that declines as we mature.

It is argued that there are very sound evolutionary reasons for this duality of intelligence. It is the role of the young to be able to rapidly absorb the wisdom of the elders; after all, any person who has managed to live to a ripe old age must surely possess the very wisdom that enabled this. Therefore, if the young wish to live to similar age, it is prudent to defer to their elders and pay attention without question.

This evolved strategy has obviously stood the test of time. Its existence is testament to the 'first watershed' mentioned earlier; yet it is predicated upon the external environment being stable, where such stability bestows relevance upon the wisdom of the elders. In stable environments, creativity and innovation that run contrary to that wisdom would be regarded as foolhardy and positively dangerous.

Yet as the environment becomes increasingly turbulent, where the question of adaptation to new conditions necessitates new strategies, our unconscious disposition to defer to our elders, and/or their unconscious disposition to maintain the status quo may be to our collective detriment. For instance, in 2007, polls showed that 95% of the Australian population were concerned about global warming; yet the Government of the day insisted that these concerns were unfounded, despite overwhelming scientific evidence to the contrary.

Human Motives

Human beings are driven by motives. Motives have to do with how behaviour gets started, is energized, is sustained, is directed, or is stopped; it is about the why of behaviour, as opposed to the how or what of behaviour (McClelland, 1987). Motives have aims or goals which orient behaviour in particular directions, leading to learned responses or habits. While other personality characteristics or situational factors may override a preferred motive in specific instances, motives are predictive of behaviour patterns in most situations and over time (Furman and Lanthier, 1996).

Stable motive dispositions are unconscious; people do not necessarily know why they act the way they do. Nor can the motive be inferred from the behaviour, since the same behaviour may be motivated by different drives (McClelland, 1987).

Three particular motives have been of interest to researchers for over half a century; they are *need for affiliation*, *need for achievement*, and *need for power* (McClelland, 1987). Each is defined below.

Need for Power: To control one's human environment. To influence or to direct the behaviour of others by suggestion, seduction, persuasion or command. To dissuade, to restrain, or to prohibit (Murray, 1938). People whose primary motive is need for power are unconsciously driven to seek positions of leadership, to acquire material possessions publicly flaunted, and to pursue status and public prestige.

Need for Achievement: To accomplish something difficult. To master, manipulate, or organize physical objects, human beings, or ideas. To do this as rapidly and as independently as possible. To overcome obstacles and to attain a high standard. To excel oneself. To rival and to surpass others. To increase self-regard by the successful exercise of talent (Murray, 1938). People whose primary motive is need for achievement are driven to excel through individual effort, to solve problems through determined application, often original, and to seek feedback on their performance, be it negative or positive.

Need for Affiliation: To draw near and to enjoyably cooperate or reciprocate with another. To please or to win the affection of another. To adhere and to remain loyal to a friend (Murray, 1938). People whose primary need is for affiliation are driven to pursue meaningful social relationships as confirmation of their acceptance.

The Origin of Motives and their Evolutionary Function

Plowman (2005) found strong support for motives having their origins in childhood, where each sibling establishes a particular psychological niche that enables it to reduce competition with siblings. Examination of differences in personality among siblings finds that up to 45% of differences in personality are attributable to non-shared environment, or experiences that each sibling has different from the other. Up to 90% of those different sibling experiences occur within the family. Growing up in the one family does not make children the same; it makes them different (Plomin and Daniels, 1987) and those differences occur in relatively systematic ways.

Eldest children are generally more conscientious and seek parental favour through acting as surrogate parents toward younger siblings; they are more responsible, conservative, and defensive. The firstborn arrives in the world needing to survive and recognizes that its resource needs are being met primarily by parents. When a younger sibling arrives, available parental investment drops significantly, without warning; this creates anxiety in the young child who explores strategies to try to reclaim the lost resources and to strengthen the probability of its own survival. The chosen anxiety-reducing strategy is to try to regain parental attention through taking responsibility and seeking to position itself advantageously in the resource flow. "I'll be safe if I can do what is expected of me, and control the resources" is the chosen strategy (Plowman, 2005); the underlying motives are need for power and need for achievement via conformity.

Middle children have broader interests, have lower self-esteem and tend to be more independent, innovative, risk tolerant, rebellious and open to experience; middle-borns also need to maximize resource flow to them. They have competition firstly from older sibling(s) who are stronger, more articulate and more assertive and later from younger siblings who are more "cuddly" and charming. Suffering from lower self-esteem, middle children frequently adopt a strategy of differentiation through broader interests and pursuits. For example, if the firstborn shows academic inclinations, the middle-born will show interests in other directions. He or she attempts to build self-esteem and reduce anxiety through independence, rebelliousness, risk-taking, innovation and personal endeavour. "I feel safe if I can prove my own competence and independence" is the strategy (Plowman, 2005); the resulting motive is achievement via independence.

The radicalism of middle children is reflected in them, as youths and adults, frequently finding their ecological niche by moving away from their environment of origin (Sulloway, 1996). Unlike older siblings with much to lose and little to gain through innovation, middle children have everything to gain and nothing to lose. Thus in businesses, middle children are less likely to be found in mature multigenerational industries (since they leave), and more likely to be found in fledgling ones (which they start), where established practice either doesn't exist or is there to be tested and varied (Plowman, Ashkanasy, Gardner and Letts, 2004). Mature industries and organizations are likely to have ‘elders’ and because of this middle children are more attracted to new enterprises that do not have elders.

Youngest siblings are commonly less ambitious, less conscientious, and more socially oriented. They are sometimes described as popular, easy going, lazy or spoilt. Lastborns, as children and later as adults, often enjoy considerable parental investment, although without the burden of parental expectations of 'success' (which is placed on firstborns). They also enjoy considerable support from older siblings, although this attention is often ambivalent. Sometimes the lastborn child is feted and cuddled like the family pet; at other times told by its older siblings that it is a nuisance and to 'go back to mum' (Plowman, 2005). It is quite possible that lastborns intuitively know, through their 'genetic memories' that, throughout history, in times of severe resource shortage, infant mortality is highest in lastborns (Pinker, 1997). This ‘genetic knowledge’ might result in the youngest child developing excellent social skills to ensure it continues to be included. "I feel safe if I can be loved" is their strategy; the resulting motive is need for affiliation.

Please note that there is not an invariant connection between birth-order and motives; the birth-order/motive relationships just described are systemic, regardless of culture, geography or ethnicity. Yet there are exceptions to these patterns, at least in the order in which they occur; factors such as domestic conflict or death or disability of a sibling can alter the patterns (Sulloway, 1996). Birth order is merely a proxy for the more important niche differentiation, where the purpose in early childhood is to establish an ecological niche that will enable the child to reduce competition with other siblings. Strategies that humans adopt are limited in number and manifest themselves as motives. Although an individual might exhibit elements of all three social motives, one will be dominant and enduring.

Motives and Social and Occupational Roles

Since motives reflect deeply unconscious needs, it is not surprising that occupational and social roles are vehicles that provide a means to having those needs met (Stahl, 1986; Plowman, 2005). At work, affiliative people spend more time interacting with others (Schmidt and Frieze, 1997) because they enjoy work for its social aspects. Affiliative people are attracted to the blue-collar and helping professions (Stahl, 1986). Since typical management jobs require the manager to be competitive, to attempt to influence others, and to make hard decisions that may hurt feelings, those with high affiliation needs are unlikely to seek out or to be successful in management or leadership roles (Stahl, 1986).

The achievement motive is commonly highest in sales, engineering research, academia, individual sport and technical and professional jobs, particularly in non-supervisory roles (Stahl, 1986). It is common among entrepreneurs, since business requires that people take moderate risks, assume personal responsibility for performance, pay close attention to feedback in terms of costs and profits, and find innovative ways to make a product or provide a service. High 'need for achievement' individuals spend a great deal of time thinking about how to do things better and they take pride in their individual contributions. Managers also have this motive, but more commonly at lower to middle levels of management, when individual contributions are more important (McClelland, 1987).

The motive that best predicts senior roles is the need for power (House, 1988; Medcof, 1985; Plowman, Ashkanasy, Gardner and Letts, 2003, 2004; Plowman, 2005). The motive profile of senior executives is frequently a combination of high need for power and low need for affiliation (Stahl, 1986). According to House (1988), the disposition towards power arises from both learned and genetic influences. It then interacts with a contextual variable, namely the opportunity to influence others.

Plowman (2005) confirmed that people with different motive preferences have different occupational or role preferences. Hence it is understandable that a whole organization or society might comprise different role clusters within which there is a degree of motive homogeneity (Figure 21.1 illustrates).

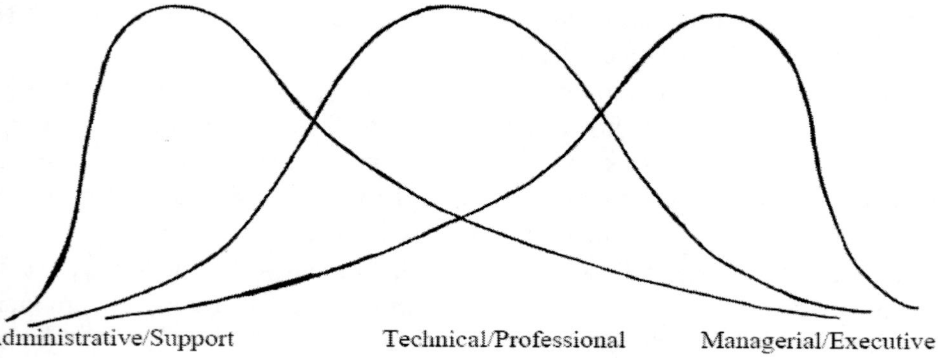

Figure 21.1. Different role clusters utilising McClelland's (1987) motives.

At the left hand end of the figure, are the administrative and support people, those for whom both work and life is largely a social activity. Here the work they do provides validation of their acceptance. Provided their social needs are met, people in these roles commonly have no great ambition to progress further up the hierarchy. In the middle are those technical and professional people who are endeavouring to make a difference through individual effort. Here the work provides a vehicle for demonstration of personal competence, the key to their sense of self-worth. At the right of the diagram are the managerial and executive people, those who would wish to influence, direct and control. Having a position of status and privilege, enjoying perks not available to those more junior, being feted by others and enjoying the patronage of those more powerful is what fuels, though seldom satiates, their needs.

Like all motives, the need for power is insatiable. Corporations gobble up smaller organisations in an unending personal pursuit by their executives for increased power. This is frequently disguised and rationalised as being in the interests of shareholders and economies of scale. Unfortunately, the larger the organization, the more distant the executives are from the realities of the operating environment and the lower the level of resilience. The world is now populated with multinational corporate giants, frequently with budgets greater than many nations, and which operate substantially in their own short-term interests.

[This chapter sets out a number of scenarios that are illustrative of the particular concept under discussion. While they are boxed, they are meant to be read in sequence with the text.]

Scenario Example

The scaling up of corporations into global conglomerates has produced benefits. For example, the mass production of motor vehicles has seen real prices fall and hence affordability increase.

In Australia in the 1950's, the average price of a motor vehicle equated to the average annual wage. Today the average price of a motor vehicle in Australia is less than half the average wage. On the other hand, global corporatisation reduces variety and choice; it also reduces corporate resilience. Ford and General Motors find themselves still producing high fuel consumption vehicles for a market that is increasingly resistant to purchasing them.

Motives and Language

Since motives are unconscious psychological drivers to satisfy particular needs, it is perhaps not surprising that each motive has its own distinct language. This is indeed the case; language is one of the vehicles commonly used to give effect to one's needs. Stop for a moment and imagine a scenario in which people, whose primary need is affiliation, talk to each other. For example, consider the dialogue of two colleagues who have just encountered each other after a long weekend away from the office. Shut your eyes while you do this, and pay particular attention to the nature of the utterances.

Now, repeat this exercise, for achievement behaviour, particularly some form of collective achievement (so that dialogue can occur). Again pay attention to the utterances of the players. Finally, repeat the exercise, imagining a scenario showing power behaviour, again paying attention to the utterances.

In relation to the three sets of utterances you have just generated, you are invited to consider the following:

1. Are the utterances subjective or objective?
2. Are the utterances generally questions, statements, or a combination?
3. What pronoun captures the essence of the utterances, the core focus?

Senge (1991) points out that learning organisations, those that demonstrate the capacity to adapt, maintain a balance between "advocacy" - the making of statements - and "enquiry" - the asking of questions. In contrast, learning disabled organisations, those with limited adaptive ability, demonstrate higher levels of advocacy (i.e., statements), and lower levels of enquiry (i.e., questions), particularly in their upper ranks. Given the suggested motive clusters in Table 21.1, those in the support ranks frequently ask subjective questions; those in the technical professional ranks frequently ask objective questions and make objective statements; while those at the top frequently make subjective statements. Most significantly, among those at the helm of an organization, association, industry and society at large, there is an absence of questions. Table 21.1 summarises responses provided in numerous exercises where workshop participants have been invited to undertake this exercise.

Table 21.1. Type of Motive and Language Attributes

Motive	Language attributes
Need for affiliation	Subjective, tentative questions, often of a somewhat personal nature that seek to engage the other party. The representative pronoun in these utterances tends to be 'we' or 'us'.
Need for achievement	More objective, where the focus is on the project or task, rather than the individuals engaged in it. Utterances tend to be a combination of rational questions and statements. The representative pronoun would be 'it' or 'that'.
Need for power	The subtext of power utterances is subjective (drawing attention to self) though often disguised as objective. Statements substantially exceed questions. The representative pronoun is 'I' or 'me'.

Most significantly, among those at the helm of an organization, association, industry and society at large, there is an absence of questions. This point is seminal to the whole chapter and cannot be overstated. Limited questions means limited enquiry; limited enquiry means the absence of critical new information which in turn leads to the failure to adapt. Those in the best position to influence an organization, an industry or a community, those who control its resources and influence its decisions are, unconsciously, commonly uninterested in critical information which is frequently available and which would enable survival and adaptation. Notice how the language of motives is totally consistent with the concepts of fluid and crystallized intelligence mentioned earlier. The 'tribal elders' are expected to know, not to ask. Similarly, the language of motives and the distribution of those motives explain how social systems are locked into the 'second watershed' mentioned at the beginning of the chapter.

Plowman (2005) interviewed representatives from (i) the administrative and support ranks, (ii) the technical professional ranks, and (iii) the executive managerial ranks across a number of organizations, public and private. All were asked exactly the same questions about the perceived level of innovation within their respective organizations. Interviews were audiotaped. Data was collapsed across the participating organizations and within the three role clusters. Two findings are critical to the thesis of this Chapter. First, respondents from the executive managerial ranks scored the levels of innovation within their organizations as being 27% higher than did the technical professional middle ranks. Further, the former used twice the number of first person pronouns than did the latter. These results suggest that not only are the executive managerial ranks less objective in their orientation, they have a distortedly optimistic view of their world, thereby substantially reducing their capacity to take timely adaptive action. And this is precisely what would be expected from an executive cohort that is generally unable to ask questions.

> Scenario Example
>
> The Intergovernmental Panel on Climate Change (IPCC) declared in 2006 that the world cannot continue to burn fossil fuels, yet mining companies, with the blessing of the Australian Government, continue to ramp up the extraction of coal, claiming that our economic prosperity depends upon it.
>
> So 'global warming' becomes Australia's biggest export. At the same time, governmental support for alternate energy sources was being wound back, when all the evidence suggests exactly the opposite was required.

In examining CEO attitudes as determinants of organization design, Lewin and Stephens (1994) determined that people with low tolerance for ambiguity and/or those driven by need for power, prefer to simplify complex problems, to deal with a minimum of information from the environment and to resist change. People with high tolerance for ambiguity prefer the converse; they are more likely to experiment, to try a variety of approaches, to seek diverse opinions, and to be comfortable with long-term planning horizons.

The literature on dispositions and motives also provides some clues for creativity and adaptability within social structures. Neurological studies reveal that norepenephrine is the neurochemical which underpins the need for power. Individuals with a high need for power tend to experience motive arousal as a feeling of strength, vigour, and energy. Physiological changes that correspond to the arousal of the need for power include increases in epinephrine and norepinephrine, which seem to be the source of the 'kick' or energized subjective state (Woike, 1994). Further the need for power inhibits creativity.

When an organization is faced with critical opportunities or threats (times when adaptation might be particularly desirable), the more that power is entrenched in the upper ranks, the slower the organization's capacity to respond. Fodor and Smith (1982) found that persons high in need for power impede both the introduction of factual information from others into group discussion and the emergence of tentative solutions to the group problem, particularly when groups are small. Management teams, councils, and committees might be good examples. As a result, the number of alternative ideas considered is fewer and the quality of decisions is lower for groups led by high power-motivated individuals (House, 1988).

> Scenario Example
>
> The invasion of Iraq in 2003 is a classic illustration of the propensity of leaders to simplify complex problems and to distort information for their own purposes. The stated reason for the invasion was to disarm Iraq of weapons of mass destruction and to end Saddam Hussein's support for terrorism, in particular Al-Qaeda.
>
> Soon after the invasion and the toppling of Saddam Hussein, President Bush declared that the war had been won. Although it was subsequently proven that there was no evidence of weapons of mass destruction, that there was no connection, until after the war began, between Iraqis and Al-Qaeda, and that the war was far from 'won', no apologies have been forthcoming from the leaders of the invading 'coalition of the willing'.

Similarly, high need for power in a group leader both minimizes cooperation and generates negative self-affect in group members (Fodor and Riordan, 1995). Yet it is positive self-affect that correlates with creativity (Isen, Daubman, and Nowicki, 1987).

> Scenario Example
>
> In 1961, Cuban exiles, supported by the US military, suffered a humiliating defeat at the Bay of Pigs whilst endeavouring to overthrow the government of Fidel Castro. Subsequent evaluation of the defeat revealed that the presence of President J.F. Kennedy in the military planning room was the principle cause for the overly optimistic endeavour. Further, the phenomenon of "group think" leads to group decision making being distorted by the desire to keep the group cohesive.

In contrast, people with a high need for achievement are constantly seeking a shorter, different, or more efficient path to a goal. They are restless innovators who avoid routine. They tend to seek out moderately challenging tasks, and look for feedback that tells them how they are progressing. They are frequently independent individuals who think things through on their own and often come up with novel solutions. However, on the negative side, McClelland (1987) reports that they are less effective in managing others and in having their good ideas implemented. Therefore they are more likely to work alone, or in small groups of similar individuals. Whilst need for achievement is a desirable characteristic of CEO's, McClelland (1987) concluded that most senior executives had a need for power that exceeded their need for achievement. In fact, need for achievement seemed to peak in supervisory to lower middle management ranks, not the senior ranks. Sulloway (2001) observes that there are two subsets of need for achievement. The first is achievement via conformity, where people strive to meet expectations set by others. The second is achievement via independence. The latter would probably fall into a classification identified by Rogers (1995) that he labelled 'innovators' and who represent only 3% of the population. These independent thinkers are unconstrained by convention and established practice. It is from their ranks that leadership during the third watershed in human evolution will need to be spawned. Paradoxically, they commonly do not reside within large mainstream social structures capable of bringing about large-scale change.

Scenario Example

It is from the ranks of these innovators that alternate energy technologies arise. Wind, solar, tidal and biofuels are all energy sources that began on the margins, in the middle of the 20th Century and which are gradually moving to more centre stage. In almost every case, this shift is occurring in the absence of support from big business and government, and in many cases, despite direct opposition. The General Motors EV1 electric car in the United States in the 1990's is a story of how government, oil companies and automobile companies conspired to remove an energy efficient technology in order to foster their own interests.

In 2005, Plowman investigated the variables that predicted whether or not a person would progress to the senior organizational ranks. Predictive characteristics included being male, being older, and having a high need for power; interestingly, level of education had no predictive value.

Scenario Example

Early in 2009, the British House of Lords held a commission of enquiry into the collapse of the Bank of Scotland, the largest corporate collapse in British history. Members of the bank's Board admitted under questioning that none of them had any formal training in banking.

Paulus and John (1998) report a perceptual bias in those with a need for power. Those who score high on 'need for power' are said to possess "an egoistic bias, a self-deceptive tendency to exaggerate one's social and intellectual status. This tendency leads to unrealistically positive self-perceptions on such traits as dominance, fearlessness, emotional stability, intellect, and creativity" (p. 1025). In partial summary, the discussion on motives suggests that: they are formed in childhood; they are unconscious and endure into adulthood; and they influence peoples' language as well as their preferred occupational and social roles.

In consequence, social structures are commonly comprised of three distinct clusters, with those with a need for affiliation occupying support roles, those with a need for achievement occupying technical/professional roles, and those with a need for power occupying senior managerial and executive roles. Executives show a propensity to make statements and an inability to ask questions, thereby inhibiting adaptation to internal or external signals.

Two Forms of Collective Behaviour

Pierce and White (1999) observe that primates, including humans, have the capacity to form different social structures, depending on the context. Hierarchical structures occur when resources are centrally located and contestable. In this context, hiearachical groups are comprised of dominant and submissive members. Those who attain and maintain dominance do so through overt acts and displays of aggression. Submissives tend to focus most of their attention on the dominants. They show little curiosity about others or about their broader physical surroundings. When external danger threatens, they cluster together and look to the dominants for protection.

In contrast, when resources are scarce and scattered, co-operative behaviour is favoured. Groups scatter more widely, keeping in contact aurally, rather than visually – hence covering a larger territory; leadership is much more fluid and keen attention is paid to the external environment, since that is the source of random reward.

Human social systems are commonly hierarchical, with multiple layers of authority. The evolutionary purpose of hierarchical systems is that they aid species survival. Males create hierarchical systems in order to showcase dominance. Females commonly choose, as mating partners, males who are located further up the hierarchy; in this way, females increase the chance of survival of the mother's genes through pairing them with dominant male genes. Thus hierarchical systems tend to be perpetuated.

Further, it is common in hierarchies for resources to be centrally located - budgets, authority and policy being examples of such. These budgets and other centralized resources are commonly controlled by the senior ranks. Promotion up the hierarchy is a function of 'pull' from above, rather than 'push' from below. Hence most attention is paid to the level above, since that is the source of one's rewards. Further, peers are competitors for those same scarce resources, so co-operation and horizontal communication in hierarchical systems is largely unrewarding (and unrewarded). In large hierarchies, most attention is directed internally, since the resources appear to be internally based and controlled. In the upper ranks, the powerful elite spend most time watching each other. In consequence, the external environment seems less salient, so external threats are less likely to be perceived.

Patterns of communication within hierarchies are particularly revealing. Since resources are more likely to come from above, there is a tendency to *ask* upwards, and to *tell* downward. Each level in the hierarchy has the tendency to filter information, forwarding upward that which is more inclined to please.

Scenario Example

There are myriad examples of leaders ignoring external environmental signals. Entering the words 'leaders ignore' into Google raises 17,000,000 examples, that many of which have enormous social, economic or environmental consequences. A small cross section of those examples follow, sufficient to illustrate the point. The San Francisco Chronicle, 29th June 2008, reports that tens of thousands of buildings are prone to collapse in the event of an earthquake, yet government and building authorities continue to ignore the risk. The UK Guardian of 24th Sept 2003 has a front page article entitled 'Why gather intelligence if our leaders deliberately ignore it?' The article points out that while government security decisions are based on intelligence, in the case of the British intervention of Iraq, it was the other way round. First came the decision to make war with Iraq and then came the search for supporting evidence. The USA adopted the same approach. Democratic presidential candidate and former general, Wesley Clark, has spoken about how he was phoned on 9/11 by "people around the White House" urging him to blame the attacks on Saddam Hussein. Never mind the lack of proof; it was the end goal that mattered. The Sowetan, a South African newspaper, reports on the 21st May 2008 how South Africa's leaders have ignored the AIDS crisis, rampant corruption, and escalating crime, whilst pursuing their own political agendas.

Hence critical strategic information, gleaned by those in the technical professional ranks, those closest to the clients, competitors, and the external environment in general, stands little to no chance of flowing upward, without considerable distortion. In this sense, hierarchy systemically blocks resilience. Herein lies the explanation of major corporate breakdowns, the warning signs for which are commonly evident to 'the workers' for some time prior to the event.

CONCLUSION

Our Collective Capacity to Adapt – Some Key Messages. This chapter began with the suggestion that there have been three watersheds in the evolution of mankind. The first occurred when instinctive solutions to life threats were replaced by the wisdom of the elders. The second watershed was when solutions provided by the elders proved inadequate for contemporary problems. The third watershed is now upon us, where even contemporary solutions are proving inadequate for contemporary problems. For problems such as global warming, rampant and unsustainable consumption, and the growing gulf between the 'haves' and the 'have-nots', new and creative solutions need to be found.

Our social structures are largely locked into the 'first watershed' where institutionalised systems of elders – codified legislation, laws, and cultural norms, all systemically serve to block reforms that need to occur at the level of the 'third watershed'. Leaders, those in the best position to help their organisations and social structures to adapt to changing external signals are those commonly least able to receive those signals, or to be interested in them.

Society continues to create social and organisational structures in a system of governance that is obsolete and largely incapable of responding to external signals.

Consider the need for the third watershed in contrast with the content of the bulk of this Chapter. On the one hand, the watershed argument says that yesterday's and today's solutions are inadequate for today's critical issues. On the other hand, the material offered suggests that most of our behaviour individually and collectively is unconsciously supportive of the first watershed. The enormity of the gulf between these two positions cannot be overestimated.

In a world that changed fairly slowly, innovation and radical thinking would not have been welcome. To challenge the wisdom of the elders derived from centuries of experience would have been sheer folly, but no longer. In our rapidly changing world, the wisdom of the elders (and here I use the terms in its broadest possible sense), and all of the social systems that give that wisdom authority and legitimacy, are increasingly unable to respond rapidly to external or internal signals.

Rather than centralised and codified wisdom, decentralised and random ideas are becoming increasingly important. These ideas are more likely to come from individuals driven by the need for achievement via independence - those people more attracted to non-hierarchical or co-operative social systems. Yet, these new adaptive and creative ideas need to be sensed, pooled, collated and acted upon and conventional hierarchical social systems are largely incapable of doing this. Instead we need grass-roots leaderless systems.

Such systems are common in nature, as well as in unstructured human social systems. Such is the wisdom of a swarm of bees, a flock of birds or a school of fish. The rules under which these leaderless collectives operate are simple, yet effective; (i) diversity of participation, (ii) independence of contributions, (iii) free comparison of ideas (peer review), (iv) time for all ideas to be considered, (v) an effective mechanism to narrow choices, and (vi) total transparency[1] (Sirowiecki, 2004). The "Web" is our contemporary leaderless technical equivalent.

If we are to survive as a species, and if the world as we know it today is to survive, then we need new systems of social governance. Old behaviours and social structures, for which we are hard-wired, and which may have been functional for millennia, are no longer adequate.

We cannot manage what we do not understand. The intent of this chapter has been to provide explanation for the sources of individual and collective behaviour and their consequences. Equipped with that understanding, we are better placed to consider more effective alternatives.

ACKNOWLEDGMENTS

1. The content of this chapter is based on my PhD research; some material in this chapter has been presented at conferences in 2006 at the Adult Learning Conference in Canberra and the Australasia Pacific Extension Network Conference in Victoria; Figure 21.1 and associated ideas were contributed to a paper prepared for the Sugar Research and Development Corporation in 2007. For further information on these papers, you can email me on ian@plowman.com.au.

2. My thanks to Robert Maxwell and Dr. Paul Wildman for helpful guidance.

---

[1] National Geographic.com/Magazine (July, 2007). "The wisdom of swarms".

REFERENCES

Buss, D.M. (1995). Evolutionary psychology. *Psychological Inquiry,* 6 (1), 1-30.

Dalmau, T. and Dick, B. (1990). *Politics, conflict and culture: concepts for cultural change.* Chapel Hill: Interchange.

Fodor, E. and Riordan, J.M. (1995). Leader power motive and group conflict as influences on leader behavior and group member self-affect. *Journal of Research in Personality*, 29, 418-431.

Fodor, E.M. and Smith, T. (1982). The power motive as an influence on group decision-making. *Journal of Personality and Social Psychology*, 42, 178-185.

Furman, W. and Lanthier, R.P. (1996). Personality and Sibling Relationships. In G.H. Brody (Ed.) *Sibling Relationships: Their Causes and Consequences* (pp. 128-140). Norwood, NJ: Ablex Publishing.

Horn, J. L., and Cattell, R. B. (1967). Age differences in fluid and crystallized intelligence. *Acta Psychologica, 26,* 107-129.

House, R.J. (1988). Power and personality in complex organizations. *Research in Organizational Behavior,* 10, 305-357.

Isen, A.M., Daubman, K.A., and Nowicki, G.P. (1987). Positive affect facilitates creative problem solving. *Journal of Personality and Social Psychology,* 52, 1122-1131.

Lewin, A. Y. and Stephens, C. U. (1994). CEO attitudes as determinants of organization design: an integrated model. *Organization Studies,* 15(2), 183-212.

McClelland, D. C. (1987). *Human Motivation*. New York: Cambridge University Press.

Medcof, J. W. (1985). The Power Motive and Organizational Structure: A Micro-Macro Connection. *Canadian Journal of Administrative Science,* 2(1), 95-113.

Miller, E.M. (1997). Could non-shared environmental variance have evolved to assure diversification through randomness? *Evolution and Human Behaviour*, 18, 195-221.

Murray, H.A. (1938). *Explorations in Personality*. New York: Oxford University Press.

Nicholson, N. (1998). How hardwired is human behaviour? *Harvard Business Review.* July/August, 135-147.

Paulhus, D.L., and John, O.P. (1998). Egoistic and moralistic biases in self-perception: The interplay of self-deceptive styles with basic traits and motives. *Journal of Personality*, *66*(6), 1025-1060.

Pierce, B.D. and White, R. (1999). The evolution of social structure: why biology matters. *Academy of Management Review*, 24, 843-853.

Pinker S. (1997). *How the Mind Works.* London: Penguin.

Plomin, R. and Daniels, D. (1987). Why are children in the same family so different from one another? *Behavioural and Brain Sciences,* 10, 1-60.

Plowman, I. (2005). *Birth order, motives, occupational role choice and organizational innovation: An evolutionary perspective.* Unpublished doctoral thesis. Fryer Library, University of Queensland, Brisbane.

Plowman, I. Ashkanasy, N., Gardner, J. and Letts, M. (2003). *Innovation in rural Queensland. Why some towns thrive while others languish.* http://www.business.uq.edu.au/display/research/Research+Reports

Plowman, I., Ashkanasy, N., Gardner, J. and Letts, M. (2004). *Innovation in rural Queensland. Why some primary industries and their associations thrive while others languish.* http://www.business.uq.edu.au/display/research/Research+Reports

Rogers, E.M. (1995). *Diffusion of Innovations* (4th ed.). New York: The Free Press.

Schmidt L. C. and Frieze, I. H. (1997). A mediational model of power, affiliation and achievement motives and product involvement. *Journal of Business and Psychology*, 11, 425-446.

Senge P. (1991). *The Fifth Discipline.* New York: Doubleday.

Stahl, M. J. (1986). *Managerial and Technical Motivation.: Assessing Needs for Achievement, Power, and Affiliation.* New York: Praeger Special Studies.

Sulloway, F. J. (1996). *Born to Rebel.* London: Little, Brown.

Sulloway, F.J. (2001). Birth-order, sibling competition and human behaviour. In H.R. Holcomb III (Ed), *Conceptual Challenges in Evolutionary Psychology: Innovative Research Strategies* (pp. 39-79). Dordrecht and Boston: Kluwer Academic Publishers.

Surowiecki, J. (2004). *The Wisdom of Crowds – Why the Many are Smarter than the Few.* London: Abacus.

Woike, B. A. (1994). Vivid recollection as a technique to arouse implicit motive-related affect. *Motivation and Emotion, 18*(4), 335-349.

Wright, R. (1994). *The Moral Animal.* London: Little, Brown.

Contributed by Tim McCrorey, Toowoomba.

Photo: High and dry at Lake Wendouree: Who's steering the boats?

In: Meltdown: Climate Change, Natural Disasters…
Editor: Kathryn Gow
ISBN 978-1-60876-153-1

Chapter 22

IT'S NOW OR NEVER: THE IMPORTANCE IN RAISING ECOLOGICAL CONSCIOUS AWARENESS OF THE WORLD'S CLIMATE CHANGE

Anna Lee Mraz Bartra
National Autonomous University of Mexico, Mexico

ABSTRACT

Currently, the world we live in is not only threatened by wars, famine, poverty and social injustice, but by a bigger problem that will affect all life around the globe, and the ecological crisis may be even worse than these current issues. Climate change is a fact that will have a great effect on every living thing. The world as we know it will cease to exist, and adaptation is a must if we are to survive as a species.

The ecological crisis is a new concern. There have been, and probably always will be, wars and poverty; but without a sustainable ecology, everything will become worse. Our planet - our home - will change so dramatically that we have to act now, or a massive destruction of natural resources will determine the future of human existence. The first step in solving any difficulty is knowing that it exists. We need information; learning about current issues, understanding the problem and where we stand on it, gives us the power to influence decisions and provide the means to solve the crisis.

In this chapter, the importance of raising ecological conscious awareness of the world's climate change is emphasised and the different effects on natural and human resources are considered, by comparing a developed country such as Australia with Mexico, an under-developed nation.

Keywords: Mexico, climate change, ecology, environmental, forests, water shortage

THE WORLD'S SITUATION AND CLIMATE CHANGE

> The struggle to save the global environment is in one way much more difficult than the struggle to vanquish Hitler, for this time the war is with ourselves. We are the enemy, just as we have only ourselves as allies. (Al Gore [1])

Climate variation is a fact, although whether or not the facts about climate change per se are correct in terms of the projected dire scenarios for the future, is a subject that is still being debated. Many scientific institutions around the globe have recorded earth's temperature variations across time. The environment is changing year by year, and while it has always done so, it is the speed of change that represents the real problem. Nature has a very delicate balance, where species exist in interdependence. No plant or animal is excluded from life's equilibrium, and this includes us, because not only are we a result of our context and environment, we are active agents in shaping it as well. We have been informed by many scientific sources that a variation in temperature will affect all life on the planet: sea levels will rise, altering tides and sea currents which will modify the seasons, and influence animal and plant life. These will subsequently change human lifestyles and the way people interact with each other.

The concentration of carbon dioxide in the atmosphere reached a record level, increasing by 0.6 per cent in 2008 alone and methane levels are rising for the first time in a decade according to new results, thus renewing fears about global climate change prospects. Researchers at the Mauna Loa observatory in Hawaii report that carbon dioxide levels in the atmosphere are now at 387 parts per million (ppm), up almost forty per cent since the industrial revolution and the highest for at least 650,000 years. The figures, published by the National Oceanic and Atmospheric Administration (NOAA), also confirm that carbon dioxide, the chief greenhouse gas, is accumulating in the atmosphere faster than expected. The annual mean growth rate for 2007 was 2.14 ppm, the fourth year in the last six to witness an annual rise greater than 2 ppm. From 1970 to 2000, the concentration rose by about 1.5 ppm each year, but since 2000, the annual rise has increased to an average of 2.1 ppm.

A review of some of the pertinent history of human sourced emissions will assist us in this discussion, commencing from the Industrial Revolution. It is evident that since that change, human beings have been exploiting the world's resources, and one result of this has been that the levels of CO_2 in the atmosphere have risen and accumulated. In other words, we are told now that we are overheating the planet and for many decades have been using too many resources. For example, if we take a simplistic view, this is the story: first we cut down the trees that live off CO_2; the trees accumulate CO_2 in their trunks by a process of photosynthesis and they return oxygen to the atmosphere. However, cutting down trees eliminates a natural oxygen provider, a water absorber and climate change mitigator, as well as a valuable asset in wildlife and human habitats.

> There is a sufficiency in the world for man's need but not for man's greed.
>
> (Mohandas K. Gandhi[2])

---

[1] M. P. Singh, *Quote Unquote: A Handbook of Quotations,* Lotus Press, 2007, p. 129.

[2] Shepard Krech, John Robert McNeill, Carolyn Merchant, Encyclopedia of World Environmental History, Routledge, 2004, p. 571.

Researchers foresee massive famine, brutal world migration waves, aggravation in climate-sensitive ailments and pollution initiated illnesses, and ultimately the destruction of life as we know it.[3]

MEXICO

Mexico is classed as an underdeveloped country. It is the fifth-largest country in the Americas in total area and the 14th largest in the world. With an estimated population of 109 million, it is the 11th most populous country in the world and the most populated Spanish-speaking nation. Mexico reflects the complexity of its history through the blending of pre-Hispanic civilizations and the Spanish culture, imparted during Spain's 300-year colonization. Exogenous cultural elements, mainly from the United States, are also an important influence on Mexico's culture.

With a vast diversity of natural environments, Mexico ranges from temperate and rainy tropical regions, to dry deserts. Its mountainous topography allows countless microclimates, which support one of the most diverse arrays of plant and animal species on the planet. However, many of Mexico's species are endangered. The human impact is enormous and the list of environmental problems is long.4

Mexico's Forests

Who leaves the pine-tree, leaves his friend,
Unnerves his strength, invites his end.

(Ralph Waldo Emerson, from poem 'Woodnotes' [5])

With one of the highest deforestation rates in the world, Mexico shamefully occupies second place in this "honour". According to Greenpeace and the Green Party in Mexico, we lose 600 thousand hectares per year, which is equivalent to four football fields every minute. If this pace continues, in six decades all forests in Mexico will have disappeared, and their loss will have an impact beyond the national level, since Mexico is ranked number eleven among nations with forests. One third of the country was once covered with tropical and temperate forests and jungles. Today, only 50% of the original forest remains. It is estimated that 14% of CO_2 atmospheric emissions are due to deforestation. Mexico has 23,000 different species of plants, of which more than half exist only in this country, and 84% of Mexico's cacti are endemic.

Near Mexico City lies The Great Water Forest, one of the most beautiful and impressive forests in the world, which was protected by the National Commission for Knowledge and Use of Biodiversity (CONABIO), because of its variety of benefits to the country. This forest makes the survival of the species that live in, or near, it possible, giving lodgings to almost

---

[3] Lester R. Brown Plan B: Mobilizing to Save Civilization (New York: Earth Policy Institute 08) in http://www.global-challenges.org/23population-issues.html.

[4] http://www.questconnect.org/mexico_environment.htm.

[5] Emerson, Ralph. In 'Woodnotes'. English Poetry III: From Tennyson to Whitman. The Harvard Classics. 1909–1914.

2% of the world's biodiversity. Moreover, it regulates climate, air and soil, produces food and other goods, and provides 75% of the water consumed in Mexico City. The Great Water Forest also supplies two very important rivers: the Lerma and the Balsas. This forest is now under enormous threat, and every year, 2,400 hectares are lost due to uncontrollable city growth, illegal logging, fires, and soil extraction. It is estimated that in 50 years time, the forest will cease to exist. As if this were not depressing enough, the Secretariat of Environment and Natural Resources (SEMARNAT)[6] has approved a highway project that will cut through the middle of it, and construction will commence in the near future.

Building this highway places the whole region in danger. Without this forest, all the cities that surround it will lose their main water supply and clean air. The habitat of thousands of endemic endangered animals will disappear. It will also lead to population expansion and an urban sprawl that could ultimately destroy the region.

Greenpeace has denounced Juan Elvira, Secretary of SEMARNAT, for approving the highway project before the legal environmental evaluation had been completed. Academic research by the Autonomous Metropolitan University (UAM) stated that 90% of the forest area is legally protected, as well as having been declared high-priority land for conservation by the CONABIO[7]. However, these laws and regulations do not seem to apply to international corporations, small private industries, or even some governmental groups, due to the fact that Mexicans are unaware of what is happening to their forests.

Mexico's Water Problems

> Thousands have lived without love, not one without water.
>
> (W.H. Auden[8])

Mexico is considered to have a water shortage. However, this does not result from the lack of rain, but rather from an unequal distribution system, and poor water management. Further, according to the Green Party, 92% of the rivers in Mexico are contaminated.

The water situation in Mexico is very delicate and multifaceted. To understand these issues, we have to look at the country's geography. Two-thirds of the nation has a semi-arid climate. It almost never rains in the north or central parts of Mexico and water resources are scarce. However, one third of the country consists of tropical forests, where rain fall is heavy. High mountains divide the north from the south, and the capital mega-city of more than 21 million inhabitants rests on a plateau.

Mexico is a very centralised country, and most of its population, industrial, agricultural and cattle production, are distributed in the driest areas, where the rate of poverty and marginalisation is the highest. Around 11 million people do not have access to water, and 15 million do not have a sewerage system.[9] This huge contrast in Mexico is due to the poor distribution of all natural and economic resources. According to the statistics of the National Water Commission (CONAGUA), water availability in the south is 7.3 times higher than in

---

[6] Secretaría de Medio Ambiente y Recursos Naturales, (SEMARNAT).

[7] Comisión Nacional para el Conocimiento y Uso de la Biodiversidad: National commission for knowledge and the use of biodiversity.

[8] W. H. Auden, British Council, Michigan, Northcote House, 1997, p.72.

[9] http://www.pronatura.org.mx/agua_mexico.php.

the north and centre of the country. Economic and population growth will keep affecting this problem, until it becomes irreparable. CONAGUA states that more than 75% of all water resources are directed towards agricultural purposes, 10% is for industrial use and the 14% that remains goes to public use. Nevertheless, due to poor maintenance in water distribution systems and a lack of water culture, more than 45% of the water that is circulated is wasted.[10]

Only 15% of the water that is distributed in Mexico City is treated. The main problem in Mexico results from the fact that both industry and the public turn fresh water into contaminated water by throwing garbage and industrial chemicals into it. In turn, the water then goes into rivers and lakes, polluting the environment and making it impossible to purify. This creates serious health problems for the population. Further, residents are not educated in terms of water use and control, so people are unaware of issues, including the importance of the forests, regulating climate and water filtration, or the costs for maintaining these ecosystems.

Mexico City was built on top of a lake, but as the city grew and the lake dried, the authorities had to come up with a new water system. The method they developed consisted of pumping water up the surrounding hills, in order to create enough pressure and get the water flowing to all parts of the city, as well as intubating all the rivers. Today these former rivers are now paved highways. Water deposits in the highlands attracted population growth. This has resulted in a widening inequity in water distribution that in turn affects all social factors. The first main source of water for Mexico City had been the Lerma system, but extensive irrigation, regional conflicts and the collapse of the subsoil, have required the metropolis to develop another system: the Cutzamala dam. Here, water is sourced from local springs that have been utilized by residents and indigenous communities for centuries. The latter are now afraid these springs will dry up, become acidic and create a drought. This has been the pattern with all other areas that have been incorporated into Mexico City's water reticulation system in the recent past.

Northern Mexico is dry, while the south is often flooded. Large portions of entire southern states are often inundated. In October 2007, the state of Tabasco went through the worst natural disaster of the last 50 years.[11] That flooding put 80% of the state under water, and one million people were left without any shelter. The causes of these disasters have been attributed to the flow of both of the biggest rivers in the country, the opening of the dams, high tides, and the destruction of the docks. However, it is believed that it was mainly due to the enormous illegal deforestation of the tropical rain forest, which left the state topsoil unprotected from run-off of storms and rainfalls. This was not reported by the media.

Mexico's Air Pollution

> There's so much pollution in the air now that if it weren't for our lungs there'd be no place to put it all. (Robert Orben[12])

---

[10] http://www.agua.org.mx/content/category/7/27/54/.

[11] René Alberto López. "Vive Tabasco la "peor catástrofe" en 50 años; 500 mil damnificados y un muerto", La Jornada, 31 October 2007.

[12] Blatt, Harvey. America's Environmental Report Card: Are We Making the Grade? MIT Press, 2004, p. 155.

Mexico City was formerly renowned for having one of the cleanest atmospheres in the world. The well known novelist Carlos Fuentes titled his first book: *Where the Air is Clear* (1959). Today that seems utterly ironic because of what the city's atmosphere has become. A thick grey layer of smog makes Mexico City one of the most polluted places in the world. The average visibility went from 100 km in the 1940s to 1.5 km today[13]. The famous volcanoes (Popocatépetl and Iztacihuatl) that were once part of an incredible landscape are today rarely visible. This is dangerous, and not only in an aesthetic sense. Levels of almost all pollutants, such as nitrogen dioxide (NO_2), often surpass international standards by two to three times. Mexico City's levels of ozone (O_3), which is dangerous to breathe, are twice as high as the maximum limit allowed by international standards for one hour a year, and this occurs several hours a day every day.[14] Imagine one of the world's largest metropolitan areas, where nearly 21 million people within the Valley of Mexico live in these conditions, yet are unaware of these facts. Even if they were aware, they would have little choice of abode, for the city grows by 3% annually and more than six million vehicles travel on the streets every day.[15]

Mexico City is set in a high altitude valley, almost completely surrounded by mountains and volcanoes both dormant and active. Further, seismic activity is frequent and the area is well known as an 'earthquake zone'.[16] Within such a geographical context of overpopulation and industrialization, the hydrocarbons, emitted from combustion engines and factories, affect the growth of trees and plants in general. This creates health problems for people living in Mexico City, causing eye irritations, persistent headaches, as well as aggravating respiratory and cardiovascular ailments leading to premature death.[17] "It's not air pollution that kills people but some people die sooner than they would otherwise," says biologist Roberto Muñoz Cruz sub-director of information and analysis at Mexico City's atmospheric monitoring system, part of the Secretaria del Medio Ambiente.[18] However, research suggests that pollutant levels and respiratory illnesses are associated.

Mexico's Endangered Species

> Natural species are the library from which genetic engineers can work.
>
> (Thomas E. Lovejoy, Time, 13 October 1986[19])

Human activity has not been compatible with nature's flora and fauna. Humans are the main problem, causing an incredible variety of plants and animals to disappear over time. In order for everything to survive and play their role in nature, which is fundamental for the conservation of living species (including our own), there has to be a certain balance. However, we human beings have not learned to live in harmony with our surroundings, and

---

[13] http://www.sbg.ac.at/ipk/avstudio/pierofun/mexico/air.htm.

[14] http://www.sbg.ac.at/ipk/avstudio/pierofun/mexico/air.htm.

[15] http://www.dcp2.org/features/47/clearing-the-smog-fighting-air-pollution-in-mexico-city-mexico-and-so-paulo-brazil.

[16] http://www.sbg.ac.at/ipk/avstudio/pierofun/mexico/air.htm.

[17] http://www.idrc.ca/en/ev-31594-201-1-DO_TOPIC.html.

[18] Michelle Hibler, The International Development Research Centre, in http://www.idrc.ca/en/ev-31594-201-1-DO_TOPIC.html, 08-12-2003.

[19] Barbara K. Rodes, Rice Odell, A Dictionary of Environmental Quotations, Michigan, Simon and Schuster, 1992, p. 18.

Mexico is far from being an exception. Endemic species such as the jaguar, grey wolves, pumas, turtles, caimans, and rare birds such as the royal eagle, face extinction.[20]

To the south, in the Yucatán peninsula alone, exist around 2,100 species of plants, and more than 2,000 vertebrate animals[21]. Because of this enormous biological variation, Mexico is recognized as one of the richest countries in biodiversity. Unfortunately, many of the plants and animals that have coexisted with us for so long are now on the verge of extinction, due to massive logging, agriculture, cattle raising, air and water pollution, illegal hunting, and trading of wild animals, among other issues.

One species of concern in particular, is the monarch butterfly, a very important species to Mexico and famous all around the world. Every year, this amazing species migrates about 4,500 kilometres from southern USA to the forests of Oyamel in the State of Michoacán in Mexico in order to reproduce, covering a huge pine forest with its colours and beauty. These creatures are endemic to Mexico's environment, as well as to the indigenous culture. In Nahuatl, a main indigenous language, butterflies are called 'papalotl'. From this word derives the Mexican word for kite: 'papalote', and Monarchs are known as "kites of the mountains".[22] However, their forests are being destroyed which consequently affects the butterfly's reproduction capabilities and its very existence.

What Is Being Done in Mexico?

> The great awareness comes slowly, piece by piece. The path of spiritual growth is a path of lifelong learning. The experience of spiritual power is basically a joyful one.
>
> (M. Scott Peck[23])

Compared to a developed country like Australia, Mexico differs economically, politically, socially and environmentally. While an underdeveloped nation does not lack the wealth and sustainability to achieve a developed country's standard, the unequal distribution of capital goods, government corruption, and years of colonization and neo-colonization by the USA seem to have determined Mexico's future. Even though there are some environmental organizations that are doing what they can, the problems are aggravated by internal migration to cities. When people find themselves with nothing to eat and basic needs not being met, they are obliged to move into the city in search of work and a better life. When it comes to choosing between feeding your children today or worrying about the future of the environment, emphasis understandably falls on the importance of immediate survival.

On the optimistic side, there are international and local organizations working on economic, social, racial and poverty issues, as well as environmental problems. Greenpeace, WWF Mexico, the Green Political Party and the National Institute of Ecology, along with other indigenous communities and student groups, are working towards addressing these problems.

---

[20] http://www.mexicodesconocido.com.mx/notas/7742-Especies-en-la-l%EDnea-de-peligro.

[21] http://especiales.yucatan.com.mx/especiales/faunaenextincion/.

[22] http://www.mbsf.org/.

[23] Hernandez, Anthony, The Enlightened Savage: Using Primal Instincts for Personal and Business Success, Morgan James Publishing, LLC, 2006, p. 266.

Fifteen years ago, Greenpeace began campaigning for the environment and raising awareness about air pollution. They demanded the implementation of public transport services, such as the Metrobús, which is now in operation. They also fought for a higher budget for communities that live in protected forest areas, and finally the Mexican Congress doubled the sum available. Greenpeace isn't the only organization working on saving forests, lakes, rivers, and wildlife. The World Wildlife Fund (WWF) granted eight Oaxaquean communities "the gift of the land", creating protected community areas in 2002, as well as publishing the book "*Voces de las montañas. Testimonios orales de la Sierra Norte*".[24] A fund has also been created for the protection of the Monarch Butterfly Sanctuary. Moreover, the construction of electricity and water distribution dams leaves the people who live in indigenous communities with no other alternative than to protest constantly against the use of their natural resources. A couple of years ago, Mazahuan women rose up against the authorities and took control of the Cutzamala dam, demanding a sustainable plan.[25]

Nonetheless, compared to a developed country, the achievements in Mexico have not been completely successful and there is still a lot more work to do.

What Can Be Done in Mexico?

Awareness is empowering. (Rita Wilson[26])

Exploitation of natural and human resources in Mexico is a real problem; and there is no way around it: people have to become aware of what is happening to their forests, lakes and rivers, their oceans, the wildlife and the social communities that depend on their environment for a healthy adequate existence. It seems unbelievable that 600 thousand hectares of pristine forest are being blatantly logged and that very few people are aware of this.

The SEMARNAT approved the highway that will destroy more than 120 thousand hectares of Mexico City's "lungs" and water resources, without even considering the annihilation of endemic wildlife, as well as the irreparable damage to communities. When the author conducted a survey among educated students from diverse backgrounds and different research areas, not one student knew anything about this massive project.

The Mexican government puts a lot of tax money into reforestation; for example, nine million trees were planted in July, 2007.[27] Reforestation is important, and planting trees is a good thing, but it is not the real solution to the problem. To plant thousands of trees does not lead to reforestation, and reforestation does not preclude deforestation. Clearly, if we are losing natural native forest at an exceedingly fast rate on one side, while only part of the money obtained is used to plant a few trees on the other, it does not prevent the problem. According to Mexico's Secretary of Environment, out of every ten trees planted, only one survives. Even if this one tree lived long enough, new forests do not interact with nature in the same way that native species do. Plantations do not absorb sufficient water, they don't mitigate climate change, and they do not sustain an adequate variety and quantity of plants and wildlife.

---

[24] Oral testimonies of the North mountain range.
[25] http://www.esmas.com/noticierostelevisa/mexico/394268.html.
[26] http://www.absolutely.net/Rita_Wilson/quote.htm.
[27] http://www.pvem.org.mx/web/index.php?option=com_frontpageandItemid=1.

It is imperative that we take care of the forest we still have; and this can only happen if the people who care about this matter speak up and inform others about the issues. The government has to change its present forest policy, and instead of replanting a few trees, use that same money to finance communities that protect and handle sustainable forestry. These are communities that take care of the forests. They prevent illegal deforestation which in turn mitigates climate change by reducing CO_2 gas emissions; and they control wild fires, while still generating eco-tourism that brings economic and social benefits to the country.

AUSTRALIA

Australia is a developed country and it has a generally prosperous, Western-styled mixed economy. Originally inhabited by aboriginals, the first European settlers arrived in the mid-nineteenth century. Formed mostly by European immigrants, the population today is just over 21.3 million, with approximately 60% of it being concentrated in and around the coastal state capitals of Sydney, Melbourne, Brisbane, Perth, and Adelaide. It has maintained a stable liberal democratic political system and remains a member of the British Commonwealth of Nations.[28]

The world's flattest and driest continent, Australia has the oldest and least hospitable soils. Only the southeast and southwest corners of the continent have a temperate climate. It is here where most of the densest population areas are found.

Although most of Australia is semi-arid or desert, it includes a diverse range of habitats, from alpine heaths to tropical rainforests and is recognised as a mega diverse country. With its extremely variable weather patterns, and its long-term geographic isolation, much of Australia's biota is unique and varied.

Australian Forests

> He who plants a tree Plants a hope.
>
> (Lucy Larcom, from 'Plant a Tree' [29])

The British Commonwealth by itself is responsible for 40% of the world's annual deforestation rate. Charlie Sherwin, director for the Victorian National Parks Association, declares Australia as one of the leading countries in deforestation. It would appear that humans have come to normalise matters that should not be taken lightly. During the time I worked with the Wilderness Society, when I attempted to talk with people about the fact that their national treasures were being ripped out of the ground for lucrative purposes, they would look at me with disdain, blink and say: "Oh, yeah, The deforestation stuff…"

It could be argued that logging of native forest occurs at an alarming rate: every year the equivalent to 85% of the size of the Canberra is logged; that is 200,000 hectares, or the equivalent of two million 1⁄4 acre suburban household blocks.[30] These are some of the richest

[28] http://en.wikipedia.org/wiki/Australia.
[29] Susan L. Rattiner, Great Poems by American Women: An Anthology, Dover Publications, 1998, p. 73.
[30] http://www.acfonline.org.au/default.asp?section_id=18.

and most diverse forests that exist in the world. According to the Wilderness Society, they are being logged, burned down and poisoned at the equivalent of twenty four football fields a day. A natural source of clean air, water and home to over 5000 unique and endangered species is being obliterated. The Australian National University research shows that unlogged native forests are the most carbon dense in the world. When these forests are destroyed, the carbon stored in the branches, trunk, roots and soil is released back into the atmosphere, contributing to ever-accelerating climate change. Every time we tear down irreplaceable trees, the chance of reducing climate change gets slimmer by the minute.

The Wilderness Society states that in Tasmania alone, 15,000 hectares of ancient native forests are being clear felled and burnt each year. For example, Tasmania's Styx Valley is home to the biggest hardwood trees in the world, many of them taller than a 25-storey building, over 400 years old and up to five metres wide at the base. They are able to store over 1200 tons of carbon per hectare. Only 18% of Tasmania's original tall old-growth forest is left.[31] Nonetheless, for big corporate groups and industries such as Gunns Limited, Australia's largest logging company and export wood chipper, these issues seem the least of their concerns. In order to make a profit, they are willing to wipe out all the natural old-growth, irreplaceable, forest that is left in Australia. Gunns Limited insists that natural eucalyptus forests have been the primary source of sawlogs, veneer logs and pulpwood in Australia since European settlement. The corporation states that it is "a firm *supporter of the continuing use of sustainable management of natural forest* as an environmentally responsible, renewable resource, now and into the future".[32] Yet, is using natural forest and replacing it with timber plantations a sustainable solution? Consider this practice of Gunns. After clear felling the forests, whatever remains of it is burned with a napalm-like substance dropped from helicopters. This "stuff" does not permit native forests to return. Moreover, the fires are one more major contributing step to global warming.

Gunns Limited is proposing to build a native forest fed pulp mill, using chlorine bleaching mill, in the Tamar Valley, an old growth forest area in Tasmania. The pulp mill has been approved for construction via inadequate, fast tracked approval processes that have ignored many of the key issues that concern the public. According to the Wilderness Society, if this pulp mill is built,[33] it will consume up to 4.5 million tons of wood from native forests and plantations annually for the next 30 years. This mill will lead to a dramatic increase in the rate of logging in Tasmania, and double the rate of total wood chipping.[34] It will also discharge 30 billion litres of pollution into Bass Strait each year, a daily effluent of 64,000 tons.[35] Independent economists have warned that the pulp mill could cost Tasmania's economy $3.3 billion, as well as hundreds of lost jobs in the tourist, fishing, and agricultural industries.

Tasmania is only one example of what is happening to Australia's native forest throughout the country. According to Bush Heritage, rates of land clearing in Queensland and New South Wales are still unacceptably high and proposals for development in northern

---

[31] http://weblog.greenpeace.org/tasmania/styx_background.html.
[32] Gunns Ltd, Bell Bay Pulp Mill, http://www.gunns.com.au/Forest/forests1.html.
[33] It appears at the time of finalising this chapter, that this pulp mill has been approved to go ahead.
[34] Gunns Ltd, Bell Bay Pulp Mill, Draft Integrated Impact Statement.
[35] Gunns' referral under the EPBC Act, April 2007.

Australia involving clearing hundreds of thousands of hectares continue. About 50% of Australia's rainforests have already been cleared.[36]

Two major contributing factors to an increase in global warming are the constant rise in deforestation around the world and the CO_2 emissions. Australian National University's research shows that given that the country already has enough plantations to meet almost all timber needs, and that every other major greenhouse gas emitter is being asked to reduce emissions, it seems absurd to continue on this path.[37] Perhaps if residents truly understood the depth of the problem and the importance these forests represent to Australia's future, they would think twice before saying: "Oh, yeah, the deforestation stuff..."

Australian Water

We never know the worth of water till the well is dry.

(Thomas Fuller 'Gnomologia' 1732 [38])

Despite living in the driest of continents, Australians are among the largest consumers of water in the world, on a per capita basis.39 Unless this problem is addressed quickly, things will get worse because only 12% of the annual rainfall over Australia results in runoff into streams and rivers, or soaks into and is retained in the ground.[40] According to the Australian Bureau of Statistics, 70% of this water is used for agricultural irrigation,[41] the rest is taken for urban and industrial purposes (about 20%), and in rural areas for stock and domestic use (5%).[42] Australia stores more water than any other country, in order to guarantee a reliable supply during periods of drought, because it is likely to vary, given that catchment conditions and rainfall patterns can be erratic. The water obtained in some areas is not very pure and, commonly, some form of chemical treatment is required to ensure that is potable.[43]

This is just the beginning of the water problem that threatens Australia's environment, ecosystems, and all living systems. Ecosystems depend completely on the surrounding lakes, rivers, marine and coastal waters. According to the Wilderness Society, in spite of the improvement that was achieved with water reforms and certain protection to wild rivers, water assets around the country are under serious pressure. For example, 80% of major rivers in the Cape York Peninsula that flow to the Great Barrier Reef violate water quality guidelines, and due to this, Queensland is losing about 7000 hectares of wet land per year because of construction[44]. Cape York's monsoonal river network is the life support system for local communities and its incomparable environments. Currently Cape York's wild rivers remain vulnerable to many of the same threats that have destroyed other Australian rivers,

[36] Bush Heritage, Creating Markets for Biodiversity, Productivity Commission, Canberra, April 2001.
[37] http://forestpolicyresearch.org/.
[38] Fuller, Thomas. Gnomologia: Adagies and Proverbs; Wise Sentences and Witty Sayings, Ancient and Modern, Foreign and British, Oxford, Printed for B. Barker, 1732, p. 273.
[39] http://www.ozh2o.com/h2use2.html.
[40] http://www.waterquality.crc.org.au/consumers/Consumersp2.htm.
[41] http://www.environment.gov.au/soe/2001/water.html.
[42] http://www.ozh2o.com/h2use2.html.
[43] http://www.ozh2o.com/h2use2.html.
[44] http://www.wilderness.org.au/articles/green-audit-qld-govt-indifference.

such as the Murray. Many rivers are dying out because of strip mining and the water extraction that comes with this activity.[45]

Australian's marine waters are another issue. These cover twice the area of the land, where another impressive vast underwater world exists, filled with canyons, undersea mountains, and the Great Barrier Reef that is home to an extensive diversity of life forms. The Great Barrier Reef includes over 2,900 reefs, around 940 islands and cays, and stretches 2,300 kilometres along the Queensland coastline. This Marine Park is 345,000 km2, which is larger than the entire area of the United Kingdom and Ireland combined.[46] Unfortunately, overfishing, mining, oil and gas drilling, and pollution are destroying this delicate ecosystem, leaving Australia without its unique marine life. Sediments and nutrients, fertilisers, pesticides, toxic chemicals, sewage, rubbish, detergents, heavy metals, and oil, run into rivers and out to the Great Barrier Reef, where they threaten plant and animal life.[47]

Australia's Air Pollution

> We're in a giant car heading towards a brick wall and everyone's arguing over where they're going to sit.
>
> (David Suzuki (Environmentalist, scientist and broadcaster)

Air pollution in Australia is relatively low, since 60% of the population lives near the coast, and because these cities are not surrounded by mountains, the air flows freely, clearing the potential pollution. Further, in a developed country such as Australia, government stipulations and regulations are more effective than in Mexico. Recently in Australia several states have adopted newer approaches to pollution control based on economic incentives. Each class of environmental or natural resource management problem has its own properties or attributes, thus the suitability and type of economic instrument must be considered on a case-by-case basis. General areas of application are the following: pollution control (discharge of solids, effluents and gases), noise and thermal pollution, physical resources, conservation reserves, and natural areas (national parks and wilderness areas).

Australian Wildlife

> Without habitat, there is no wildlife. It's that simple.
>
> (Wildlife Habitat Canada[48])

Australia is one of the countries in the world with the greatest biodiversity. Its habitat holds more than one million species of plants and animals, many of which are endemic, Including 140 species of marsupials, such as koalas, wombats, and the Tasmanian devil. The last is now on the extinction list and is found only in some parts of the Tasmanian wilderness. More than 750 species of birds have been recorded in Australia, 350 of which are found

[45] http://wildrivers.org.au/updates.
[46] http://www.wwf.org.au/ourwork/oceans/gbr/.
[47] http://www.reef.crc.org.au/discover/threats/index.html.
[48] http://www.whc.org/EN/home.htm.

nowhere else in the world. Among them are the kookaburra, the rainbow lorikeet and fairy penguins. There are also fifty-five different species of macropods native to Australia.[49] Moreover, about 85% of flowering plants, 84% of mammals, more than 45% of birds, and 89% of inshore freshwater fish are unique to Australia.[50]

Australians are very concerned for the future of their native animals. Fifty-four species have disappeared, 22 animals are currently critically endangered, 127 species are endangered, and 200 animals are considered in a vulnerable state.[51] Many animals have not been seen in Australia for a long time, and are probably extinct, such as the Tasmanian tiger, or the lesser bilby.[52] Their extinction is usually a result of the fact that most Australian fauna depend on a specific food or habitat, and if these disappear, the species do not survive. Further, many native animals are killed by the introduction of predators such as cats, dogs and foxes. One example would be leadbeater's possum[53], which has not been seen since the 1960's. Introduced non-predator animals are also a cause of extinction; these include cattle, sheep, rabbits and deer that eat the food needed by native animals, thereby reducing the latter's numbers.

There is one little creature in particular whose story is worth noting. It is the Tasmanian devil. As indicated earlier, now found only in Tasmania, it captures people's attention and mine when I heard about its sad plight. European settlers evidently named it "devil", because when they hunted at night, their screeches and growls frightened the men. The Tasmanian devil is quite the opposite of what the name conjures up in one's mind: it is in fact a relatively small, very cute, furry animal. Sadly, in May 2008, the Tasmanian devil was listed as "endangered" under Tasmania's Threatened Species Protection Act 1995.[54] Tasmanian devils with large facial tumours, have been sighted at 63 different locations across more than 60% of Tasmania's mainland. This fatal disease was first detected during 2008, and it is estimated that the impact will result in as much as a 50% decline in the devil population.[55] Affected high-density devil populations undergo up to 100% mortality in 12–18 months.[56]

What Is Being Done in Australia?

> Raising awareness is the key and getting people to ask the right questions - only they can decide what they do from there.
>
> (James Gibbs[57])

Some Australians have understood the profundity and importance of this environmental crisis and they began uniting in groups such as The Wilderness Society, Green Peace, Bush Heritage, WWF-Australia, and Conservation Volunteers. Other organizations, government

---

[49] http://www.australia.gov.au/Our_Natural_Environment.
[50] http://www.dfat.gov.au/facts/env_glance.html.
[51] http://www.environment.gov.au/cgi-bin/sprat/public/publicthreatenedlist.pl?wanted=fauna.
[52] http://www.environment.gov.au/cgi-bin/sprat/public/publicthreatenedlist.pl?wanted=fauna.
[53] http://www.kidcyber.com.au/topics/Austendangered.htm.
[54] http://www.tassiedevil.com.au/disease.html.
[55] Byrnes, M. 2007. "Battle to save Tasmanian devil from extinction. Retrieved" on 15 March 2007.
[56] DPIWE. 2005. Devil Facial Tumour Disease - Update June 2005.
[57] http://www.ntsad.org/TOW/NATIONAL_TAY-SACHS_AWARENESS_MONTH%20_9_5_08.asp.

parties, communities and individuals are also working very hard to help stop global warming, massive deforestation, pollution, water waste, and animal extinction.

Conserving the Australian environment is a gigantic task that requires collective action to stop climate change by protecting forests and rivers, as well as terrestrial and marine wildlife. Such groups undertake scientific research, put a lot of their time into community awareness raising, and serve as environmental advocates. They work with elected leaders to make sure that they are aware that preserving forests is one of the quickest, easiest and cheapest ways of fighting climate change and protecting endangered species, as well as helping to protect freshwater resources and marine environment.

Raising awareness and uniting against the injustice that is occurring can make a real difference. In July 2003, 5000 people marched in the Tasmanian forests, the major shareholders of Gunns Limited were lobbied, and groups were formed that raised awareness about the impact of deforestation. A year later, 15,000 people marched through Hobart's streets in order to stop Styx Valley deforestation. Today, Styx Valley is a protected national forest.

What Can Be Done in Australia?

> The first step toward change is awareness. The second step is acceptance.
>
> (Nathaniel Branden[58])

Even though a lot of effort has been put into saving forests, lakes and rivers, there is still much to do. Motivated individuals need to start by talking about the problem with friends, family, neighbours, classmates and co-workers. Knowing about something, and letting other people know, makes a difference. Human networks are so interconnected that they can create a waterfall effect. We may only represent one single drop of water, but once we start uniting, others will follow, making change possible in the future.

All of us can start to think about doing more. All non-government organizations depend on a certain amount of financial support from their members in order to keep campaigns going. In spite of the global financial problems, one simple way to help Australia's environment and wildlife is to actively support them. This will bring an end to unrestrained instances of land clearing and most native forest logging – a guaranteed, immediate and cost-effective way to stop carbon pollution – and create more protected areas to give habitats and wildlife a better chance to cope with climate change. However, not everything can be resolved by contributing funds.

> If we are willing, change can start at home. We can save energy by using appliances with 5-star energy ratings and turn them off at the wall. We can change over to compact fluorescent light globes, and turn lights off when we don't need them. We can install solar water heating and choose accredited Green Power electricity suppliers. Driving less is also important, as every 5 kilometres driven creates 1.5kg of carbon dioxide. We can reduce consumption by reusing and recycling. By buying local and buying less, we can cut down on packaging. We can try to save in every way possible. Australians create tons of greenhouse gases to cool and

[58] Sanders, Hugh B. The Subconscious Diet: It's Not What You Put in Your Mouth; It Is What You Put in Your Mind! Liberation Press, 2005, p. 24.

heat their homes, so we need to turn down the air conditioning and the heaters. We can plant trees, because each tree can absorb a ton of carbon dioxide during its lifetime. We can fill our gardens with water-wise native plants. Finally, the most important thing for us to do is to raise awareness. The more people know about environmental degradation and climate change, the faster, cheaper and easier it will be to address them.

CONCLUSIONS

> Look deep into nature, and then you will understand everything better.
>
> (Albert Einstein [59])

The ecological crisis will strike and affect all living things on the planet, as is already happening. Yet, even though change will occur everywhere, it will not have an equal effect around the world. People will have to adjust differently, according to the place they live in, their needs and the resources they have. Mexico as an under-developed country will go through the ecological crisis at a very different level from Australia, which has a much higher economic development level. The gap between all the developed and the underdeveloped countries will widen. This will be a real threat to those with the least amount of natural and human resources. Social class disparity will widen and racial and gender disadvantage in the world will increase. Some members of society will have too much, and others too little.

In a place like Mexico, which does not have the infrastructure, the laws, or the resources to adjust to an ecological crisis, the problem will worsen. In Australia, where things tend to happen as expected, it is likely that major action will be taken to prevent famine, migration struggles, water wars, and massive death tolls.

The challenge is a rapid transition of the political, economic, and social organization in order to live within our environmental means, while preserving and enhancing our general wellbeing. This can be done not only in the developed countries, but in an under-developed country like Mexico as well. A greater effort has to be made, and it has to be done fast. It really is now or never.

REFERENCES

Auden, W.H. (1997). *British Council.* Michigan: Northcote House.

Blatt, H. (2004). *America's Environmental Report Card: Are We Making the Grade?* MIT Press, p. 155.

Martín, B., Saurí, D. (2002). *El cambio global.* Barcelona: Rubes.

Rubens, B. (2007). *Regímenes internacionales y politicas de cambio climático* in *El clima.* Madrid: Popular.

Bulat, S. (2007). *Planeta Frito,* Barcelona: Urano.

Byrnes, M. (2007). *"Battle to save Tasmanian devil from extinction" in http://www.alertnet.org/thenews/newsdesk/SYD321428.htm Retrieved 15 march* 2007

[59] Edwards, Andres. David Orr. The Sustainability Revolution: Portrait of a Paradigm Shift, New Society Publishers, 2005, p. 113.

Durban. (2007). *Declaration for carbon commerce, ¡La justicia climática, ahora!* in *El clima*, Madrid: Popular.

Edwards, Andres. (2005). *The Sustainability Revolution: Portrait of a Paradigm Shift.* Gabriola Island, BC: New Society Publishers, p. 113.

Emerson, R. (1909). In 'Woodnotes'. *English Poetry III: From Tennyson to Whitman*. The Harvard Classics.

Fuller, T. (1732). *Gnomologia: Adagies and Proverbs; Wise Sentences and Witty Sayings, Ancient and Modern, Foreign and British*. Oxford, Printed for B. Barker.

Gurría Di-Bella, M. (2007). *Lecciones de étice ambiental,* Cholula: Universidad de las Américas Puebla.

Hernandez, A. (2006). *The Enlightened Savage: Using Primal Instincts for Personal and Business Success*. United States: Morgan James Publishing, LLC.

Hernandez-Garduño E., Perez-Neria J., Paccagnella, A.M., Pina-García, M., Munguía-Castro, M., and Catalan-Vazquez. (1997). Air Pollution and Respiratory Health in Mexico City. *J. Occup. Environ. Med, 39*(4), 299-307.

Holland-Cunz, B. (1996). *Ecofeminismos*, Madrid: Cátedra.

Martín, L. and Maximiliano, M. (2005). *Êl negocio del agua a nivel mundial,* in *Le Monde diplomatique: El agua y el futuro del mund.* Santiago: Aún creemos en los sueños.

López, R. A. (2007). "Vive Tabasco la "peor catástrofe" en 50 años; 500 mil damnificados y un muerto." *La Jornada*, 31 October.

Rattiner, S. L. (1998). *Great Poems by American Women: An Anthology*. Mineola, N.Y.: Dover Publications.

Rodes, B. K. and Rice, O. (1992). *A Dictionary of Environmental Quotations*. Michigan: Simon and Schuster.

Van Ypersele, J-P. (2007). *La principal injusticia de los cambios climáticos,* in *El clima*. Madrid: Popular.

Sanders, H. B. (2005).*The Subconscious Diet: It's Not What You Put in Your Mouth; It Is What You Put in Your Mind!* California: Liberation Press.

Krech, S., McNeill, J.R., and Merchant, C. (2004). *Encyclopedia of World Environmental History*. New York: Routledge.

Singh, M. P. (2007). *Quote Unquote: A Handbook of Quotations*. New Delhi: Lotus Press.

Sweet, L. (2003). *Citizen Participation and Environmental Management*, Return to Focus on Connections Between Health and the Environment: *IDRC Bulletin,* August, 12.

Xanic, A. (2008). Algo verde bajo el Sol, in *Expansión*, July 7.

Further Reading on the Internet

Agua, *...desperdician más?,* Source: CONAGUA Comisión Nacional del Agua, http://www.agua.org.mx/content/category/7/27/54/

Agua en México (Water in Mexico), *Homepage*: http://www.aguaenmexico.org/

ATL (The water portal), *Homepage*: http://www.atl.org.mx/

Australian Conservation Fund, *Homepage:* http://www.acfonline.org.au/default.asp?section_id=2

Barclay, Eliza. *Clearing the Smog: Fighting Air Pollution in Mexico City, Mexico, and São Paulo, Brazil*, http://www.dcp2.org/features/47/clearing-the-smog-fighting-air-pollution-in-mexico-city-mexico-and-so-paulo-brazil, June 23, 2007.

BBC News, *World water crisis map,* http://news.bbc.co.uk/hi/english/static/in_depth/world/2000/world_water_crisis/default.stm

BBC World, *En imágenes: inundaciones en Tabasco,* http://news.bbc.co.uk/hi/spanish/photo_galleries/newsid_7072000/7072865.stm

Bush Heritage, *Land Clearing and Its Impacts,* http://www.bushheritage.org.au/natural_world/natural_world_land_clearing

CONABIO. 2006. *Capital natural y bienestar social.* Comisión Nacional para el Conocimiento y Uso de la Biodiversidad, México. http://www.conabio.gob.mx/

Consumer's Guide to Drinking Water, *About water and Australia,* http://www.waterquality.crc.org.au/consumers/Consumersp2.htm, May 2006.

E ambiental, Homepage: http://www.eambiental.org/index.php?option=com_frontpage andItemid=288

E ambiental, *Conservación:* http://www.eambiental.org/index.php?option=com_contentand task=sectionandid=48andItemid=204

Ecomunidades, *Homepage*: http://red-ecomunidades.blogspot.com/

EcoPortal, *Homepage*: http://www.ecoportal.net/content/view/full/42724

Greening Australia, *Working for our unique landscape,* http://www.greeningaustralia.org.au/our-solutions/land

Greenpeace, *Bosques y selvas in Mexico:* http://www.greenpeace.org/mexico/campaigns/bosques-y-selvas-de-m-eacute-x

Greenpeace, *Tala ilegal en el Gran Bosque de Agua:* http://www.greenpeace.org/mexico/fungames/animations/tala-ilegal-en-el-gran-bosque?MX_BTRACK=banner_tala

Greenpeace, *Por los hijos de Calderón*, http://porloshijosdecalderon.org/index.html

Greenpeace and The Wilderness Society, *The Styx Valley:* http://weblog.greenpeace.org/tasmania/

Greenpeace and The Wilderness Society, *The Styx Valley:* http://weblog.greenpeace.org/tasmania/styx_background.html

Gunns Limited, homepage: http://www.gunns.com.au/

Gunns Limited, *Natural Forests,* http://www2.gunns.com.au/Forest/forests.html

Iniciativa Mexicana de Aprendizaje para la conservación, *Cuánta agua tiene México*, http://www.imacmexico.org/ev_es.php?ID=16646_208andID2=DO_TOPIC

Instituto Nacional de Ecología, *El enfoque multimedios y su aplicación en la regulación ambiental en México,* http://www.ine.gob.mx/ueajei/publicaciones/gacetas/244/regulacion.html, Thursday 21 august 2007.

Las noticias México, *Produce Cutzamala una de las aguas más puras,* http://www.lasnoticiasmexico.com/184280.html

Legorreta Jorge, *Las Cuencas Externas,* Ecología:Agua, http://www.planeta.com/ecotravel/mexico/ecologia/97/0897agua2.html

Mexico Forestal, *Homepage:* http://www.mexicoforestal.gob.mx/index.php

Monarch Butterfly Sanctuary Foundation, *Homepage*: http://www.mbsf.org/

Olguín, Rosa Maria. *Mujeres mazahuas mantienen un bloqueo en la entrada de la planta potabilizadora del Sistema Cutzamala, en Villa de Allende, Estado de México,* in Es mas, source Noticiero Televisa: http://www.esmas.com/noticierostelevisa/mexico/394268.html, Sep 27, 2004

Portal Forestal, *La Commonwealth es responsable del 40% de la deforestación anual,* http://www.portalforestal.com/index.php?option=com_contentandtask=viewandid=137andItemid=30, Friday, 05 de Octubre de 2001

ProNatura, *El agua en México*, http://www.pronatura.org.mx/agua_mexico.php

Ruth Rodríguez, *Alertan sobre daños de contaminación en desarrollo infantil*, in http://www2.eluniversal.com.mx/pls/impreso/noticia.html?id_nota=114847andtabla=Nacion, Monday 23 de august 2004.

Reyes Mariusa, México: los desatinos del agua, BBC Mundo

http://news.bbc.co.uk/hi/spanish/science/newsid_4402000/4402020.stm, November, 3 2005.

Save the Tasmanian Devil, *Homepage*: http://www.tassiedevil.com.au/disease.html

Sainz Santamaría, Jaime. Mariana Becerra Pérez, *Los conflictos por agua en México: avances de investigación*, Instituto Nacional de Ecología, http://www.ine.gob.mx/ueajei/publicaciones/gacetas/389/conf_agua.html

SEMARNAT. 2005. *Informe de la situación del Medio Ambiente en México. Compendio de estadísticas ambientales*. México, D.F. http://www.semarnat.gob.mx

Sosa Iván, *Sufre falla el sistema Cutzamala*, from Reforma, in Iniciativa Mexicana de Aprendizaje para la conservación, http://www.imacmexico.org/ev_es.php?ID=38859_201andID2=DO_TOPIC June 07 2007.

Susan Briggs, Sandra O'Malley and Kate Corbett, *Water problems could get 'even worse',* Herald Sun, November 03, 2006.

http://www.news.com.au/heraldsun/story/0,21985,20694411-5005961,00.html,

Tirel, Magali Iris, *Mazahuas y guerrerenses, unidos en defensa de los ríos*, La jornada ecológica, September 25, 2006, http://www.jornada.unam.mx/2006/09/25/eco-d.html

The Wilderness Society, *Homepage:* http://www.wilderness.org.au/

The Wilderness Society, *Climate Change:* http://www.wilderness.org.au/campaigns/climate-change

Tribunal Latinoamericano del agua, *Piden a Tribunal cancelar el sistema Cutzamala,* http://www.tragua.com/es/index.php?option=com_contentandtask=viewandid=47andItemid=34

UNESCO, *Water resources,* http://www.unesco.org/water/wwap/index_es.shtml

United States Environmental Protection Agency, *Homepage*: http://www.epa.gov/Border2012/issues.html

Urrutia, Alonso and René A. López, *La crisis en Tabasco ya afecta a un millón de personas: Granier,* La jornada, November 2nd, 2007;

ttp://www.jornada.unam.mx/2007/11/02/index.php?section=sociedadandarticle=035n1soc,

Villarreal, Lilia Ruiz, *Especies en la línea de peligro,* in Mexico Desconocido, http://www.mexicodesconocido.com.mx/notas/7742-Especies-en-la-l%EDnea-de-peligro, April 2005.

Wikipedia, *Inundación de Tabasco y Chiapas de 2007,* http://es.wikipedia.org/wiki/Inundaci%C3%B3n_de_Tabasco_y_Chiapas_de_2007

WWF Australia, *Great Barrier Reef:* http://www.wwf.org.au/ourwork/oceans/gbr/

Contributed by Geoffrey Frost, Tasmania.

Photo: A Tasmania Tiger on the Endangered Species list.

In: Meltdown: Climate Change, Natural Disasters... ISBN 978-1-60876-153-1
Editor: Kathryn Gow

Commentary

HOW DID WE BLOW IT ON CLIMATE CHANGE?[1]

Stephan Lewandowsky
University of Western Australia

You ask a toddler if they would rather have a lolly now, or that big shiny new toy in a few minutes. You can bet that the toddler will take the lolly now and forgo the delayed bigger reward.

You ask a nine-year-old if they want a new computer game right now, or a snazzy new laptop a week from now and many of them will happily wait a week for the bigger prize.

This ability to forgo an instant fleeting pleasure for a greater reward later is called “delay of gratification” and it has been identified by psychologists as a sign of maturity and adulthood. As we grow up, we become capable of weighing the costs and benefits of present and future actions and we learn to consider our long-term interests.

In fact, psychologist Walter Mischel, at Columbia University, New York, has shown that our ability to delay gratification during childhood predicts our social competence and academic abilities as adults. The more we learn to regulate our desires as children, the more competent we are as adults.

Accordingly, most adults who buy a new car will willingly pay for an oil change every six months, because they realise that the small cost incurred now will prevent the much greater and unavoidable expense of a new engine at some future point. We might hedge and delay or grumble, but we will eventually get that next service done before too long.

This fact is so obvious that if we came across a corporate fleet manager who is running all their cars without maintenance, just to improve the bottom line for a year or two, we would shudder at their economic mismanagement. And their boss would probably fire them.

The same simple, clear and inescapable logic applies to climate change. In fact, this logic has applied to climate change for several decades now, as most of my fellow scientists know all too well.

Twenty or so years ago, we could have dealt with climate change for the price of an oil change. Ten years ago, the price had gone up and it would have cost us a new engine. Right now, as revealed by the United Nation's top scientific panel in a report last week, we are in

[1] First published in The West Australian 22nd November 2007; permission given to reproduce in this text.

for the cost of a new car. And if we do nothing for another five years, our planet will undergo unstoppable remodelling with us no longer in the driver's seat.

In light of those facts, seeking to expand an economy by unabated continued reliance on fossil fuels is, simply put, infantile. It is only the toddler who wants the lolly now, only the nine-year old who cannot wait a week for a laptop and only the fleet manager who makes a buck by saving on the oil change who would think that our country's record on climate inaction represents good economic management.

Anyone who services their car regularly will recognise that our leaders' persistent refusals to tackle climate change represents a case of serious economic mismanagement.

As a consequence of this mismanagement, the issue that is facing us now is not just the pain over the rapidly increasing costs of dealing with climate change.

Instead, the issue is also for people in their 20s to decide whether they want the Great Barrier Reef to be devastated by coral bleaching before they get to see it.

The issue is for people in their 30s to decide how they will explain to their children that 80 per cent of Kakadu wetlands disappeared on their watch.

The issue is for people in their 40s to decide how they will cope with the rise in food prices, after the northern boundary of the Wheatbelt has inexorably moved south.

The issue is what to do when entire countries will be threatened by rising sea levels.

The issue is how to hear the total price tag of $US20 trillion ($US20,000,000,000,000) which represents current estimates of the cost of climate inaction for the remainder of the century.

And the issue is for all of us to think of answers when our grandchildren ask us why we did not force our politicians to act more like adults and make them look after our future, the way we know how to look after our new cars.

How Could We Live without These?

The Nature o' the world
the Trees high in the sky
on the bueaty o' the world;
It's birds, animals and trees.
How could we live without These?

Earth is Nature.
we must respect Nature.
with a lot o' care. Ohhhhhhhhh
ohhhh oh oh oh oh!

The Nature o' the world
the Trees high in the sky
on the bueaty o' the world
It's birds, animals and trees.
How could we live without These?

We must respect Nature by Picking up Rubbish
and Recyalbes.
Also respect Nature by Taking your bike to school
or bus or walk.

The Nature o' the world
the Trees high in the sky
on the bueaty o' the world.
It's birds, animals and trees.
How could we live without These?

(Jordy Roberts, Gold Coast, Australia)

[Editor's note: Author is boy, aged 6; he has set this piece to music.]

SAVE OUR PLANET

Majestic pillars of strength in a verdant curtain
Buttressed bastions denoting their stability
And yet we know their future is uncertain
Threatening to become sterile aridity.

High above the canopy of shade
Protecting the symbiotic life beneath
A food chain in a delicate balance made
By a Creator who now looks on in grief.

Millenniums of gradual evolution
Produced our planet's precious "gold"
To be plundered via greed and prostitution
For economic gain: to be sold.

Our forests appeal for consideration
Will no one listen to their piteous plea?
Save their souls from extermination
So earth remains for posterity.

And as I labour to pen these words
Acres of forest are being destroyed
Insects, trees, animals and birds
No longer there to be enjoyed.

International cooperation
Is necessary to save the day
Rigid rules and legislation
Are all that will stop the world's decay.

Elizabeth Tindle

(Elizabeth Tindle, Brisbane, Australia)

INDEX

D

E

F

G

H

J

Q

R

S

T

U

Y

Z